Whales and Dolphins

Whales and Dolphins

Cognition, Culture, Conservation and Human Perceptions

EDITED BY

Philippa Brakes

AND

Mark Peter Simmonds

London • Washington, DC

First published in 2011 by Earthscan

Earthscan Ltd, Dunstan House, 14a St Cross Street, London EC1N 8XA, UK
Earthscan LLC, 1616 P Street, NW, Washington, DC 20036, USA

Earthscan publishes in association with the International Institute for Environment and Development

For more information on Earthscan publications, see www.earthscan.co.uk
or write to earthinfo@earthscan.co.uk

ISBN: 978-1-84971-224-8 hardback
978-1-84971-225-5 paperback

Typeset by JS Typesetting Ltd, Porthcawl, Mid Glamorgan
Cover design by John Yates
Title page illustration: humpback whale by Mark Peter Simmonds

A catalogue record for this book is available from the British Library

Library of Congress Cataloging-in-Publication Data

Whales and dolphins : cognition, culture, conservation and human perceptions / edited by Philippa Brakes and Mark Peter Simmonds.
p. cm.
Includes bibliographical references and index.
ISBN 978-1-84971-224-8 (hb) – ISBN 978-1-84971-225-5 (pb) 1. Cetacea–Psychology. 2. Cetacea–Behavior. 3. Cetacea–Conservation. 4. Human-animal relationships. I. Brakes, Philippa. II. Simmonds, Mark P. III. Title.
QL737.C4W44145 2011
599.5'15--dc22
2010047848

At Earthscan we strive to minimize our environmental impacts and carbon footprint through reducing waste, recycling and offsetting our CO_2 emissions, including those created through publication of this book. For more details of our environmental policy, see www.earthscan.co.uk.

Printed and bound in the UK by CPI Antony Rowe.
The paper used is FSC certified.

For Skye and her generation

Contents

List of figures and boxes *ix*
Foreword *xi*
Acknowledgements *xiv*
Contributors *xv*
List of acronyms and abbreviations *xxvi*

1 Why Whales, Why Now? 1
Philippa Brakes

PART I: WHALES IN HUMAN CULTURES

2 Impressions: Whales and Human Relationships in Myth, Tradition and Law 9
Stuart Harrop

3 Whales of the Pacific 23
Viliamu Iese and Cara Miller

4 The Journey Towards Whale Conservation in Latin America 29
Miguel Iñíguez

5 Whales and the USA 37
Naomi A. Rose, Patricia A. Forkan, Kitty Block, Bernard Unti and E. C. M. Parsons

6 Whales in the Balance: To Touch or to Kill? A View of Caribbean Attitudes towards Whales 47
Nathalie Ward

7 The British and the Whales 56
Mark Peter Simmonds

8 Whales in Norway 76
Siri Martinsen

9 Of Whales, Whaling and Whale Watching in Japan: A Conversation 89
Jun Morikawa and Erich Hoyt

10 A Contemporary View of the International Whaling Commission 100
Richard Cowan

PART II: THE NATURE OF WHALES AND DOLPHINS

11 The Nature of Whales and Dolphins 107
Liz Slooten

12 Brain Structure and Intelligence in Cetaceans 115
Lori Marino

13 Communication 129
Paul Spong

14 Lessons from Dolphins 135
Toni Frohoff

15 Highly Interactive Behaviour of Inquisitive Dwarf Minke Whales 140
Alastair Birtles and Arnold Mangott

16 The Cultures of Whales and Dolphins 149
Hal Whitehead

PART III: NEW INSIGHTS – NEW CHALLENGES

17 Whales and Dolphins on a Rapidly Changing Planet 169
Mark Peter Simmonds and Philippa Brakes

18 From Conservation to Protection: Charting a New Conservation Ethic for Cetaceans 179
Philippa Brakes and Claire Bass

19 What is it Like to Be a Dolphin? 188
Thomas I. White

20 Thinking Whales and Dolphins 207
Philippa Brakes and Mark Peter Simmonds

Index *215*

List of Figures and Boxes

Figures

I	Solitary bottlenose dolphin, Kent, UK	7
3.1	Spinner dolphins, Fiji	25
3.2	A humpback whale comes into view as Uto ni Yalo, a traditional Pacific *waka*, is being sailed by the Fiji Islands Voyaging Society	26
4.1	Sperm whale in Grytviken factory, South Georgia	31
4.2	Sei whales caught in southwest Atlantic in the 1930s	32
4.3	Watching southern right whales in Peninsula Valdes, Argentina	34
4.4	Southern right whale watching in Patagonia	35
6.1	Flensing of a humpback whale on Semple Cay in Bequia, Saint Vincent and the Grenadines	48
6.2	'Let dolphins swim free for all generations to see' – educational poster sponsored by WDCS, UNEP, the Caribbean Environment Programme, Caribbean Conservation Association and the Marine Education and Research Centre	53
7.1	Northern bottlenose whale in the Thames, 2006	57
7.2	An unknown whaler in the mouth of a whale about to be flensed at a land station	59
7.3	Francis Buckland applies 'medicine' to a porpoise – an illustration from his book	62
7.4	'Dave' the bottlenose dolphin and friends off Folkestone in 2007	67
7.5	Whale on the beach at Stonehaven, 1884	73
9.1	A whale-watching lookout on Chichi-jima in Ogasawara, Japan	96
9.2	Whale watching on Chichi-jima in Ogasawara	97
9.3	Whale watching on local fishing boats in Tosa Bay, Shikoku Island	98
II	Humpback whale spyhops to get a better view above the surface	105
11.1	Highly social group of sperm whales	108
11.2	Hector's dolphins playing with seaweed	111
11.3	Another Hector's dolphin investigating seaweed	111
11.4	Female Hector's dolphin jumping out of the water close to the photographer's boat, providing the dolphin with an excellent view of the humans on board	112

12.1 Sagittal (sideways) images of cetacean brains at four stages of evolution along a timeline showing approximate Encephalization Quotient and some major structures 117
12.2 Illustrations of the modern human brain and bottlenose dolphin brain showing the location of the primary visual and auditory cortex in each 123
13.1 The closest communication among orcas occurs between mothers and their offspring, here A42 (Holly) with her 2008 baby (A88) 130
13.2 Orca A42 (Holly) with her youngest daughter (A88) and eldest son (A66) 130
15.1 A headrise and breath for a nearly vertical and very curious dwarf minke whale near swimmers 141
15.2 A close binocular examination for the author (Alastair Birtles) by a young female dwarf minke ('Strumpet') 143
III Minke whale surfaces for air 167
17.1 Marine conservation efforts are sometimes too little, as well as too late 176
20.1 'The whale' as portrayed in *Natural History Scripture* 207

Boxes

7.1 The main international agreements and domestic laws affecting cetaceans in UK waters 70
20.1 Marine scientists petition to the IWC delivered June 2010 211
20.2 Declaration of Rights for Cetaceans: Whales and Dolphins 212

Foreword

The unbelievably huge whale swam close to the surface, undulating, parallel to our boat, her calf close beside her. The gentle rhythmic sound of her breathing brought a sense of deep peace. I was aware of her ancient knowledge, carried in her genes through millions of years of blue whale evolution. That day out on the ocean provided one of the most memorable of the great store of memories that I have been gifted with throughout my 75 years. Dolphins came to us also on that same day, some 2000 our captain estimated. Suddenly they were all around us, hunting. Wherever we looked there were dolphins, racing through the water, intent on their prey. It was breathtaking. Just as suddenly, they were gone. And then, a few hours later, there they were again, enchanting us with their vitality, their energy. This time they were surely playing as they dived under the boat, torpedoed to the surface, zigzagged. Playing for the sheer joy of life, as do the chimpanzees when the ripe fruit is all around and their bellies are comfortably full.

My fascination with the denizens of the ocean was first triggered by Jules Verne's *20,000 Leagues Under the Sea.* For several years I dreamed of learning about the beings that lived in the mysterious world under the water. It was at that time that I read *Moby Dick* as one of my school books. I couldn't sleep at night. I used to lie imagining what it must be like – being pursued by killers, knowing that if you come up for air you will be most cruelly harpooned. Yet knowing too that you could not forever put off the increasingly desperate need for air. And it has become far worse: in those days there was no sophisticated technology that can track a whale's movements from afar. Today there is truly no escape for a hunted whale.

I have never lost my love for the ocean and her creatures, my fascination with whales and dolphins, even though my path took me into a very different world in the rainforests of Africa. And this interest in the cetaceans has only intensified as we learn more and more about their complex lives, their cultures, their communication. Who can fail to be moved when they listen to recordings of the songs of the humpback whales?

That they have emotions and a capacity for suffering I have never doubted. When I was a child I had a wonderful teacher from whom I learned a great deal about animals, their intelligence and their emotions. That knowledge stood me in good stead when, after a year in the field studying chimpanzees, I was admitted to

Cambridge University in England to do a PhD in ethology in 1961. Although I was thrilled to have this opportunity, I was also a bit apprehensive as I had never been to college at all. Imagine my shock when I was told I had done everything wrong. It was not scientifically acceptable to talk about chimpanzees having personalities and their own individuality, nor the capacity to think, and I certainly could not ascribe to them emotions. All these things were unique to the human animal. To suggest otherwise was to be guilty of that worst ethological crime: anthropomorphism, attributing human-like behaviours to other-than-human animals. Fortunately, despite my scientific lack of know-how, I was able to stick to my convictions, mainly, I am sure, because of the lessons I had learned from that great teacher of my childhood. And that was my dog, Rusty.

Given the similarity in chimpanzee and human DNA and brain anatomy, why should it be surprising that chimpanzees should be capable of intellectual abilities once considered unique to humans? Why would there not be similarities in emotional expressions? And the differences between individuals in behaviour as well as appearance were so obvious that it was clearly ridiculous to deny them. The 1960s saw the burgeoning of field studies of a variety of creatures with complex brains and correspondingly complex social behaviour. It became increasingly clear that reductionist explanations for many of these behaviours were simply not appropriate.

As this book points out, cetacean social behaviour and cognition is a comparatively new field, largely due to the physical limitations associated with studying them in a watery environment, so different from our own. But with incredible determination and dedication, and through skilful use of some of the latest technology, cetacean researchers have provided many fascinating and extremely significant insights into the lives of their subjects. There can now be no reasonable doubt that these animals are highly intelligent, and have extraordinarily complex social behaviour, rich communication patterns and cultural traditions that vary between groups of the same species. Nor can it be denied that they show emotions and are capable of altruism, caring for sick and dying companions. This book shares also many stories of the sometimes extraordinary relationships that have been observed between cetaceans and humans.

A story that attracted a good deal of attention concerned a whale who had become hopelessly tangled in fishing lines and the volunteers who went to her aid, cutting through the lines. Some wondered how she would react when she was finally free – with a small slapping of tail or fluke she could easily have killed them. As the last line dropped off she dived down deep – but then came up and swam to each in turn as though thanking them, before swimming to freedom. One of the team said that after looking into her huge eye as he cut the rope from her head, he will never be the same again.

The book also discusses the use of dolphins in entertainment, pointing out the cruelty of capturing these denizens of the open sea, destroying their families, maintaining them in hopelessly inadequate captive conditions, and teaching them inappropriate 'tricks'. One of the most moving accounts of the effect of unnatural confinement is provided by Alexandra Morton in her book *Listening to Whales.*

Whales and Dolphins: Cognition, Culture, Conservation and Human Perceptions is a very important book. It makes a compelling case for scientists, conservationists and animal welfare groups to combine to develop a new approach to the conservation of cetaceans. An approach that takes into consideration the various environmental threats such as decrease in fish stocks, chemical and noise pollution, the navy's use of low-frequency sonar, the effect of global warming on ocean habitats, collisions with shipping and, of course, hunting – for food, for 'research' or for the live animal trade. And one that combines not only concern for the species, but also for the individual and his or her social group and culture.

We must be grateful to Philippa Brakes and Mark Peter Simmonds, and to all who contributed to this book, for it results in a clarion call for action. Whales and dolphins are ancient and wonderful sapient and sentient beings. How would we be judged by our great, great grandchildren and all unborn generations if, knowing what we do, we do not fight to prevent their extinction? The whales and dolphins need and deserve our help – now, before it is too late.

Jane Goodall, PhD, DBE
Founder – the Jane Goodall Institute
UN Messenger of Peace
www.janegoodall.org

Acknowledgements

We would like to thank Claire Bass for having faith in this project from the outset, the World Society for the Protection of Animals (WSPA) and the Royal Society for the Prevention of Cruelty to Animals (RSPCA) for their assistance and financial support, and the Whale and Dolphin Conservation Society (WDCS) for allowing us the time to complete this project. Many WDCS colleagues helped in the production of this book (although some may not have even known they were doing so) by providing inspiration and advice on issues, and also being patient with us in the final stages of this volume. We thank Chris Butler-Stroud in particular for his support and advice. Several people helped to review the chapters and we are particularly grateful for the keen eyes and sharp minds of Sue Fisher, Naomi Rose, Nicky Kemp and Guy Harris.

We would also like to thank the following for kindly providing photographs or illustrations and appropriate permissions: Terry Whittaker, Walter Roli, Miguel Iñíguez, Paul Spong, Steve Dawson, Erich Hoyt, Nathalie Ward, Brendan Beirne (Rex Features), Paul Brown (Rex Features), Lori Marino, Nardi Cribb, Saras Sharma (Fiji Fisheries), Matt Curnock, Alastair Birtles, Duncan Murrell, Rob Lott and Aberdeen Art Gallery & Museums Collections.

Thanks also go to IPSOS MORI for allowing citation of the results of their opinion poll.

Contributors

Editors

Philippa Brakes
The Whale and Dolphin Conservation Society, Brookfield House, 38 St Paul Street, Chippenham, Wiltshire, SN15 1LJ, UK
Philippa Brakes is a marine biologist, specializing in marine mammal welfare and the ethical issues associated with our interactions with cetaceans and their environments. Philippa has served as an expert on cetacean welfare issues and whaling policy with the New Zealand government delegation to the International Whaling Commission (IWC), as well as an informal adviser to other government and non-government delegations, as a lecturer in Zoological Conservation Management, as Marine Advisor to the RSCPA and as the Curator of a British Zoological Gardens. She is also the author of numerous reports, papers and conference presentations on marine mammal welfare and ethics. Philippa's academic focus is the evolution of the existing conservation paradigm towards a more sophisticated approach that integrates conservation, animal welfare and the inherent value and interests of the individual into conservation and protection policy. Her interest is the evolution from traditional genetic and geographic boundaries towards protecting lower orders of organization, such as cultural units, and recognizing the conservation value of the individual and smaller social groups to encourage the advancement of a novel, scientifically and ethically sound conservation approach. Philippa currently serves as a Senior Biologist with WDCS. She has recently become a mother and this book is dedicated to her daughter.

Mark Peter Simmonds
The Whale and Dolphin Conservation Society, Brookfield House, 38 St Paul Street, Chippenham, Wiltshire, SN15 1LJ, UK
Mark Peter Simmonds is an environmental scientist specializing in the problems facing marine mammals in the 21st century. His research interests include the effects of chemical pollution, noise and climate change. He is also involved in field studies in UK waters and the development of conservation policy. Mark is currently the International Director of Science at WDCS. Before joining WDCS

in 1993, he was a researcher at Queen Mary and Westfield College, University of London, and then Senior Lecturer in Environmental Sciences and Reader in Wildlife Conservation at the University of Greenwich. He has continued to teach on various university courses, including regular contributions at the universities of Bristol and Bangor. He has been the Chair of the UK's Marine Animal Rescue Coalition since 1989. He has also been a member of the Scientific Committee of the IWC since 1994, and in recent years has also been part of the UK delegation to the IWC. Mark has authored over 200 papers, articles and letters in scientific and popular periodicals – mainly on marine environmental themes. He has contributed to a number of books and wrote *Whales and Dolphins of the World*, first published by New Holland in 2004. He is also a wildlife artist.

Chapter Authors

Claire Bass
World Society for the Protection of Animals, 5th Floor, 222 Grays Inn Road, London, WC1X 8HB, UK
Claire Bass manages the marine mammal programmes at WSPA, focusing on the welfare problems of cetacean hunting, sealing and the keeping of cetaceans in captivity. She holds an MSc in Marine Environmental Protection from the University of Wales, which led her into the field of marine mammal conservation and protection in 2003. Claire previously worked as a campaigner and investigator for the Environmental Investigation Agency, where she undertook field-based projects to gather information on marine mammal exploitation, using the results in scientific reports to lobby governments for better protection. She has attended meetings of the IWC annually since 2004, where she has been a member of the UK delegation, providing advice to the UK and other pro-conservation governments on issues relating to whaling and whale conservation and welfare.

Alastair Birtles
Minke Whale Project, Tourism, School of Business, James Cook University, Townsville, Queensland 4811, Australia
Alastair Birtles is Senior Lecturer in Environmental Management and Ecotourism at the School of Business, James Cook University in Queensland, Australia. He grew up in the UK, studied Zoology at Oxford University and obtained a PhD in Marine Biology at James Cook University. He has nearly 40 years of experience in tropical research and teaching, including over ten years of marine biology and zoology in Great Barrier Reef environments. Alastair's interests include ecotourism, environmental management, integrated coastal zone management and ecologically sustainable tourism. He is leader of the 20-year Minke Whale Project (conducting 15 years of field studies since 1996) and other marine wildlife tourism projects (featuring sharks, whales, dolphins, turtles, dugongs, groupers and scuba diving);

also tourism in protected areas (especially World Heritage Wet Tropics Rainforest and the Great Barrier Reef), visitor management, interpretation and Aboriginal tourism – all aimed at enhancing tourist experiences, minimizing impacts and developing more ecologically sustainable management practices. He has worked extensively in developing countries, supervised 100 honours and postgraduates from 25 countries (currently supervising eight PhD students) and has produced many publications and natural history films.

Kitty Block
Humane Society International, 2100 L Street NW, Washington, DC 20037, USA
Kitty Block is the Vice President of the Humane Society International (HSI). She oversees efforts by HSI in numerous international treaties and agreements, including the Convention on International Trade in Endangered Species of Wild Fauna and Flora, the IWC and the Inter-American Tropical Tuna Commission. She is a member of the Maryland and District of Columbia Bar Associations. Prior to joining HSI, Block was an associate at Hirschkop & Associates. She served as a law clerk with the Public Corruption Unit of the US Attorney's Office at the US Department of Justice.

Richard Cowan
Tunbridge Wells, Kent, UK
Richard Cowan CBE was born in Edinburgh in 1949. He was educated at the Royal Blind School in Edinburgh, the College for the Blind in Worcester and then at St John's College Cambridge, where he read economics. He joined what was then the Ministry of Agriculture, Fisheries and Food in 1971 and spent most of his career negotiating with others in the European Community on the management of markets for milk products, beef and lamb. In 2000, he was appointed a Commander of the Order of the British Empire for services to the beef and sheep sectors. In May 2001 he joined the Ministry's Fisheries Directorate in which, among other things, he served as UK Commissioner to the IWC until his retirement in March 2010.

Patricia A. Forkan
Humane Society International, 2100 L Street NW, Washington, DC 20037, USA
Patricia A. Forkan is the Humane Society of the United States (HSUS) Senior Envoy to the Obama Administration. She has been involved in the protection of whales and dolphins for over 30 years. She has been a non-governmental organization (NGO) representative at the annual meetings of the IWC and has also served as a member of the US delegation. She was also a member of the US delegation to the Law of the Sea negotiations and successfully worked to enhance protection for marine mammals in that treaty.

Toni Frohoff
TerraMar Research and Learning Institute, 27 W. Anapamu St., Suite 336, Santa Barbara, CA 93101 USA
Toni Frohoff is a consultant, author, and Director and Faculty Affiliate of TerraMar Research and Learning Institute (www.TerramarResearch.org). She obtained her Master's at Texas A&M University and received her PhD from The Union Institute Graduate School in 1996. Since the 1980s she has specialized in the behaviour of marine mammals, with a distinctive emphasis on psychology, communication, stress and wellbeing of cetaceans in the context of human interaction and anthropogenic activities. She conducted the first studies of dolphins in captive swim programmes and dolphins interacting with human swimmers in the wild. Her research is unique in that she has studied the influence of human interaction on a diversity of cetacean species (beluga whales, orcas, bottlenose and spotted dolphins, and baleen whales) in a complex range of conditions (solitary dolphins and groups; in captivity and in the wild). Frohoff has developed 'interspecies collaborative research' methodologies that are expanding to include baleen whale and elephant psychology and cognition. Near her home in Santa Barbara, Frohoff and her students are studying the behavioural ecology and psychology of coastal dolphins. She has authored the books, *Dolphin Mysteries: Unlocking the Secrets of Communication* (Yale University Press, 2008) and *Between Species: Celebrating the Dolphin-Human Bond* (Sierra Club Books/UC Press, 2003) and contributed to almost 20 others. She lectures internationally, including at the TED Global 2010 conference in Oxford. Her work on wildlife psychology and wellbeing has contributed to the implementation and revision of legislation protecting marine animals in over a dozen countries.

Stuart Harrop
Durrell Institute of Conservation and Ecology, School of Anthropology and Conservation, University of Kent, Canterbury, Kent, UK
Stuart Harrop holds the Chair of Wildlife Management Law at the University of Kent and is the Director of the Durrell Institute of Conservation and Ecology within that university. As such he teaches and researches in the field of the human relationship with the conservation of nature. His preferred emphasis within that subject is to seek to promote the welfare of animals and to protect human traditional practices while simultaneously conserving nature. He has written widely on animal welfare, whaling, human traditional practices along with human rights and the environment and presented papers on these subjects at conferences in diverse locations around the world ranging from Sumatra, Argentina, Peru, Mexico and the US to Oxford University and the Zoological Society of London within the UK. He has also advised the UK government, the European Union and the United Nations Food and Agriculture Organization (FAO) and other international institutions on aspects of these subjects. Prior to joining the University of Kent, Professor Harrop held a number of posts including: Director, Legal Services for

the RSPCA, and Director, Legal Services for the London Stock Exchange. He is also a natural history photographer.

Erich Hoyt
Whale and Dolphin Conservation Society, North Berwick, Scotland, UK
Erich Hoyt has written or co-written 18 books and hundreds of magazine articles on whales, dolphins, as well as ants, other insects, wild plants and other subjects. He is currently Senior Research Fellow with WDCS in the UK, and Head of WDCS's Global Critical Habitat Marine Protected Area Programme. For the past decade, he has jointly directed the first killer whale (orca) study in eastern Russia (in Kamchatka), an international collaboration with Russian scientists. He has lectured and worked on conservation and scientific projects in Japan, Russia, Indonesia, Taiwan, Iceland, Mexico, Chile and the Caribbean. He has also taught as a visiting lecturer at the Ohio State University, the University of Edinburgh, and Massachusetts Institute of Technology (MIT). His first book, *Orca: The Whale Called Killer* (Firefly) is still in print after 30 years and was recently translated into Japanese. His other books have been translated into Chinese, German, Danish, Spanish, French, Italian, Dutch and other languages. Erich is a charter member of the Society for Marine Mammalogy, as well as a long-time member of the European Cetacean Society. In 2006, he was invited to be a member of the International Union for Conservation of Nature's (IUCN) Species Survival Commission – Cetacean Specialist Group, and in 2010, the IUCN World Commission on Protected Areas (WCPA). He is also a member of the IUCN High Seas Task Force.

Viliamu Iese
Pacific Center of Environment and Sustainable Development (PACE-SD), Faculty of Science, Technology and Environment (FSTE), The University of the South Pacific, Laucala Campus, Suva, Fiji
Viliamu Iese is a Research Fellow for the Pacific Centre for Environment and Sustainable Development at the University of the South Pacific. He has been involved in various biodiversity, food security, agricultural and conservation projects in Samoa, Tuvalu, Fiji and the Federated States of Micronesia throughout his studies and working life. His research in Tuvalu as a Marine Environment Officer is of particular relevance as he coordinated the first systematic surveys on large marine species (turtles, sharks, whales and dolphins) to be run in this country. The Department of Environment, Tuvalu, and Department of Conservation, New Zealand, NZAID project focused on cetaceans, sharks, rays and turtles and included both standard scientific methodology as well as traditional knowledge surveys. Viliamu has presented his work at a variety of scientific symposia, undertaken specialized training in cetacean and turtle research survey techniques, and has attended numerous regional conservation conferences and meetings. He is also the chairperson for the Climate Change Working Group in the Pacific Nature

Conservation Roundtable. Viliamu holds Bachelor's (Biology and Education), Postgraduate Diploma and Master's (Biology (Plant conservation genetics)) degrees from the University of the South Pacific and intends to undertake his PhD work on Pacific conservation issues in the near future.

Miguel A. Iñíguez
Fundación Cethus and Whale and Dolphin Conservation Society Latin America, Potosí 2087, (B1636BUA), Olivos, Prov. Buenos Aires, Argentina
Miguel A. Iñíguez was born in Buenos Aires, Argentina. He is the President and Founder of Fundación Cethus, which is based in Argentina, and has been involved in cetacean work for 25 years, including field studies on Commerson's dolphins, orcas, bottlenose dolphins, Peale's dolphins and southern right whales in Patagonia. Since 1998, he has also specialized in the development of responsible whale watching, working on capacity building for this activity along the Latin American coast. Miguel is the author of *Orcas de la Patagonia Argentina* (Propulsora Literaria, 1993) and *Toninas overas, los delfines del fin del mundo* (Zagier & Urruty, 1996). He is also the author and co-author of numerous scientific papers and popular science articles. He is an Associate Professor at both the Universidad Nacional de Costa Rica and the Universidad Marítima Internacional de Panamá. Miguel has also lectured and led courses for the Asociación Balaena, Universidad Complutense de Madrid, Spain; Universidad de Murcia, Spain; Universidad Nacional de Costa Rica; Universidad de Buenos Aires, Argentina; Universidad Nacional de la Patagonia Austral, Argentina; and the Universidad Marítima Internacional de Panamá. He has acted as a consultant to the WDCS since 1991 and, from 2002 to date, he has been part of the Argentine delegation to the IWC, including acting as the Alternate IWC Commissioner in recent years for his country.

Arnold Mangott
Minke Whale Project, Tourism, School of Business, James Cook University, Townsville, Queensland 4811, Australia
Arnold Mangott is based at the School of Earth and Environmental Sciences and Business of James Cook University. He grew up in Austria, where he started his degree in biology at the University of Vienna. In 2003 he moved to Australia, where he finished his BSc in Marine Biology at James Cook University, Townsville. Arnold became involved in dwarf minke whale research and the JCU Minke Whale Project when Alastair Birtles supervised his Masters degree in 2004 in which he studied the management problems of dive boats interacting with minkes on short daytrips to the Great Barrier Reef. He then undertook a PhD study on dwarf minke whale behaviour from 2006 to 2008 and submitted his thesis in 2009. During his studies, Arnold was involved in several other research projects, including studies on the sustainable management of tourism associated with whales, dolphins, sharks, turtles, dugongs and corals. He has been a member of the International Relations Committee of the Society for Marine Mammalogy since 2008.

Lori Marino
Department of Psychology, 488 Psychology and Interdisciplinary Sciences Bldg, 36 Eagle Row, Emory University, Atlanta, GA 30322, USA
Lori Marino is Senior Lecturer in Neuroscience and Behavioural Biology and affiliate of the Center for Ethics at Emory University and co-founder and Executive Director of *The Aurelia Center for Animals and Cultural Change, Inc.* Her research interests include the evolution of brain, intelligence and self-awareness in cetaceans (dolphins and whales) and other species, human–nonhuman animal relationships, and animal welfare/rights and ethics. She is the author of over 80 publications in the areas of cetacean neuroanatomy and brain evolution, comparative behavioural ecology and evolution in cetaceans and primates, and the ethical dimensions of human–nonhuman relationships. In 2001, she and Diana Reiss published the first definitive evidence for mirror self-recognition in bottlenose dolphins in the *Proceedings of the National Academy of Sciences*. She is an active scholar-advocate for cetaceans and other animals and a founding signatory of the Declaration of Rights for Cetaceans. She has also published several methodological critiques of dolphin-assisted therapy and dolphin–human interaction programmes. She teaches animal intelligence, brain imaging, animal welfare and other related courses.

Siri Martinsen
NOAH, Osterhausgate 12, 0183 Oslo, Norway
Siri Martinsen is a veterinarian, educated at the Norwegian School of Veterinary Science, with farm animal behaviour as her student research project. She has worked with 'NOAH – for animal rights (Norway)' for 20 years; 14 years as director and spokesperson. The last two years she has also worked with WSPA (on the Norwegian whaling campaign) and the International Fund for Animal Welfare (on the international sealing campaign). For the last 20 years she has been the Norwegian representative of InterNICHE (International Network of Individuals and Campaigns for Humane Education), being the first veterinarian to graduate without the use of animal experimentation in Norway. She has published a number of scientific papers on the issue of alternatives to animal experiments and also authored several hundred articles and published letters about a wide range of animal welfare issues.

Cara Miller
Institute of Marine Resources, The University of the South Pacific, Laucala Campus, Suva, Fiji
Cara Miller has been involved in cetacean research for the last decade, including documenting spinner dolphins in critical resting habitat in Fiji and undertaking some of the first cetacean diversity surveys in Papua New Guinea. Her undergraduate studies focused on animal behaviour and marine biology, whereas her Master's degree in Applied Statistics examined population models and field survey design for marine species. A four-year project investigating habitat and abundance of a

small resident population of bottlenose dolphins in the Gulf of Mexico served as her PhD research. Presently Cara is the Pacific Islands Program Leader for WDCS International, a Visiting Research Fellow at the University of the South Pacific in Fiji, and also has a research appointment at Flinders University in South Australia. Her work in the Pacific Islands has given her the exciting opportunity to collaborate with many researchers and staff across this region, become engaged in local capacity-building efforts, the chance to conduct ongoing research surveys, and also to progress cetacean conservation under the Convention of Migratory Species Memorandum of Understanding for the Conservation of Cetaceans and their Habitats in the Pacific Islands Region.

Jun Morikawa
Rakuno Gakuen University, Sapporo, Japan
Jun Morikawa is a Professor in the Department of Regional Environmental Studies at Rakuno Gakuen University in Sapporo, Japan and a visiting research fellow at the University of Adelaide, Australia. He is the author of *Japan and Africa: Big Business and Diplomacy*, published by Hurst in 1997, and specializes in Japan's relations with the South, especially development and overseas aid. He is also the author of *Whaling in Japan: Power, Politics and Diplomacy* (Hurst and Columbia University Press, 2009). This book focuses on the gap between the political myths and reality of Japan's whaling policy and focuses on seldom discussed aspects of the political and decision-making structures that support it. It also examines how Japan has used diplomacy and aid gradually to expand support for its whaling policies at the IWC.

E. C. M. Parsons
Department of Environmental Science & Policy, George Mason University, 4400 University Drive, Fairfax, Virginia 22030, USA
E. C. M. Parsons is an Associate Professor in the Department of Environmental Science and Policy at George Mason University, and is a Research Associate with the University (of London) Marine Biological Station in Scotland and the Smithsonian Conservation Biology Institute in Virginia. He has been a member of the Scientific Committee of the IWC since 1999 and is currently the Marine Section President, and sits on the Board of Governors, for the Society for Conservation Biology. Dr Parsons has published over 100 scientific journal articles, book chapters, reports and papers, and is currently writing a textbook on marine mammal biology and conservation.

Naomi A. Rose
Humane Society International, 2100 L Street NW, Washington, DC 20037, USA
Naomi A. Rose is the Senior Scientist for HSI, where she oversees marine mammal protection issues. She has been a key player in the international debate on the welfare of marine mammals in captivity and has been a member of the Scientific Committee of the International Whaling Commission since 2000. Dr Rose has

authored or co-authored several popular and scientific articles, as well as chapters in several books. She lectures annually at several universities.

Liz Slooten

Department of Zoology, University of Otago, P.O. Box 56, Dunedin, New Zealand

Liz Slooten is an Associate Professor in the Department of Zoology, University of Otago, in New Zealand. She studies the biology of Hector's dolphins and sperm whales. Together with a team of other staff, postdocs and graduate students, she is working on projects on a range of human impacts (including fishing, tourism and marine mining) on marine mammals (including Hector's and bottlenose dolphins, sperm and right whales, fur seals and sea lions). These projects include line transect surveys, acoustic surveys, photo-ID, ecological and behavioural studies. A particular focus is research to determine the effectiveness of current and potential protection measures. For example, Liz's research was instrumental in the creation of a protected area for Hector's dolphins around Banks Peninsula (east coast of New Zealand's South Island). Long-term research on the effectiveness of this sanctuary has shown that the boundaries need to be extended north, south and offshore. In 2008, protection was extended north and south. Discussions are ongoing to achieve better protection for Hector's dolphins further offshore. Liz is especially interested in assessments of risk and uncertainty for resource management decisions. In her spare time she enjoys sailing and other outdoor activities.

Paul Spong

P.O. Box 510, Alert Bay, BC, V0N 1A0, Canada

Paul Spong is co-director of OrcaLab, a land-based whale research station on Hanson Island in British Columbia, Canada, which operates under the philosophy of 'learning without interference' (www.orcalab.org). Paul acquired a PhD in physiological psychology from UCLA in 1966. He began studying dolphins and orcas in 1967, initially in captivity, then in the wild. His insights soon led to his involvement in the save-the-whales movement during the 1970s, when he participated in Greenpeace campaigns against commercial whaling. Following the 1982 decision of the IWC to impose an indefinite moratorium on commercial whaling, Paul returned to full-time research. Since then, his work has focused on the long-term life history of the 'Northern Resident' community of British Columbia orcas, and on the protection of orca habitat. He is also involved in the development of technology that connects people to the natural world via the internet, and in 2000–2006 participated in a successful experiment in virtual experience called 'Orca-Live' (www.orca-live.net). Paul is joined, in all aspects of his work, by his research partner and wife, Helena Symonds. They live and work on Hanson Island and at a second operations base in nearby Alert Bay on Cormorant Island.

Bernard Unti
Humane Society International, 2100 L Street NW, Washington, DC 20037, USA
Bernard Unti is Senior Policy Adviser and Special Assistant to the CEO and President of the HSUS, and a historian of animal protection in the US. He is the author of *Protecting All Animals: A Fifty-Year History of The Humane Society of the United States* (Humane Society of the USA, 2004) and several public policy articles on animal welfare as a social concern. He is currently working on a history of concern for animals in the US before World War II.

Nathalie Ward
Eastern Caribbean Cetacean Network, Bequia, Saint Vincent and the Grenadines, West Indies
Nathalie Ward is the Director of the Eastern Caribbean Cetacean Network (ECCN). She has been employed at the US National Marine Sanctuary Program at Stellwagen Bank National Marine Sanctuary (SBNMS) since 2002, wherein she developed the Marine Mammal Sister Sanctuary Program between the SBNMS and the Santuário de Mamíferos Marinos de la República Dominicana. Ward has a Masters of Science in Cultural Anthropology and Doctoral degree in Environmental Studies from Antioch University. She has studied the behavioural ecology of humpback whales in the US Gulf of Maine and the Caribbean since 1978, where she conducted research on aboriginal subsistence whaling with the Bequia Humpback Whale Fishery from 1984 to 1990. Ward was employed as Biology Faculty at Boston University's Marine Program for ten years, where she established the marine mammal undergraduate and graduate programmes and taught courses in marine mammal biology, anatomy and policy as well as an annual field course in Dominica, West Indies. Ward is a marine educator who has worked with the United Nations Environment Programme's (UNEP) Specially Protected Areas and Wildlife (SPAW) Programme since 1989. She developed SPAW's marine mammal educational materials and has served as UNEP consultant in the development and drafting of the Marine Mammal Action Plan (MMAP) for the Wider Caribbean Region. Since 2005, she has conducted marine mammal stranding training workshops in the French, Dutch and Lesser Antilles. Currently, her marine policy work focuses on the management and conservation of trans-boundary marine mammal species. She has authored scientific publications, field guides, environmental curricula and children's books.

Thomas I. White
Center for Ethics and Business, Loyola Marymount University, Los Angeles, CA 90045, USA
Thomas I. White is the Hilton Professor of Business Ethics and Director of the Center for Ethics and Business at Loyola Marymount University in Los Angeles, California. Professor White received his doctorate in philosophy from Columbia University and taught at Upsala College and Rider University in New Jersey before

moving to California in 1994. His publications include five books (*Right and Wrong, Discovering Philosophy, Business Ethics, Men and Women at Work* and *In Defense of Dolphins*) and numerous articles on topics ranging from 16th-century Renaissance humanism to business ethics. His most recent research has focused on the philosophical implications – especially the ethical implications – of scientific research on dolphins. His book on this topic (*In Defense of Dolphins: The New Moral Frontier*, Blackwell Publishing, 2007) addresses the ethical issues connected with human–dolphin interaction – for example, the deaths and injuries of dolphins in connection with the human fishing industry and the captivity of dolphins in the entertainment industry. Professor White is a Fellow of the Oxford Centre for Animal Ethics and a Scientific Advisor to the Wild Dolphin Project, a research organization studying a community of Atlantic spotted dolphins in the Bahamas. He served as US Ambassador to the United Nations' Year of the Dolphin Program in 2007–2008.

Hal Whitehead
Department of Biology, Dalhousie University, Halifax, Nova Scotia, B3H 4J1, Canada
Hal Whitehead is a University Research Professor in the Department of Biology at Dalhousie University. He holds a PhD in Zoology from Cambridge University in England. His research focuses on social organization and cultural transmission in the deep-water whales, but he also works on their ecology, population biology and conservation. Field work is mainly carried out in the North Atlantic and South Pacific Oceans from a 12-metre sailing boat. He has developed statistical tools and software for analysing vertebrate social systems, and uses individual-based stochastic computer models to study cultural evolution, gene-culture co-evolution and mating strategies. Hal co-edited *Cetacean Societies: Field Studies of Whales and Dolphins* (University of Chicago Press; 2000) and has written *Sperm Whales; Social Evolution in the Ocean* (University of Chicago Press, 2003) and *Analyzing Animal Societies: Quantitative Methods for Vertebrate Social Analysis* (University of Chicago Press, 2008).

List of Acronyms and Abbreviations

ASCOBANS	Agreement on the Conservation of Small Cetaceans of the Baltic, North East Atlantic, Irish and North Seas
BDMLR	British Divers Marine Life Rescue
CITES	Convention on International Trade in Endangered Species of Wild Fauna and Flora
ECCN	Eastern Caribbean Cetacean Network
EQ	Encephalization Quotient
FAO	United Nations Food and Agriculture Organization
GBA	Grupo Buenos Aires
GBR	Great Barrier Reef
HSI	Humane Society International
HSUS	Humane Society of the United States
ICR	Institute of Cetacean Research
ICRW	International Convention for the Regulation of Whaling
ILO	International Labour Organization
InterNICHE	International Network of Individuals and Campaigns for Humane Education
IUCN	International Union for Conservation of Nature
IWC	International Whaling Commission
LOS	Law of the Sea
MARC	Marine Animal Rescue Coalition
MIT	Massachusetts Institute of Technology
MMAP	Marine Mammal Action Plan
MMPA	Marine Mammal Protection Act
MPA	marine protected area
MWP	Minke Whale Project
NEPA	National Environmental Policy Act
NGO	non-governmental organization
OSPAR	Oslo/Paris Convention (for the Protection of the Marine Environment of the North-East Atlantic)
PCB	polychlorinated biphenyl
RMP	Revised Management Procedure

RSPCA	Royal Society for the Prevention of Cruelty to Animals
SAC	special area of conservation
SBNMS	Stellwagen Bank National Marine Sanctuary
SPAW	Specially Protected Areas and Wildlife
UDAW	Universal Declaration on Animal Welfare
UN	United Nations
UNCLOS	United Nations Convention on the Law of the Sea
UNEP	United Nations Environment Programme
WCPA	World Commission on Protected Areas
WDCS	Whale and Dolphin Conservation Society
WSPA	World Society for the Protection of Animals
WWF	World Wide Fund for Nature

1

Why Whales, Why Now?

Philippa Brakes

INTRODUCTION

There can be little doubt that we are at a pivotal point in human history: the global commitment and cooperation required to abate and mitigate climate change will be of an unprecedented scale. In rising to the considerable challenges of tackling climate change, and other environmental threats, we must also consider how our action – or inaction – is likely to influence the lives of every other species, in every habitat on the planet. At the end of the first decade of the 21st century we face a convergence of crises: on the one hand, dramatic changes in our environment threaten to have an influence on every organism on the planet, and on the other, a large-scale global economic crisis is likely to influence our ability – and perhaps willingness – to tackle these pressing problems.

Why, then, should we turn our attention to cetaceans, an entire order of mammals that includes the baleen whales, beaked whales, sperm whales, dolphins and porpoises (some 83 species in total, with new additions still occasionally being identified)? As a group of animals they are fascinating and compelling. But, there are already many books about cetaceans and there are many other threatened, vulnerable and endangered species. The answer is arguably threefold.

First, whales remain today, as they did in the 1960s and 1970s, an icon for the environmental movement; a motivating emblem of what could be lost forever if we do not act swiftly to protect these remarkable animals and their habitats. In this role, they are not only ambassadors for their own species, but also for entire marine ecosystems and, potentially, for the biosphere as a whole.

Second, scientific understanding of the social and cognitive complexity of some of the species with which we share the planet, including whales and dolphins, is evolving dramatically. Through a miraculous twist of fate, the modern era of environmental uncertainty has coincided with a period of accelerated growth in our understanding of the lives of some of these mysterious ocean dwellers, many of whom have previously evaded our best efforts to assess the more complex details of their existence. Modern techniques and sheer determination from a committed community of cetacean researchers has resulted in the collection of longer-term and more comprehensive datasets. From these we can now interpret new understanding of the lives of our marine mammal cousins and what this means for their conservation and individual welfare.

This new knowledge provides the third element of our argument: our improved understanding of the lives of some of these animals reveals that they have some qualities shared with the primates, including ourselves, and this too arguably demonstrates that we need to pay them special attention.

Whales may also act as vital sentinels for some environmental changes, particularly those associated with ocean acidification and climatic and oceanographic changes. By contrast, some of their smaller coastal-dwelling relatives, whose habitats differ greatly from those in the deep oceans, may also be important indicator species, residing as they typically do at the apex of food-webs in coastal areas where human activities are frequently most intensive.

This volume aims to illuminate some of the secret lives of some of these illusive ocean dwellers, and here we draw together scientific insights and opinions from experts in a variety of fields to present a unique view of whales in the 21st century. Throughout this volume runs the theme of 'culture' – how whales and dolphins are viewed in various human cultures and also how unique cultures have been identified in certain groups of whales and dolphins. Understanding more about the biological and social complexity of the lives of cetaceans heralds an increased sense of responsibility for protecting their habitats and ecosystems, as well as protecting these sentient animals as individuals. In light of new research, novel questions can now be posed in scientific terms. For example, how does the removal of a key individual or individuals influence the wellbeing of the rest of the group and can this have an influence on the long-term survival of the population or social subunit?

For many people the conservation ethic often seems to dominate the fact that some nonhuman species can suffer. But the relevance of animal welfare, not just to the suffering of individuals, but also in relation to key conservation questions, is beginning to gain recognition. Conservation and animal welfare both have scientific and ethical components. Science can help identify and quantify problems, and then we make ethical decisions informed by the data that science provides. The multidisciplinary field of animal welfare is now being championed through an initiative led by a number of governments inside the United Nations (UN), where support is being garnered for a Universal Declaration on Animal Welfare (UDAW).[1]

CETACEANS AND HUMAN CULTURES

Leading experts in cetacean research have been brought together in this volume to provide authoritative descriptions of investigations into the complex lives of cetaceans. This new information is then considered within the context of how various human cultures view cetaceans and whether new information on issues such as cetacean intelligence, culture and the ability to suffer, warrants a significant shift in global attitudes.

We also describe how the ocean, coastal and estuarine habitats of cetaceans are under threat from a variety of factors and examine the paradox that the iconic status of dolphins and our inherent fascination with these intelligent animals has been the impetus for the development of a global cetacean captivity industry, dedicated to holding these animals captive for our entertainment.

Our journey begins with an overview of human relationships with cetaceans in myth, tradition and law and then moves to a tour through some of the cultures of the Pacific Islands, Latin America, US, Caribbean, UK, Norway and Japan. These chapters compare the cultural values, myths, traditions and history of the relationship of these various peoples with cetaceans and examine how views and attitudes are enshrined in modern domestic policy, politics and legislation. Among the issues considered are, inevitably, our use of whales, which includes not only whaling but also whale watching.

The political entrenchment of certain government policies towards cetacean protection, or cetacean utilization, is also discussed along with the attitudes towards cetaceans in cultures where protection is a key theme, borne out through legislation, but where significant threats still exist from other factors such as fishing and pollution.

WHALES AND SCIENCE

The second part of the book focuses its attention on some of the recent advances in science which provide remarkable insights into the lives of these animals, particularly in relation to cognitive function. A great deal of cunning and patience has been required to unravel some of the mysteries of the lives of cetaceans from often only momentary glimpses at the surface or brief sound recordings.

Most cetacean species, specifically the toothed whales, also predominantly 'see' the world through sound; thus there is a challenge inherent for us in understanding an aquatic way of life dominated by sound. The baleen whales do not use echolocation in the same way as the toothed whales, but sound is still of enormous importance to these huge mammals, which have the capacity to communicate over hundreds, perhaps thousands, of kilometres.

However, perhaps one of the greatest challenges to studying cetaceans, beyond the practical difficulties involved in obtaining the relevant data, is interpreting what specific behaviours may mean in terms of an individual's psychology. Why, for example, do whales regularly and often repeatedly leap clear of the water (a behaviour known as breaching)? Is it sheer exuberance and joy; are they trying to rid themselves of parasites as some suggest; are they stunning fish or other prey; or are they using their entire bodies to communicate a message to other whales and dolphins in the vicinity? Could the answer lie in a mixture of these explanations and be context specific?

In considering cetacean behaviour and intelligence here, we start with a journey into the complex brain structure of dolphins and learn more about dolphins' ability to 'recognize themselves' – a trait associated with self-awareness – and their use of tools. We learn more about the recent and significant discovery of spindle neurons in the brains of certain cetacean species – these neurons were previously believed only to be found in the brains of humans and other primates.

In these chapters, we then explore cetacean communication to provide some useful insights into the possible significance of cetacean vocalizations and other forms of communications. The roles that specific individuals play in a group, or cetacean society, are also discussed in relation to the importance of such research to conservation and welfare issues. For example, are there specific individuals within a cetacean society that are responsible for identifying critical habitat for feeding, or for leading a baleen whale migration? For many of these illusive species these are very difficult questions to answer.

This section includes a contribution from Hal Whitehead, one of the world's leading authorities on culture in cetacean societies. Growing evidence suggests that some cetacean species can pass information from one generation to the next through a form of cultural transmission or cultural learning. The existence of 'cultural units' indicates that the genetic and geographic boundaries that we have traditionally used to define cetacean populations may need to be revaluated and that we need to protect much lower levels of organization than species in order to protect some of these unique cultures, which are, in fact, component parts of the species.

Finally, we explore, more broadly, the nexus between differing human cultural perspectives and the growing body of cetacean research, to ask: what does this mean for the long-term future of cetacean populations and cultures? Whether lauded or vilified by humans, predicting the future for cetaceans in view of increasing threats from climate change, ocean acidification and other industrial activities is an enormous challenge. It is likely that the collapse of certain resources, such as fisheries, may bring even greater pressures to bear on cetacean populations, through prey depletion and, potentially, through hunting as cetaceans become increasingly considered as an alternative protein source, or – more remarkably – as a source for omega oils (and other food supplements) or biofuels.

In the closing chapters of this book, we delve deeper into ethical questions related to appropriate levels of protection for cetaceans, reflecting on the findings of recent research that provide insights on the capacity for these animals to suffer.

Where do we draw the line for protection: at the species, population, cultural group or individual level? Does science now demonstrate unequivocally that these animals are sentient, sapient, intelligent beings that are as worthy of individual protection as the great apes?

Acknowledgements

I would like to thank my co-editor Mark Simmonds for his unending energy and enthusiasm in helping to bring this book together and to specifically thank Chris Butler-Stroud and WDCS for providing the time and space to allow this venture to come to fruition. WSPA and the RSPCA also provided invaluable funding and support for this endeavour.

Note

1 Draft text for the UDAW can be viewed at http://s3.amazonaws.com/media.animalsmatter.org/files/resource_files/original/en_draft.pdf

Part I

Whales in Human Cultures

Figure I *Solitary bottlenose dolphin, Kent, UK*

Source: Terry Whittaker

2

Impressions: Whales and Human Relationships in Myth, Tradition and Law

Stuart Harrop

Introduction

The predicament of whales and whaling provides a focus for various distinct perspectives and a battleground for a range of epistemic groups that are shaped by multiple influences: myths, traditions, practices, ethics, laws and frameworks of knowledge. These viewpoints are guarded and reinforced and provide the basis for validation of each group's political stance within the IWC and other regulatory forums dealing with the predicament of whales. In this chapter samples of these perspectives are examined as they relate particularly to controversial aspects of polar whaling, where a battle has been fought for some years now, albeit with sporadic skirmishes rather than constant fighting. A key area for conflict is the granting by the IWC of aboriginal subsistence exemptions for whaling. Specific questions may be extracted from the political haze that beleaguers whaling politics. These include the need to define the meaning of 'aboriginal' in the context of the loss of traditions and the developmental transitions of indigenous people and the need to determine the extent to which traditional whaling may cross over from the status of 'subsistence' to 'commercial'. By examining aspects of the root perspectives and influences that relate to these questions it is hoped that some meaning may be discerned that may contribute to shortening the long days spent debating the fate of the great whales that are the subject of aboriginal subsistence quotas.

Whale Origins

Instinctively human communities may have elevated the whales above the fish because of our mammalian proximity and embodied this relationship within their legends and mythology. Similarly, and perhaps ironically, given the recent history of our treatment of whales, contemporary repositories of culture, such as Hollywood blockbusters, often impute whales with the ability to feel and ultimately to suffer on a par with humans and occasionally endow them with superhuman abilities.

Human communities across continents and through time have created a colourful mosaic of whale stories, many of which link the origin of whales to those of humans. These are often neglected vehicles of knowledge, relegated to possessing literary or historical significance only, and yet this 'wisdom of the mythtellers' (Kane, 1998) may contain pointers to assist us to understand the contemporary relationship between humans and whales and thus move forward in the resolution of some conservation dilemmas.

With our contemporary capacity to track genetic development, we have recently learned much about the complex evolutionary trajectory of whales. First, there was a vast wave of adaptations and metamorphoses as early animals made the slow transition from the sea. Next we see them reaching the complexity of mammals. Then, drawn back to the source, and perhaps never far from it, some 50 million years ago a cetacean ancestor (now believed to be common also to the hippopotamus) followed a very different track and returned again to the lagoons that ultimately led to a marine existence (Ursing and Arnason, 1998; Jonathan et al, 2003). Beyond this, the first cetaceans began to spread into the greatest wilderness on Earth. They went on to become superlative in size, utterly at home in the whole range of the oceans and, for some representatives, adapted to become masters of the depths (Evans, 1987). The ancestral link between humans and whales lived not so long ago in the context of the entire history of life on Earth but in the very distant past in human terms. Nevertheless, even landlocked human communities possess cultural traditions replete with cosmological and other mythical allusions that suggest some sort of co-evolution.

An examination of mythology reveals themes and images that might hold clues to the perspectives of some proponents of contemporary whaling. The rich variety of stories range from meta-myth describing the human relationship with ancient forces of creation and cosmological principles, through to anecdotal tales that have arisen through locally specific traditions. Many myths identify whale species with people or deities in human form, whether through a permanent metamorphosis or through forms of transient shape-shifting. In the Peruvian Amazon local people believe that some deaths by drowning are caused by the shape-shifting 'boto' (the river dolphin, *Inia geoffrensis*) who is believed to be capable of transforming selectively either into the shape of a beautiful man or a woman, depending upon the target and thereby luring men and women into the river to their deaths (Blackburn, 2002). In China, the Yangtze River dolphin, the 'baiji' (*Lipotes vexillifer*), now functionally extinct, was regarded as the embodiment of a reincarnated princess thrown into the Yangtze River to her death because she refused to marry a man chosen for her. The theme of whales deriving from women who were drowned because of perceived folly in marriage is carried further in the well-known Inuit legend of Sedna, whose severed fingers were believed to be the source of all cetacean species.

For some communities, living at the extremes of human existence in the northern polar regions, their mythical traditions possess greater intensity and detail as whale species assume archetypal qualities and foundational positions in culture. They also tend to possess a red-blooded quality that is sometimes far removed from our contemporary attitudes but reflects the reality of life in a harsh world. We can immediately appreciate why, for example, many totemic artefacts were used, as part of the Native American coastal hunting tradition, to protect against the hostile deity represented by the killer whale (Fauconnet, 1959). However, in other cases myths may comprise a vehicle to give us deeper insights into the gulf between current attitudes by pro-whalers deriving from a long history of living in regions at the edge of human survival capability and the so-called 'protectionists' who derive their culture from a less hostile environment.

The challenges of life in the inhospitable polar worlds and the attitudes required for survival are graphically described in the myth of Sedna. The Inuit people both revere and fear the goddess Sedna and the tale of the genesis of whales is a doorway into understanding aspects of northern polar views on whale hunting. As a young Inuit girl, Sedna became the lover of a bird spirit and her disapproving father tried to hide her away from this shape-shifting being. While he was attempting to do so in a boat on the ocean, the spirit raised a terrible storm and, in order to appease the spirit, Sedna's father decided to sacrifice his daughter and threw her into the waters (an act that mirrors the fate of the Princess of the Baiji). Sedna tried to climb out of the sea and her father cut off her hands and fingers as she desperately gripped the side of the boat. The separate parts became the seals, walruses and whales. Consequently, in this depiction of tragedy, cruelty and ambivalent moral approaches to the value of human life, the body of Sedna generated all of the sea mammals that were essential to the survival of the Inuit. She also became the divinity of the sea who, in Inuit beliefs, governs everything in that watery world beyond human domination (Fauconnet, 1959). This myth provides us with a glimpse of the inner world of the Inuit and reveals aspects of their animist perspective (Ingold, 2000). The spirit of Sedna has, by the sacrifice of her body, given to the Inuit the very species that became crucial to their survival. Moreover, the dispersal of her essence has secured that the same spirit now pervades both humans and the marine mammals (De Castro, 1998) thus removing the polarized distinction between animals and humans and generating, thereby, a different ethical perspective to that which prevails in our society.

This equality of the Inuit and the animals they hunt eradicates any need for separate concepts of human or animal rights, based as they are on the perceived differential in value of species. Further, it leaves little room for contemporary concepts of animal welfare that have evolved to deal with the differential treatment of animals and humans within the commanding and dominating position of humans within modernist cultures. To the Inuit, Sedna's sacrifice is repeated and perpetuated in the death of each whale just as much as it is in the death of each human within their communities. Thus, not only did the hunters kill their prey for the purposes of survival but also in this harsh world anthropologists record that some Inuit communities embraced euthanasia of the old (and in some cases children) in their groups to ensure that the communities did not grow beyond the capacity of the scarce natural resources around them. This was not done without compassion or respect but through an essential and culturally embedded sense of service to the community (Morse, 1980; Ponting, 1991).

Necessarily, communities in polar regions developed extensive traditions in relation to key species that they relied upon for their subsistence needs. Indeed there are apocryphal tales and ethnographies describing the manner in which Inuit hunters would deal with their prey and ensure that every part of a whale was used for subsistence and other needs of their communities. Taboos, regulating the manner in which whales were dealt with in the hunting process, were also important mechanisms to secure a sustainable relationship with one of the few species that could be depended upon to support life in the ice deserts. But this requirement of a 'sustainable' relationship did not derive from contemporary concepts of sustainability that now trip off the tongue of policy-makers, but came from a reverence for the great source of spirit that was seen to be common to humans and whales as reflected in the myth of Sedna.

Indeed, we should avoid notions of 'sustainability' in our interpretation of indigenous motives to respect the prey they hunted and the ecosystems that supported them. The Inuit in the North, and others such as the Yagan in the South, were few in number – living as they were on the edge of the planet's capacity to support them. In order to survive, they had to ensure that the hunt would continue and that every part of the prey animal was properly used. The animals hunted by them were difficult to capture, hard to locate and some were seasonally available only on their migration routes. The risks involved in hunting were extreme. There is evidence that virtually all the food of ancient northern hunting peoples, and many other key materials for life support, were derived from a mixed use of whale, seal and caribou. Other evidence suggests that adaptability was also required in that the very use of this mix of northern mammals related also to climatic variations and perturbations (Coltrain et al, 2004). Respect for the prey species was not, therefore, an emotional luxury but a necessity deriving from the slender line between survival and failure. Humans and animals were both sacrificed for the benefits of wider-community imperatives in the polar world. As a contrast, in other more forgiving and biodiverse locations in the world, some nomadic indigenous peoples living in circumstances of very low population density were able to exhaust resources through intense harvesting and then merely travel on to other areas, without ever considering sustainability in a world that gave them everlasting sustenance (Posey and Balee, 1989; Meggers, 1996).

Not all indigenous peoples living at the polar extremes hunted whales. The Yagan people lived as far south as any peoples on Earth have ever lived prior to industrialization – at the very tip of South America where the mountains of the Andes sink into the ocean and become a scattering of island peaks. They were also one of the most marginalized of peoples: treated with contempt by other groups of indigenous peoples ranging south of their territories, along with the European colonizers. They were forced to feed on whatever was available to them in this harsh environment. The last full-blooded member of the Yagan records how her people lived virtually naked in the cold, sleeping in wigwams and moving from island to island in canoes. Fish and shellfish were key components of the diet but she also recalls how whale meat, derived from stranded animals, was seen as a particular delicacy. She recapitulates how, in a dream, she encountered a sperm whale spirit in a particular bay not far away from the sleeping community. On waking she and her people canoed to the location to find a beached sperm whale rotting in the bay. This was regarded as a special gift of choicest food to the Yagan (Stambuk, 1986).

With this perspective in mind as the root of the concept of aboriginal subsistence whaling, few could reasonably condemn the traditional hunting practices of the polar peoples as they are described in historical ethnographies. Indeed this right to persist in carrying out such traditional practices wins some support in international law and policy beyond the IWC. Thus comparable aboriginal subsistence exemptions are prescribed in other international laws as exceptions to the general prohibitions on taking endangered species (see, for example, the Convention on the Conservation of Migratory Species of Wild Animals and the Agreement on International Humane Trapping Standards (Harrop, 1998)). Instruments such as the International Labour Organization (ILO) Convention 169 (ILO 169) and the United Nations Charter for the Rights of Indigenous Peoples, dealing with human rights, also recognize subsistence practices and facilitate the maintenance of the customs and practices of indigenous peoples within their traditional territories. These instruments also acknowledge a key issue that has a bearing on the traditional whaling debate. Thus Article 7.1 ILO 169 does not contemplate restricting indigenous traditional practices to museum set-pieces but instead ensures that indigenous peoples:

> *have the right to decide their own priorities for the process of development as it affects their lives, beliefs, institutions and spiritual well-being and the lands they occupy or otherwise use, and to exercise control, to the extent possible, over their own economic, social and cultural development.*

This prescription has some relevance to a number of issues that confront whaling regulation today in the context of exemptions for aboriginal subsistence.

First, this instrument, concerned exclusively with the status of humans rather than animals, acknowledges that tradition is not static. Indeed as soon as tradition is relegated to a museum or a showpiece for visiting tourists, it ceases to be tradition and becomes a record of frozen moments within the long evolution of tradition. The immediate implication is that traditional whale-killing practices can quite properly move with the times: the spear and the arrow can be replaced by rifles and mechanized harpoons. Second, this provision also makes it clear that moving from a subsistence lifestyle to one governed by commercial activities is a valid right for indigenous peoples in terms of their self-determined development route, although this departure is not necessarily within the concept of the aboriginal subsistence exemption as contemplated by the IWC. There are also other issues that arise in the aboriginal subsistence exemption debate within the IWC that are not dealt with or resolved within the ILO text, which it must be pointed out, is a separate stipulation of international law and does not necessarily rule on the development and definition of such terms within the IWC. The unresolved issues concern, among others, the extent to which a dead tradition can be resurrected or the extent to which a tradition that has only a tenuous link with the cultural heritage of indigenous peoples may be properly comprised in an aboriginal subsistence exemption.

When is a Tradition Not a Tradition in the IWC?

Subsistence lifestyles are vanishing rapidly as globalization takes hold of even the remotest corners of the wild. The prevalent economic paradigm does not necessarily support the perpetuation of traditions designed for small, isolated communities that previously had no scope or reason to accumulate capital and were restricted to subsistence living. As commercial markets have extended into every corner of the globe, it is now inevitable that subsistence activity must be linked into the financial structures of the developed world. Consequently the fine line of distinction between commercial whaling and subsistence whaling may now be utterly obscured. Further, within this obscurity it is becoming increasingly difficult to identify a direct subsistence need because of the integration of communities within wider social and economic structures.

However, the IWC seems to have avoided direct entanglement with the most sensitive areas of the debate concerning the concept of aboriginal subsistence exemptions. In the words of its current website, rather than restricting the definition, the approach has been ad hoc, and this state of affairs will continue until a comprehensive management scheme for this area can be finalized.[1] This latter process will necessarily be controversial and, indeed, previous attempts to capture the meaning of the aboriginal subsistence exemptions have never been formally accepted. The IWC's ad hoc Technical Committee Working Group on Development of Management Principles and Guidelines for Subsistence Catches of Whales by Indigenous (Aboriginal) Peoples defined the necessary activity in 1982 as being:

> *for purposes of local aboriginal consumption carried out by or on behalf of aboriginal, indigenous or native peoples who share strong community, familial, social and cultural ties related to a continuing traditional dependence on whaling and on the use of whales.*

As Reeves (2002) points out, the same committee also sought to define 'local aboriginal consumption as nutritional, subsistence and cultural requirements' and 'trade in by-products of subsistence catches'. Thus there is some evidence of an attempt by the IWC to draw the line between commercial activities and subsistence by emphasizing first the local community's needs and then permitting the incidental by-products to be sold into the wider commercial world. In reality the IWC goes much further than this, as Reeves (2002) emphasizes by pointing out that Greenland's hunting is market-led rather than driven by a subsistence need since there is a distinct correlation between the price of whale meat and the volition to carry out hunting. Further, the central purpose in the Greenland hunt is not to provide incidental income to support a subsistence society's external needs, but instead is geared to providing income within the wider, prevailing economic structures.

This early attempt by the IWC to define the scope of the exemptions also seeks to link the aboriginal subsistence whalers to a 'continuing traditional dependence on whaling'. The Greenland tradition might in some way be regarded as a modern adaptation of an ancient tradition but it has altered beyond recognition from the original subsistence practices in that it is entirely based on introduced technology, is aimed at different target species and appears to be commercially driven (Reeves, 2002). Consequently, the Greenland activities have only a slender link with the prescribed continuing traditional dependence.

However, there are some instances of this traditional dependence in extant whaling activities that trace their origin to antiquity and yet remain relatively unchanged. An example of this is the bowhead whale hunting carried out in Alaska, which has a clear, documented and ancient lineage, retaining many unchanged aspects of the hunt and appending only a few contemporary additions to enhance the chances of a kill.

Tradition need not derive from antiquity in order to be *continued*; but how far do we need to go back to discover the root of the tradition? A definition of 'tradition' is elusive but there may be some consensus on its dynamic and trans-generational nature and in this latter respect the early IWC definition's reference to a 'continuing traditional dependence' implies some demonstrable transmission through generations, but the application of this principle by the IWC appears to have been inconsistent. In one case, small-scale subsistence exemptions were granted by the IWC to the Caribbean islands of St Vincent and the Grenadines permitting local, traditional whaling. However, these traditions were not carried out by the indigenous Carib people who inhabited these islands before the arrival of the conquistadors. Instead they were comparatively newly evolved practices carried out by the descendants of slaves, many of whom were stolen from landlocked countries in Africa that would have had little traditional relationship with whaling or indeed with practices of any kind that related to the harvesting of the oceans. In this instance there is evidence of continuing traditional dependence, albeit in the form of a relatively new tradition that has passed through the hands of a minimal number of generations.

However, the case of the Makah, who live in the extreme northwest of continental USA, is much more remarkable in demonstrating that a tradition can die altogether and yet still be *continuing* for the purposes of the IWC. Although the Makah certainly have a history of whaling, this activity ceased altogether more than 70 years prior to the applications to the IWC for aboriginal subsistence exemptions in respect of this tribe. Any sense of *continuing* dependence had vanished. Indeed, in order to carry out the proposed activities the Makah had to reaquire the appropriate hunting techniques, relearn techniques of whaling boat construction and even regain an appetite for whale meat (Miller, 2002). Although, as has been discussed, there is a legal right for indigenous peoples to self-determine their own future, the decision to grant an aboriginal subsistence exemption to this community was made exclusively within the complex tangle of IWC whaling politics since there simply had not been time in this case for any inter-generational transmission of this new tradition.

By approaching the subject of aboriginal subsistence exemptions in an ad hoc manner the IWC seems to have moved far from its original intention of having a clear line drawn between commercial whaling and strictly traditional subsistence activities. It may have also expanded the scope or grounds for controversy. With such a state of affairs and the nature of some of the decisions described herein, other whalers, such as the Japanese and Norwegians, could argue for parallel exemptions that may be, in some instances, on more solid ground. However, there are issues of scale here. Aboriginal exemptions should comprise only small-scale activities with little impact on whale populations, in sharp contrast to the large-scale, clearly commercial, operations of the Japanese and the Norwegians. In addition, these large-scale commercial activities also face strong opposition bolstered by the increasing evidence of the nature and complexity of cetacean social structures, the ability of whales to suffer and the concomitant growth in importance and strength in the whaling debate of a strong protectionist lobby comprising both animal welfare campaigners and conservationists.

Contemporary Myth, the 'Protectionists' and the Reinvention of Tradition

The IWC came into being soon after World War II and was initially established as a cartel of industrial whaling nations regulating, according to the preamble to the International Convention for the Regulation of Whaling, 'the orderly development of the whaling industry'. Thereafter it gradually reinvented itself to become an instrument primarily concerned with conservation and more recently a rare breed of international instrument that also engages with animal welfare issues (Harrop, 2003a). However, the text of its founding instrument remains anachronistically concerned with the industrial activities of commercial whaling. Prior to the 1982 whaling moratorium this commercial activity, in which the British played a key part (see Simmonds' perspective in his UK chapter in this volume), maintained a slaughter that may seem nothing less than relentlessly brutal from our current perspective. A British observer, who worked in the industry as it declined in the 1950s and writing from the perspective of the whaler, recorded his compassion and sense of wonder for the whales that died in great numbers. Nevertheless he also qualified these observations by expressing a lack of regret for the killing and some of the extremes that are certainly regarded as anathema today (such as the killing of pregnant females and calves). At times, however, despite his pro-whaling perspective, he transmits to the reader a vision of the darker side of operations within industrial whaling as he describes a view of these activities as 'a vision of mediaeval hell' (Ash, 1964).

The same author also mentions subtle cultural differences within whaling nations. Thus the market for whale products differed substantially between Europe and Japan. He notes that whereas Europeans had an innate prejudice against eating whale meat, in Japan it was regarded as a delicacy. He possessed an uncharacteristic taste for whale meat himself and regretted that 'in England it [was] used as the main component of many canned foods for dogs and cats' (Ash, 1964). The ignominious fate of many great whales condemned to end as pet food graphically demonstrates the difference between subsistence whaling and commercially driven whaling. Whereas subsistence whaling was characterized by respect, the raw commercial volition proceeded relentlessly so long as some market existed, however reprehensible. Ironically, the very owners of the pets fattened by whale meat may well have become financial supporters of key welfare NGOs that passionately fight to end all whaling today. The fact that 'animal lovers' were feeding their pets from what is now generally regarded in the UK as a cruel destruction spree, may derive from the historic differential between the welfare treatment of wild and domestic animals in the UK.

Animal welfare NGOs have had a long record of popularity in the UK at least dating back to Queen Victoria's patronage of the RSPCA. However, despite these long-standing animal welfare credentials, the UK was still at the vanguard of industrial whaling 50 years ago. Whereas, welfare laws to protect domestic animals have been enacted for over 100 years in the UK, wild mammals received no direct statutory protection from cruelty until the 1990s. This had a direct impact on whale protection and perhaps also influenced public attitudes to the plight of the whale. In the legal case of Steel and Rogers ((1912) 106LT 79) a whale was washed up on a beach and onlookers took great pleasure in cutting large chunks out of the living whale. In a private prosecution this abhorrent practice was found to be quite legal and indeed, had not other legislation come into force to protect whales because of their unfavourable conservation status, this form of cruelty would have still been legal less than 20 years ago.

Nonetheless, the position is different today and wild mammals are protected from cruelty in a number of respects (Harrop, 1997). Moreover, the larger welfare NGOs are active in the field of wild animal welfare and a number of other NGOs devote themselves entirely to this subject. In terms of lobbying, welfare and conservation lobbyists do not always agree and to an extent they derive their origins from separate communities. Conservationists typically concern themselves with the perpetuation of species and biodiversity, and not the individual animal. By contrast, the priority of the animal welfare campaigner is to protect individual animals from suffering irrespective of their conservation status. From this perspective, whatever the population status of the species, the suffering of a hunted minke whale is no less important than the suffering of a blue whale. When conservation planners require the eradication of an alien species that is threatening endemic species, the conservation planners may find themselves at odds with proponents of animal welfare. (As was the case, for example, where it was proposed to cull introduced hedgehogs to save nesting birds in northern Scottish islands). However, conservation and welfare NGOs seem to find enough common ground to join forces on some campaigns and are often indistinguishable when working together to argue against the resumption of commercial whaling or the international trade in whale products (Harrop, 1997). This may be because, first, many of the arguments concerning the conservation status of some hunted whales are scientifically controversial and thus heavily debated by opposing conservation scientists and, second, there may be no way to effectively secure a clean kill of a whale, necessarily calling into play the welfare arguments to assist the embattled anti-whaling conservationists (Brakes et al, 2006).

The joint approach has ensured that there is a strong European, Australasian, Latin American and US lobby, among others, which, prior to the recently increased membership of the IWC, secured a powerful voice in the IWC meetings. The main force of the lobby is aimed at commercial whaling and does not often stray into the realm of aboriginal subsistence exemptions because, inter alia, conservation NGOs have policies that are generally supportive of the traditional practices of indigenous peoples (Harrop, 2003b).

In the past this strong voice within the IWC meetings, resulting from the combined force of two lobbying communities, may have also had the effect of polarizing the opposition and thus galvanizing the pro-whaling movement in a number of ways. The Japanese delegates to the IWC have expressed considerable frustration in the welfare context and they have been particularly concerned that, bearing in mind the time taken in meetings to discuss the humaneness of their high-tech methods of whaling, there should also be a proportionate discussion of the often haphazard killing methods used by some indigenous peoples in their subsistence exemptions (Harrop, 2003a). This reaction to the anti-whaling lobby may, however, go deeper in the context of polar whaling. For whaling nations in the northern region, the strength of the global protectionist lobby has resulted in a certain amount of pro-whaling solidarity and even a reactive reinvention of tradition. In the latter respect, Brydon (2006) argues that there has been a political demonization of anti-whaling in Iceland and a concomitant nationalist reaction that has reinvented previously dubious tradition.

One of the antagonists to return fire is Finn Lynge, who for some time was Greenland's member of the European Parliament and is particularly vociferous in his reply to the animal welfare lobby, which has had substantial influence in the European Parliament. Lynge argues that historically the hunter–gatherer was marginalized and maligned by the growing strength of the agriculturalists who ultimately coalesced into powerful city-states. To emphasize his point, he deploys mythical material and draws upon the distinction between Jacob and Esau in the biblical account in Genesis whereby the hunter–gatherer Esau was tricked out of his birthright by Jacob, his younger agriculturalist brother. Lynge argues that the welfare lobby and the whale protectionists come from the same trickster's lineage and that this ancient archetypal event is merely carrying on in a contemporary context. He builds the case further not only by linking contemporary animal welfare and animal rights movements to the 'Jacob' camp, but also by arguing that concepts of financial accumulation, alien to hunter–gatherer communities, have also been used through the years to isolate them and destroy their traditional way of life. Lynge's words are expressed almost as a call to fight back (Lynge, 1992).

Kalland (2009) is another commentator who appears to be fighting back in response to the anti-whaling movement, in this case to protect Norwegian whaling. He argues that the so-called 'protectionists' (a term that might include any lobbyist, whether from the animal welfare camp or not) have created, with the help of Hollywood and other contemporary repositories of culture, a totemic concept – a *super whale* – that, for the non-expert who has access only to this contemporary myth, becomes more real than the actuality of the biological whale. Thus, if we accept this logic, we have a new myth within this archetype that is imbued with a plethora of magical capacities whereby the whale feels as we feel, is at least as intelligent as a human or is a god that can liaise with extraterrestrials on behalf of humanity. In her chapter in this book, Martinsen argues that claims of this nature are inventions of the Norwegian pro-whaling campaign developed to maintain waning support for Norway's commercial whaling activities. Moreover, whether or not blockbuster images of whales endowed with remarkable powers have or have not persuaded some to support anti-whaling campaigns, the improbable implication of the super whale concept is that the many respected scientists around the world who argue for a full cessation of whaling activities do so by uncharacteristically relinquishing all scientific rigour.

SYNTHESIS

This chapter has examined the perspectives of a number of the epistemic communities who focus their concern on whales and whaling. The anti-whaling lobby powerfully amalgamates conservation arguments linked with a plea for compassion deriving from various levels and qualities of science in addition to popular belief. It is argued that the lobby itself has created its own myth in the nature of a 'super whale' that feeds support from popular imagination.

The pro-whaling lobby is made up of diverse proponents. Although not extensively dealt with herein, the Japanese seek to assert a right to harvest the sea to satisfy their need for food in a highly populated but mountainous country with sparse agricultural space. Thus Japanese whaling, which can never meet these needs, may be a symbolic activity. Norway, Iceland and the Faroe Islands, in a wave of perceived nationalism, may have generated a nationalist response to the anti-whaling lobby thereby reinventing a tradition of whaling or promoting an otherwise waning activity. Greenland operates in a strange limbo of subsistence and commerce, and at the further extreme there are true traditions, still maintained although occasionally interrupted, which derive from hunting practices that date back to antiquity and were designed to keep small polar communities alive in harsh conditions.

The longest-standing polar traditions appear to operate within a form of animist perspective comparable to worldviews embodied in the myth of Sedna. That worldview provides us with an alternative foundation for respect and an ethic of sustainability. It also assists us to appreciate why some peoples still wish to hunt whales as part of the persistence of their culture. But can this cultural approach, which is far removed from the so-called rational epistemological perspective, be of use in discerning a way forward in the wider whaling debate? There have certainly been instances when conservation strategies have demonstrably benefited from the employment of traditional 'non-rational' strategies (Feit, 1998), just as there are other examples of traditional practices carried out by communities that would not assist us (Meggers, 1996). However, we now face supervening factors that are so important that the debate and the politics that surround it may seem petty in comparison.

Whether or not we appreciate the problem from within the mythic cosmologies of the aboriginal peoples, the omens are discernible for all to see since they are written in the now overwhelmingly clear predictions about the fate of life on Earth. We need to read the signs carefully and honestly and then make our choices in a world that has been heavily depleted of resources and that is facing a future beleaguered by unpredictable and dramatic climate change perturbations. As the oceans transform into marine deserts, it may now be that there is no longer any justifiable case for hunting a whale. First we need to be clear, in the context of precaution and a future that is becoming increasingly precarious, that we can accurately predict a sustainable removal from a species and that our hunting will not have other repercussions throughout the ocean ecosystems. Second, if this can be achieved we should ask ourselves why we need to hunt them. Can we honestly declare that this is necessary for food security and resilience, either at the national and thus commercial level or for local subsistence? Can we even say, if our hunting is based on tradition, that in the face of this uncertain future our tradition still calls for the death of the whale whether as a sacred act or otherwise? All in all the impeccable choice may well be to leave the whales free to live out their last days.

Note

1 See www.iwcoffice.org/conservation/aboriginal.htm

References

Ash, C. (1964) *Whaler's Eye*, George Allen & Unwin Ltd, London

Blackburn, T. (2002) 'Behaviour and ecology study of river dolphin species in the Peruvian Amazon', practical research project DI512, Durrell Institute of Conservation and Ecology, University of Kent, Canterbury, Kent

Brakes, P., Butterworth, A., Simmmonds, M. and Lymbery, P. (eds) (2006) *Troubled Waters: A Review of the Welfare Implications of Modern Whaling Activities*, World Society for the Protection of Animals, London

Brydon, A. (2006) 'The predicament of nature: Keiko the whale and the cultural politics of whaling', *Iceland Anthropological Quarterly*, vol 79, no 2, pp225–260

Coltrain, J. B., Hayes, M. G. and O'Rourke, D. H. (2004) 'Sealing, whaling and caribou: The skeletal isotope chemistry of Eastern Arctic foragers', *Journal of Archaeological Science*, vol 31, no 1, pp39–57

De Castro, E. V. (1998) 'Cosmological deixis and Amerindian perspectivism', *Journal of the Royal Anthropological Institute*, vol 4, no 3, pp469–488

Evans, P. G. H. (1987) *The Natural History of Whales and Dolphins*, Christopher Helm, London

Fauconnet, M. (1959) 'Mythology of the two Americas', in *New Larouse Encyclopaedia of Mythology*, Hamlyn, London

Feit, H. A. (1998) 'Self-management and government management of wildlife: Prospects for coordination in James Bay and Canada', in R. J. Hoage and K. Moran (eds) *CULTURE The Missing Element in Conservation and Development*, National Zoological Park Smithsonian Institution, Kendall/Hunt, Iowa, pp95–111

Harrop, S. R. (1997) 'The dynamics of wild animal welfare law', *Journal of Environmental Law*, vol 9, no 2, pp287–302

Harrop, S. R. (1998) 'The agreements on international humane trapping standards: Background, critique and the texts', *Journal of International Wildlife Law & Policy*, vol 1, no 3, pp387–394

Harrop, S. R. (2003a) 'From cartel to conservation and on to compassion: Animal welfare and the International Whaling Commission', *Journal of International Wildlife Law & Policy*, vol 6, pp79–104

Harrop, S. R. (2003b) 'Human diversity and the diversity of life: International regulation of the role of indigenous and rural human communities in conservation', *The Malayan Law Journal*, vol 4, pp xxxviii–lxxx

Ingold, T. (2000) *The Perception of the Environment: Essays on Livelihood, Dwelling and Skill*, Routledge, London

Jonathan, H., Geisler, J. H. and When, M. D. (2003) 'Morphological support for a close relationship between hippos and whales', *Journal of Vertebrate Paleontology*, vol 23, no 4, pp991–996

Kalland, A. (2009) *Unveiling the Whale: Discourses on Whales and Whaling*, Berghan, Oxford, UK

Kane, S. (1998) *Wisdom of the Mythtellers*, Broadview Press, Ontario

Lynge, F. (1992) *Arctic Wars: Animal Rights Endangered Peoples*, Dartmouth College, University Press of New England, Hanover, New Hampshire

Meggers, B. J. (1996) *Men and Culture in a Counterfeit Paradise*, Smithsonian Institution Press, Washington, DC, and London

Miller, R. J. (2002) 'Exercising cultural self-determination: The Makah Indian tribe goes whaling', *American Indian Law Review*, vol 25, no 2, pp165–273

Morse, B. W. (1980) 'Indian and Inuit family law and the Canadian legal system', *American Indian Law Review*, vol 8, pp199–257

Ponting, C. (1991) *A Green History of the World*, Penguin, London

Posey, D. and Balee, W. (1989) *Resource Management in Amazonia: Indigenous and Folk Strategies*, Advances in Economic Botany Volume 7, New York Botanical Garden Press, New York, NY

Reeves, R. R. (2002) 'Origins and character of aboriginal subsistence whaling: A global review', *Mammal Review*, vol 32, no 2, pp71–106

Stambuk, P. M. (1986) *Lakutaia le kipa, Rosa Yagan, El ultimo eslabon*, Andrés Bello, Santiago de Chile

Ursing, B. M. and Arnason U. (1998) 'Analyses of mitochondrial genomes strongly support a hippopotamus-whale clade', *Proc. R. Soc. Lond. B*, vol 265, pp2251–2255

3

Whales of the Pacific

Viliamu Iese and Cara Miller

> *May it be. Let the canoe of you and me turn into a whale … let it leap on and on over the waves, let it go, let it pass out to my land.* — Melanesian seafarer call for help to Qat, their ancestor-spirit hero (Montgomery, 2004)

The first time I saw dolphins, I was swimming with my blind great grandmother on the beach at Lepa in Samoa. I was five years old. I clearly remembered how their dorsal fins emerged above the waves about 20 metres from us. My blind great grandmother with a perfect sense of hearing told me that they were *nunua*, which means dolphin in Samoan. She started calling loudly to them, 'Jump dolphins, jump!'. To my amazement, they did jump, spin around and land with a splash. They were responding to the old lady's call by showing off their acrobatic skills. Although she couldn't see them, my great grandmother gave a cheer of joy. I then asked her whether we speak the same language as dolphins. She explained to me that they are our ancestors who had passed on to the next world. The joy of seeing the dolphins and my great grandmother's cheer stays in my memory all these years. However, as a five-year-old Christian boy, I never understood my great grandmother's explanation until 25 years later when, as an Environment Officer in Tuvalu, I found myself in charge of coordinating a project on cetaceans and other large marine species. As part of this project I was required to interview locals about their cultural connection to whales and dolphins. When I met islanders from Niutao Island in Tuvalu they told me that they too considered whales and dolphins as their ancestors and protectors on the high seas.

During my interviews I was told that the forefathers of the people of Niutao had the ability to call whales and dolphins, and were reported to have used them as a means of transportation to neighbouring islets of Tuvalu. One of the clans also told me about a much longer journey – claiming that their Maori ancestor Tinilau had arrived in Niutao after travelling on the back of a whale all the way from New Zealand. During a later trip to New Zealand I noted with much curiosity that the (now famed) whale rider from Aotearoa had the same name. In addition, Niutao islanders shared with me that when they found a stranded dolphin or whale on the beach they considered this a message from their ancestors that one of the elders should join them. Indeed just a few days after a stranding, a prominent elder passes on to the next world. The Tuvaluans also related real stories of how pods of dolphins and whales had at times protected relatives lost at sea. In one instance they told me about a group aboard a wayward fishing vessel that was accompanied by dolphins until they eventually made landfall in neighbouring Wallis and Futuna. In Fiji I've also come across similar stories. In the Tailevu district on the main island of Viti Levu there is a pod of dolphins that come close to shore to rest during the day. When I was visiting this area I sat around the kava bowl one evening with some villagers from nearby Silana and they told me that the local word for dolphin is *babale* and that of all the reefs in the area, they only choose one of them for their daytime resting: Moon Reef, or Makalati as it is known locally. People of this area believe that these dolphins are their traditional guardians, and that they protect this reef. Furthermore, legend says that when someone passes away their spirit is launched from a point on the main island from where they dive into the water and become dolphins – spending the rest of their days at Makalati.

Some Pacific islands also have a history of utilizing whales and dolphins in some way. Fijians highly regard the sperm whale tooth (*tabua*) as a cultural item of chiefly exchange for reconciliation, dowry for marriage and strengthening alliances. Derrick (1950) explains the worth of the sperm whale tooth as 'the price of life and death, the indispensable adjunct to proposals (whether of marriage, alliance, or intrigue), requests and apologies, appeal to the gods, and sympathy with the bereaved'. Nowadays *tabua* is still exchanged during weddings, funerals, birthdays and negotiations. However, whaling activity in Fiji is banned and so only teeth from stranded sperm whales add to the large number of *tabua* in regular and active circulation among Fijians. A drive hunt for small cetaceans has been practised in Malaita of the Solomon Islands group for many years (Reeves et al, 1999). A majority of dolphins taken in this hunt (termed 'porpoise hunts') are apparently long-beaked oceanic species, including spinner, pantropical spotted, striped, common and rough-toothed dolphins (Dawbin, 1972) (see Figure 3.1). The primary objective of these hunts is to obtain teeth that are used as local currency and for collars or headbands and necklaces, as well as to obtain meat. Risso's dolphin has also been taken on occasion, but due to their lower tooth count they were apparently of lesser value to the Malaitans (Dawbin, 1966). Tonga was the most recent Pacific nation to ban whaling (in 1978) (Anonymous, 1981) but has now transformed this interest in whales into a major tourism industry with associated economic growth. Several other countries have also developed strong whale-watching industries based on migrating humpback whales, including New Caledonia, French Polynesia and the Cook Islands (O'Connor, 2008). However,

Figure 3.1 *Spinner dolphins, Fiji*

Source: Cribb/WDCS

the small nation of Guam is by far the most profitable cetacean tourism location in the Pacific, bringing in more than US$16 million per year (approximately 75 per cent of the total revenue for this industry in the region).

Some Pacific whales are also 'used' when they are away from our waters and feeding in the Southern Ocean during the Austral summer (Figure 3.2). 'Scientific' whaling by Japanese whaling boats in Antarctica take minke, fin and possibly soon humpback whales. These whales are the same ones that migrate into Pacific waters for breeding and calving. There are of course lots of politics at play here, but it seems to me that the animals themselves have been forgotten. What about those whales that are killed for the name of science? Do they have an opinion? What about our ancestors? Do they have a say? This reminds me of the short conversation I had with a Japanese fisheries officer who was in Tuvalu to assist the production and processing of fish for sale. I asked him why his country continues to hunt whales. He closed his eyes and explained their culture of eating whale meat. I replied to my Japanese friend by telling him the stories from my great grandmother and the Niutao islanders. I ended our conversation by telling him that 'Yes, it is his traditional right to eat whales and dolphin meat, but what about my traditional right to conserve my ancestors? What about my ancestors' right to roam the Pacific Ocean free of fear of being caught in a net or being harpooned, killed and eaten?'. I also asked the basic question to fishermen and Niutao islanders in Tuvalu:

Figure 3.2 *A humpback whale comes into view as Uto ni Yalo, a traditional Pacific* waka, *is being sailed by the Fiji Islands Voyaging Society*

Source: Sharma/Fiji Fisheries

whether they want to kill whales for science and food. They were all against it. It is interesting to note though that sometimes these same fishermen find themselves in competition for the same resources. Several times I was told about the regular incidence where spinner dolphins jump and take the flying fish just before it enters the fishermen's scoop net. However, most of these fishermen simply accepted that this was part of fishing. They were not disappointed or angry at the dolphins but would instead chose to turn off the light and speed away as far as possible before resuming their activities. However, I did hear complaints from some commercial fishermen and fisheries officers, although I always saw it from another angle. The big questions that came to my mind were: 'Who is actually doing the depredation?

Is it the cetacean, other large marine species or is it man?'. I would think about the whales and dolphins that are naturally living in the ocean, relying on their environment for survival, and how they must feel when man comes along and takes away their food. We feel offended if someone steals our resources from our backyard, so I'm sure they do too. I think that at times we are only thinking of ourselves and forgetting the welfare of the other living things around us. The local fishermen in Tuvalu also explained to me that they are not scared of whales when they see them. Instead they prefer to move closer to them because they know the whales and tuna always swim together. This knowledge has been passed down from their forefathers for many years. Traditional knowledge related to cetaceans in the small island of Nuitao is obviously strong.

My experiences have shown me the strong cultural connection to whales and dolphins by Pacific Islanders who inherently live and rely on the ocean and the marine species within it. Thinking about them all takes me back to my blind great grandmother's soft voice when she called the dolphins to jump and then cheered as she heard their splashes. It makes me picture how those whale riders roamed the Pacific Ocean as they travelled from island to island; it makes me imagine the whales and dolphins of the high seas protecting fishermen when they had no hope of surviving; and it makes me pause as I wonder about the ancestral spirits in Silana taking their final jump to Makalati where they would swim and spin around joyously forever. Science and reasoning may contradict these stories but as a Pacific Islander, they are part of us. We are defined by myths and legends for they are part of our culture and who we are. Our Pacific culture is based on respect, caring and sharing. We always show great respect to our elders, special care for the young and disadvantaged, and willingly share with our neighbours. This same sentiment is sustained for our deep respect for our ancestors who have gone before us. Perhaps the Pacific Islander connection to cetaceans is outlined best in the Christian principle of 'Love thy neighbour as thyself' or more broadly to do unto others as what you would want others to do to you.

References

Anonymous (1981) 'Kingdom of Tonga report of the preliminary survey of humpback whales in Tongan waters July–October 1979', in *Report of the International Whaling Commission*, vol 31, Impington, UK, pp204–208

Dawbin, W. H. (1966) 'Porpoise and porpoise hunting in Malaita', *Australian Natural History*, vol 15, no 7, pp207–211

Dawbin, W. H. (1972) 'Dolphins and whales', in P. Ryan (ed) *Encyclopedia of Papua and New Guinea*, Melbourne University Press, in association with the University of Papua New Guinea, Melbourne, pp270–276

Derrick, R. A. (1950) *A History of Fiji*, vol 1, Government Press, Suva, Fiji

Montgomery, C. (2004) *The Shark God: Encounters with Ghosts and Ancestors in the South Pacific*, University of Chicago Press, Chicago, Illinois

O'Connor, S. (2008) *Pacific Islands Whale Watch Tourism: A Region Wide Review of Activity*, IFAW, Sydney

Reeves, R. R., Leatherwood, S., Stone, G. S. and Eldredge, L. G. (1999) 'Marine mammals in the area served by the South Pacific Regional Environment Programme (SPREP)', South Pacific Regional Environment Programme (SPREP), Apia, Samoa

4

The Journey Towards Whale Conservation in Latin America

Miguel Iñíguez

Cetaceans, in particular orcas (*Orcinus orca*), are part of the cultural wealth of the native people of Latin America, who refer to them in many legends and beliefs. The people of Nazca, Peru painted orcas in their natural environment as a sacred symbol of power, courage and fertility. They even built temples in their honour, where they were worshipped until the 1st century BC (Hoyt, 1990).

There is some evidence that in historical times, natives from Latin American used the meat of stranded whales as a source of food and sometimes they organized banquets when they found a stranded animal. Three examples are provided below that illustrate this theory.

In southern Argentina, the Selkman, canoe Indians of Tierra del Fuego, told the story of a hunter and his three dogs. Despite the dogs' great hunting success, the hunter complained about their catches and was very dissatisfied. Kwonyipe, the powerful man of the Selkman, tired of hearing his complaints; transformed the hunter into Kshamenk (orca in the Selkman language) and expelled him to the sea, turning him into a hunter of whales. Since then, whenever the Selkman find a whale stranded on the coast, they worship it, and are grateful for the food that Kshamenk provides (Gusinde, 1977).

Along the coast of Santa Catarina, Brazil, archaeologists have found prehistoric artefacts made of whalebone (Palazzo et al, 2007). However, there is no evidence that whales were hunted by the Brazilian natives. It seems instead that stranded whale meat was utilized by them and the bones are likely to have been from these stranded whales.

In Baja California, Mexico, some indigenous native groups may occasionally have eaten meat of gray whales (*Eschrichtius robustus*) that had died and been washed ashore (Urbán et al, 2003). However, the use of cetaceans by the indigenous people changes drastically after the Europeans arrived in the Americas, bringing the whaling industry with them. Three species were the main target of whaling operations in Latin American waters: southern right whales (*Eubalaena australis*), gray whales and sperm whales (*Physeter macrocephalus*). Whale hunting first started in 1603 when two Basque whaling ships from Bayonne, France, settled on the coast of Brazil. For a century they hunted southern right whales along the coast of Bahia using the same techniques they had developed in Europe, where mothers with calves were the main targets. In contrast to European practices, whale meat was used to feed the slaves who worked the shore stations and sugar mills.

The first large-scale whaling activities in Latin American waters began after the US War of Independence. After 1775, whaling activities under the British, American, French, Spanish and Portuguese flags spread quickly to the southern hemisphere. This change in whaling operations required longer, better-planned voyages and improvement in the design and seaworthiness of whaling ships. From 1765 to 1812, the whaling fleet operated mainly on the Brazilian Bank, an extended area between Abrolhos Bank in Brazil and Trinidad Island. Later they established themselves from Rio de Janeiro, to the east of Rio de la Plata and even as far south as Patagonia. Between 1772 and 1792 it was reported that 1700 sperm whales were killed and between 1772 and 1812, 30,000 southern right whales were caught in these waters. Most of the whalers came from the US and UK (Richards, 1994). Despite being protected by an international agreement since 1935, southern right whales were hunted until 1973 off the coast of Santa Catarina state (27°S). Brazil expanded its whaling operation to other species of whales and the last whaling station, located in Costinha, Estado de Paraíba and operated by the Companhia de Pesca Norte do Brasil (COPESBRA), a subsidiary of the Japanese company Nippon Reizo Kabashiki Kaisha, ceased its hunting operations in 1985 and was finally closed in 1987, once the IWC moratorium on commercial whaling came into force (Palazzo and Carter, 1983).

In Uruguay, a whaling factory was established by the Real Compañía Marítima in Punta del Este in 1789. They hunted southern right whales until 1806, when the factory was burned down during the British Invasions. In 1823 a British whaler was allowed to hunt whales in Bahía Maldonado; however, the numbers of whales he killed remains unknown (Costa et al, 2005).

On the Pacific coast, hunting activities were recorded for Chile and Peru and also ended when the commercial whaling moratorium was implemented. As seen in other parts of the region and the rest of the world, whale populations were severely depleted by commercial hunting.

Even though southern right whales were the main target species of the 18th and 19th centuries, another species chosen by the whalers was the sperm whale (see Figure 4.1) , for which there were important whaling grounds in the southeast Pacific. This wonderful creature provided the oil that lit the streets and lubricated the machines of the industrial age. The Galapagos Islands of Ecuador were quite productive in this trade from 1790 to 1810 and sperm whale products were still

Figure 4.1 *Sperm whale in Grytviken factory, South Georgia*

Source: Courtesy of Walter Roil, www.instantespatagonicos.com

in use up to perhaps 1840, but were progressively abandoned thereafter, when the whales became scarcer (Shuster, 1983). The experience of the whalers in those days is well described in *Moby Dick* by Herman Melville, who was inspired to write his novel by his own experience of the whaling industry and the history of the whale ship *Essex*, an old whaling vessel from Nantucket, sunk by an enraged sperm whale 1500 nautical miles west of the Galapagos (Philbrick, 2000).

The gray whale hunt was one of the best documented (Urbán et al, 2003). The first documented catches of gray whales in Baja California, Mexico, are from the winter of 1845–1846 in Bahía Magdalena, when two US whaling vessels caught approximately 32 whales. During a period of 29 years (1845–1874), 3500 gray whales were hunted in Mexicans waters. In the 20th century, Norwegian whalers took at least 200 gray whales in this area from 1913 to 1929. In 1965, the Mexican Secretariat of Fisheries proposed the opening of a whaling station to harvest gray whales off the Baja California coast. However, as a result of international pressure, the government withdrew the proposal. The gray whale population was severely depleted following the whaling operations. In 1971, the Mexican government established the first marine protected area (MPA) in Laguna Ojo de Liebre to protect the prime gray whale mating and calving lagoon. In 1988, this was

Figure 4.2 *Sei whales caught in southwest Atlantic in the 1930s*

Source: Courtesy of Walter Roil, www.instantespatagonicos.com

expanded to include Laguna San Ignácio and part of Laguna Guerrero Negro, all of which together became El Vizcaino Biosphere Reserve.

Since the creation of the first MPA of Laguna Ojo de Liebre, Mexico in 1971, 71 MPAs for whales, dolphins and porpoises have been established in the region (Hoyt, 2005). Whale watching in Latin America is managed in more MPAs than in many other areas of the world (Hoyt and Iñíguez, 2008).

On 16 November 1904, the whale catcher *Fortuna* (sailing under the flag of Argentina even though all the crew were Norwegians), and two supply ships anchored in the bay of Gritviken, South Georgia, commenced modern commercial whaling in the Antarctic region (Walløe, 2005). In the six decades that followed, around 2 million whales were hunted for commercial purposes in the southern hemisphere (Clapham and Baker, 2002) (see Figure 4.2). This unparalleled slaughter pushed the whaling nations to regulate the activity. Many species of whales were depleted almost to extinction. In 1929, the Argentine international lawyer José Leon Suarez proposed to the League of Nations that a sanctuary for whales should be established in the Antarctic to protect whale populations, in particular blue whales (*Balaenoptera musculus*), fin whales (*Balaenoptera physalus*) and humpback whales (*Megaptera novaeangliae*), which were extensively hunted in the southern hemisphere, mainly around the South Georgia area where nine land whaling stations operated. In 1931, the League of Nations created the Geneva Convention for the Regulation of Whaling, the parties to which adopted worldwide protection for right and gray whales in the mid-1930s. However, regulation was unsuccessful, and in 1946 the International Convention for the Regulation of Whaling was agreed by 15 whaling nations, including Argentina, Brazil, Chile and Peru. It created the International Whaling Commission (IWC) in 1948.

For decades Latin American countries allowed the use of their flags of convenience for whaling fleets. Fortunately for the future of whale stocks, their position has more recently shifted to consolidate a conservation stance with respect to whales, following the implementation of the commercial whaling moratorium. Nowadays, most countries in the region work hard to protect whales stocks, understanding that a living whale has more value than a hunted whale.

64 species of whales, dolphins and porpoises, representing 75 per cent of the 86 known cetacean species are found in Latin American waters (Hoyt and Iñíguez, 2008). This cetacean biodiversity is used as a source for whale-watching activities and generates important socioeconomic benefits for the coastal communities of the region. Mexico was the first destination for whale watching in Latin America. The first trip was in 1970 with ships that came from San Diego (USA) to Baja California. In the 2006/2007 season, 885,679 people went whale watching in Latin America, spending $278.1 million in total. From 1998 to 2006, the activity increased at an average rate of 11.3 per cent per year with most of the whale watching taking place in Argentina, Brazil, Mexico, Costa Rica and Ecuador (see Figures 4.3 and 4.4).

The commitment of the region to the conservation of whales was finally consolidated in 2005, when Latin American IWC representatives established a group to coordinate efforts and develop common positions within the IWC and other international organizations and forums. It was designated under the name of 'Grupo Buenos Aires' (GBA). From its beginning, the GBA established a common position among the Latin American countries that proposed the non-lethal use of cetaceans through research and whale-watching activities. The GBA recognized the importance of the whale-watching industry for the local communities along Latin American coasts. Reinforcing its conservation strategy, in December 2007, the GBA established the 'Latin America Cooperation Strategy for the Conservation of Cetaceans', which promotes the development of responsible whale watching, supports the establishment or implementation of areas for the protection and

Figure 4.3 *Watching southern right whales in Peninsula Valdes, Argentina*

Source: Miguel Iñíguez

conservation of cetaceans and strengthens regional capabilities in the management and conservation of these species. Nowadays, the GBA plays an important role in discussions within the IWC and in the conservation of Latin American whale stocks.

Latin American people recognize the importance of protecting whales, dolphins and porpoises. Old and new threats risk the survival of cetaceans in this region. Many of these species are endemic to Latin American waters:

- boto, *Inia geoffrensis*;
- tucuxi, *Sotalia fluviatilis*;
- Guiana dolphin, *Sotalia guianensis*;
- Peale's dolphin, *Lagenorhynchus australis*;
- Commerson's dolphin, *Cephalorhynchus commersonii*;
- Chilean dolphin, *Cephalorhynchus eutropia*;
- Franciscana (*Pontoporia blainvillei*) (one of the most critically endangered of all cetacean species);
- vaquita, *Phocoena sinus*.

Figure 4.4 *Southern right whale watching in Patagonia*

Source: Miguel Iñíguez

The challenge for Latin American governments now is to protect these species from threats and to take the measures necessary to secure the future of cetaceans in our waters.

REFERENCES

Clapham, P. J. and C. S. Baker. (2002) 'Whaling, modern', in W. F. Perrin, B. Würsig and J. G. M. Thewissen (eds) *Encyclopedia of Marine Mammals*, Academic Press, San Diego, pp1328–1332

Costa, P., Praderi, R., Piedra, M. and Franco-Fraguas, P. (2005) 'Sightings of southern right whales, *Eubalaena australis*, off Uruguay', *LAJAM*, vol 4, no 2, pp157–161

Gusinde, M. (1977) *Folk Literature of the Yamana Indians: Martin Gusinde's collection of Yamana narratives*, Johanes Wilbert (ed), University of California Press, Berkeley, CA

Hoyt, E. (1990) *Orca: The Whale Called Killer*, Camden House Publishing, Buffalo

Hoyt, E. (2005) *Marine Protected Areas for Whales, Dolphins and Porpoises: A World Handbook for Cetacean Habitat Conservation*, Earthscan, London

Hoyt, E. and Iñíguez, M. (2008) *The State of Whale Watching in Latin America*, WDCS, Chippenham, UK, IFAW, East Falmouth, EEUU, Global Ocean, London

Palazzo, J. T. Jr and Carter, L. A. (1983). 'A caça de baleias no Brasil', Associação Gaucha de Proteção ao Ambiente Natural, Porto Alegre, Brazil

Palazzo, J. T. Jr, Groch, K. R. and Silveira, H. A. (2007) 'Projeto Baleia Franca, 25 anos de Pesquisa e Conservação, 1982–2007', Coalizão Internacional da Vida Silvestre IWC/ Brasil, Imbituba, Brazil

Philbrick, N. (2000) *In the Heart of the Sea*, HarperCollins, London

Richards, R. (1994) *Into the South Seas: The Southern Whale Fishery Comes of Age on the Brazil Banks 1765 to 1812*, Projeto Baleia Franca – IWC/Brasil, Paremata, Brazil

Shuster, G. W. (1983) 'The Galapagos Islands: A preliminary study of the effects of sperm whaling on a specific whaling ground', *Rep. Int. Whal. Commn.*, Special issue, vol 5, pp81–82

Urbán R. J., Rojas-Bracho, L., Pérez-Cortés, H., Gómez-Gallardo, A., Swartz, S. L., Ludwig, S. and Brownell, R. L. Jr (2003) 'A review of gray whales (*Eschrichtius robustus*) on their wintering grounds in Mexican waters', *J. Cetacean Res. Manage*, vol 5, no 3, pp281–295

Walløe, L. (2005) 'The early development of Norwegian Antarctic whaling', in Institute of Cetacean Research (eds) *Learning from the Antarctic Whaling*, Report and Proceedings, International Symposium Commemorating Centennial of the Antarctic Whaling, Institute of Cetacean Research, Tokyo, pp10–30,

5

Whales and the USA

*Naomi A. Rose, Patricia A. Forkan, Kitty Block,
Bernard Unti and E. C. M. Parsons*

The US is a whaling nation. Despite its present support for cetacean conservation and opposition to cetacean exploitation, historically the country had a major whaling industry. More recently, the continuing claims of aboriginal whalers have often been in conflict with the US leading the effort to 'Save the Whales'. Other cetacean exploitation was born in the US too; the captive dolphin show was an American invention and millions of dolphins died in the American tuna fishery in the eastern tropical Pacific before the 1980s. Although the US was once the driving force behind the global shift from killing the great whales to protecting them, in recent years it has ceded this leadership to Australia, New Zealand and the UK, among others, an evolution with a complicated and uneven trajectory.

From Whale Exploitation to Whale Protection

The American whaling industry arose from a frontier nation's utilitarian view of wildlife, in an era before plastics and petroleum made whale products obsolete. In the second half of the 20th century, Americans' appreciation of the aesthetic value of whales grew, leading to the creation of NGOs devoted to their protection, the passage of marine mammal legislation, greater investment in marine mammal research, an increase in whale watching and growing concern for cetaceans expressed in popular culture (Lavigne et al, 1999).

From Herman Melville's *Moby Dick* in 1851 to Earth Day in 1970, US policy transitioned from consumptive use of whales to placing greater value on protecting them (Forestell, 2009). Earth Day, in fact, was a 'game-changing' event. Long-established conservation organizations and newly minted environmental groups conducted rallies and marches all over the US. The media's coverage of these events strongly influenced political leaders.

Groundbreaking US laws were passed during this period. The National Environmental Policy Act of 1969 (NEPA) instituted new requirements for civil society involvement and environmental impact analyses (42 USC §4321 *et seq.*). The Marine Mammal Protection Act of 1972 (MMPA) (16 USC §1361 *et seq.*) embraced the 'precautionary principle' as no US law had before. It dictated the protection of all marine mammals, regardless of their conservation status, in order to maintain them as 'significant functioning elements in the ecosystem of which they are a part' (16 USC. §1361(2)). With these and the 1973 Endangered Species Act (16 USC §1531 *et seq.*), the US had a formidable legislative arsenal with which to defend cetaceans. In 1970 the National Oceanic and Atmospheric Administration was created with a mandate (among other things) to protect marine mammals.

Starting in the 1950s, hundreds of thousands of dolphins were dying each year in purse seine nets in the tuna fishery in the eastern Pacific Ocean, a fishery dominated by the US. The public outcry at this carnage grew substantially through the 1960s, and this, along with concerns about the deliberate killing of whales, seals and polar bears, led the US government to draft the MMPA to protect marine mammals from harmful human activities. This statute prohibits not only killing or injuring marine mammals, but also harassing them. Although the definition of 'harassment' has been a matter of debate over the decades, it is nevertheless a critical progressive element of the statute.

Making it illegal to disturb or disrupt the lives of these animals implies that their welfare merits concern. In the MMPA's preamble, it is clear that the US Congress was responding not only to scientific advice on the importance of being precautionary when seeking to protect species that are difficult to study, but also to the public's expressed concern for these iconic ocean creatures. It states that 'marine mammals have proven themselves to be resources of great international significance, esthetic [sic] and recreational as well as economic' (16 USC §1361(6)). The inherent value of marine mammals was, for the first time, held equal to their economic value.

In 1971, researchers reported that humpback whales (*Megaptera novaeangliae*) sing complicated songs that change over the course of time (Payne and McVay, 1971). Scientists were now studying cetaceans not to improve harvest yields but to explore the natural history of these little-known species (Forestell, 2009). Kenneth Norris, who was among the early pioneers of this 'new' cetacean science, published *The Porpoise Watcher* in 1974. Along with *The Mind of the Dolphin* (Lilly, 1967) and *Mind in the Waters* (McIntyre, 1974), this engaging natural history book contributed to the change in how Americans perceived cetaceans. Once creatures to kill for profit, whales were now animals to watch and enjoy.

Indeed, whale watching began off the coast of California in the 1950s but became increasingly popular in the 1970s, expanding to the east coast and around the world (Forestell, 2009). Today it is a global, multi-billion dollar industry (Hoyt, 2002). By the 1990s, Americans believed that whales should be protected, not exploited. A 1999 survey found that a large majority of Americans objected to most whaling, with more than 60 per cent objecting to commercial whaling 'under any circumstance' (Kellert, 1999). Indeed, some 70 per cent of Americans objected to killing whales on moral grounds (Kellert, 1999).

Nowhere was the changing attitude of the American public reflected better than in popular culture. In the 1960s, Hollywood produced the movie and then the television series *Flipper*, which followed the exploits of a plucky Florida dolphin and the family he adopted. In the 1970s, the portrayal of dolphins in American film turned darker. The theme of the movie *Day of the Dolphin* expressed concern for the ethics of using intelligent animals in warfare. In the 1990s, *Free Willy* imbued a generation of young American children with an activist zeal sufficient to send its real-life protagonist, the Icelandic orca Keiko, home after almost 15 years in captivity. And in 2009, the American-produced *The Cove* raised international public awareness of the cruelty inherent in some live-capture methods.

Yet the first exhibition of captive cetaceans was in the US, when showman P. T. Barnum put a beluga whale in a sideshow tank. The modern 'dolphin show' was invented by Marine Studios (now Marineland of Florida) near the town of St Augustine in 1938. By the 21st century, the number of American dolphinariums was in decline (Rose et al, 2009). Although American support for exhibiting live dolphins and whales remains relatively high, the public now deems it acceptable only if there is a strong education and conservation component to the exhibit (Kellert, 1999).

THE GOLDEN YEARS IN THE INTERNATIONAL ARENA

In 1972, the US sent a delegation to the United Nations Conference on the Human Environment held in Stockholm, Sweden. This delegation led the effort to adopt a resolution calling for the protection of whales, specifically recommending a ten-year moratorium on commercial whaling. American NGOs used the Stockholm meeting to launch the 'Save the Whales' movement, popularizing whale images via murals on buses and papier-mâché models.

The chair of the US delegation in Stockholm was also sent to the annual meeting of the IWC that year. President Richard Nixon declared that his Administration placed a 'high priority... [on] the protection of whales' (Woolley and Peters, 2009). At the same time, the US Congress passed resolutions supporting a ten-year whaling moratorium.

However, at the 1972 IWC meeting, the proposal for a moratorium was soundly defeated. The next year it was again rejected, although by a smaller margin. According to *Time Magazine*, 'the US pushed hard for a ban on all whaling. The result: the most rancorous conference in the IWC's 27-year history' (Anonymous, 1973). When both Japan and the Soviet Union objected to modest quota reductions adopted at the 1973 meeting, it set off protests among the growing environmental movement.

American NGOs mobilized the 'Save the Whales Coalition'. A week prior to the 1974 IWC meeting, the Coalition held a press conference to declare a boycott of Japanese and Soviet goods. This public uprising was not lost on US officials. In 1973–1974, the IWC quota for all whale species was 45,673 (Nowak, 1999). By 1976 it had been reduced substantially and a better management scheme adopted. The US drove these developments.

The US also wielded the Pelly Amendment to the Fisherman's Protective Act (22 USC §1971 *et seq.*). Under this provision, the US secretaries of commerce and interior can certify a country for undermining international conservation agreements; certification carries the threat of trade sanctions. The commerce secretary certified Japan and the Soviet Union early in 1974. Both countries voted later that year for reduced quotas, which was seen as a direct response to the pressure from Pelly certification. The US also used the Pelly Amendment to bring countries into the IWC who were whaling outside its jurisdiction. For example, the Commerce Department threatened to end South Korea's access to US fisheries if it did not join the IWC, which it eventually did.

The US also played a pivotal role in the fight to protect whales at the negotiations for the Law of the Sea (LOS) Treaty. The US delegation, without any real assistance from other governments, drafted and promoted a marine mammal section in the treaty; the final result of these negotiations was Article 65 (UN, 1982). Only a few sentences, it nevertheless stands as a remarkable achievement because it carves out the possibility of nations totally protecting marine mammals and not requiring their 'use'. Its adoption was a high water mark in US efforts to protect cetaceans.

However, by 1972 there was growing alarm that Alaskan natives were killing too many endangered bowhead whales (*Balaena mysticetus*), leading the IWC to vote to set a zero quota for bowheads in 1977. The US delegation abstained. The US had a clear conflict, asking for higher aboriginal quotas on an endangered species while pushing for lower commercial quotas; this tension was also reflected in the NGO community, which saw the cooperation fostered by the Save the Whales Coalition erode. Then the Alaskan natives filed a lawsuit in October 1977 against the zero quota, an opening salvo in their fight to regain their bowhead quota.

Aboriginal Subsistence Whaling and the US

The lawsuit sought to force the secretary of state to file a formal objection to the zero quota for bowheads. The US delegation had refused to do so (no doubt recognizing it would make the US as worthy as Japan or the Soviet Union of Pelly certification). The government eventually won the case, with the court ruling that the Alaskan natives did not have standing to interfere in a foreign affairs decision by the executive branch of government (Miller, 2002). Despite winning in court (or perhaps because of it), the US government continued to support the Alaskan native whale hunt, going to an interim session of the IWC in December 1977 to ask for special consideration of the bowhead quota. The quota was secured, but only after another contentious meeting.

US lobbying at the IWC eventually led to three expert panels in 1979, which considered the issues related to aboriginal subsistence whaling: wildlife science; nutritional needs of aboriginal peoples; and cultural aspects of subsistence hunting (Gambell, 1991). In 1980, the US submitted a report (which became known as a 'needs statement'), which addressed the historical, cultural and nutritional aspects of the bowhead hunt, with the goal of quantifying the number of whales the Alaskan natives needed (Gambell, 1991). By 1981, an IWC working group had agreed on a definition of aboriginal subsistence whaling that included a requirement to demonstrate nutritional need and influenced all needs statements thereafter. The working group's definitions were guidelines only, but were accepted by all IWC member nations.

At the 1996 IWC meeting, the Makah tribe of Washington State submitted a needs statement for gray whales (*Eschrichtius robustus*) as part of a campaign to revive a traditional whale hunt that had ended in 1926 with this species' commercial extinction. Many NGOs and several IWC member nations opposed the Makah proposal from the outset because it failed to demonstrate a nutritional need (given that the Makah had not eaten whale meat for 70 years). The proposal failed to win support and was withdrawn; the US then came back in 1997 to forge a deal behind closed doors with the Russian Federation.

The subsistence needs of Russia's Chukotkan natives had been recognized years previously, so the two countries bypassed the IWC's, by then established, procedure of approving aboriginal whaling proposals by consensus (a procedure the US had worked hard to establish nearly 20 years earlier), instead signing a bilateral agreement to share the gray whale quota. This deal blurred the whaling categories, weakening the distinction between aboriginal subsistence and commercial whaling.

The whaling nations considered the US machinations on behalf of the Makah to be hypocritical. The situation came to a head at the 2002 IWC meeting in Shimonoseki, Japan, where aboriginal quotas were up for renewal. Subsistence quotas are usually reconsidered every five years and had always been adopted by consensus. Angered that its proposal to reopen commercial whaling was again roundly defeated, on its home turf no less, Japan retaliated by voting against the proposed bowhead quota. It persuaded many of its allies to follow suit, and the quota failed to garner the necessary number of votes. This defeat (the bowhead quota was eventually adopted at a special interim meeting) seriously unnerved the US delegation, as it was no doubt meant to. The delegation openly began seeking a compromise to accommodate the Japanese push to reopen commercial whaling.[1]

The US National Oceanic and Atmospheric Administration was so determined to legitimize the Makah proposal that it put the Alaskan native hunt at risk and violated both NEPA and the MMPA, as several successful NGO lawsuits between 1998 and 2002 demonstrated. It can be argued that at the IWC, the US delegation established a dangerous precedent of member nations acting unilaterally, or bilaterally, to recognize aboriginal needs and to determine without IWC oversight which groups were eligible to share aboriginal quotas.

As noted earlier, from 1979 to 1981, the US delegation diligently worked to ensure rigorous standards for aboriginal subsistence whaling at the IWC, clearly delineating it from commercial whaling. More than 20 years later, it basically reversed this achievement. In 2004, the US delegation spearheaded a change in the definition of aboriginal subsistence whaling at the IWC; the new definition (thereafter serving as the guideline for subsistence proposals) was far weaker than the original, as it no longer included the requirement to demonstrate nutritional need.

The Last 30 Years – a Mixed Bag

Facing the IWC's practice of renewing aboriginal quotas every five years or so, the US developed a resistance to initiating and leading action on controversial issues at the IWC. From 1977 onward, the delegation would add its name as a sponsor or vote the 'right way' on behalf of conservation, but would not stick its neck out too far lest the US lose the bowhead quota again. Ten years after the Stockholm meeting called for a ten-year moratorium, on 23 July 1982, the IWC voted to end all commercial whaling indefinitely. The US did not initiate this proposal; it was put forward by the tiny island country of the Seychelles. Only a few years earlier this lack of leadership from the US would have been unthinkable.

However, the US delegation did help to draft the moratorium language that was the focus of the vote, inserting wording that helped to keep an indefinite moratorium on the table rather than one with a set time limit. The US also played a significant role in convincing undecided governments to support the moratorium. US policy on commercial whaling did not change; it simply became secondary to the need to preserve quotas for the Alaskan natives and later the Makah.

Throughout the 1980s and into the 1990s, the pro-whale faction at the IWC (whose majority status waxed and waned as both sides of the whaling debate worked to bring allies into the forum) held onto the moratorium, but with only tepid assistance from the US. Yet despite their gradual retreat from the front lines of the whale wars, there were some high notes for the Americans, at home and abroad.

The debate regarding the impact of human-caused sound (for example, from military sonar, seismic airguns used to explore for oil and gas deposits, and shipping) on marine mammals did not begin in the US but it has grown most heated there. Evidence is growing that certain acoustic technologies used in the marine environment can injure and harass marine mammals, particularly deep-diving cetaceans – under certain circumstances these technologies can kill. Many sound producers have attempted to take short cuts when navigating US legal requirements and, when NGOs have taken legal action, the US judicial system has, to a large extent, ordered them to be more diligent in assessing the risks.

However, litigation involving military sonar went all the way to the US Supreme Court (Winter v. Natural Res. Def. Council, 129 S. Ct. 365, 378, 2008), where the justices ruled that national security takes precedence over environmental interests. The 'marine noise' debate continues unabated, in the US and elsewhere.

The Pelly Amendment saw considerable action during this era: Japan was certified in 1988, 1995 and 2000 for engaging in lethal 'scientific' whaling. In 1986, Norway was certified for violating the moratorium on commercial whaling. Norway had objected to the moratorium in 1982, but less than a month after the Pelly certification, Norway announced that it would suspend commercial whaling after the 1987 season. This decision was seen as a direct consequence of the certification. When Norway was again certified in 1990, for taking minke whales after the majority of IWC members had expressed opposition to Norway's research proposal, this time the country paid no attention. With trade sanctions never imposed, the whaling nations no longer feared Pelly certification.

The US Congress has been a major champion for whales. Over the years, resolutions have affirmed the position that commercial whaling has no place in the modern world and that the US should 'work to strengthen the IWC as the indispensable organization for safeguarding ... whale stocks' (House of Representatives, 1990). The Senate adopted a resolution calling for the US to 'support the permanent protection of whale populations through the establishment of whale sanctuaries in which commercial whaling is prohibited' (Senate, 1998). A later House resolution called for the government to take strong action against Norway and Japan for their whaling activities (House of Representatives, 2001).

More recently, a Senate resolution was introduced that supported the IWC's new Conservation Committee (Senate, 2005). And the House unanimously passed a resolution that called on the US delegation to work to uphold the moratorium, oppose attempts to resume commercial whaling, close loopholes in the existing Convention and focus more fully on conservation (House of Representatives, 2008).

The US Congress also amended the MMPA several times after its initial passage. In 1994, a major change established a transparent regulatory regime to address fisheries that entangle marine mammals in gear. Entanglement in industrial fishing gear, such as gill and drift nets, is the major known source of human-caused marine mammal mortality (Read et al, 2006). The MMPA was one of the first legislative efforts in the developed world attempting to bring this unintentional killing under scrutiny and control, and the management requirements have caused a significant decline in US fisheries entanglement, especially of cetaceans (Read et al, 2006). Some thousands of marine mammals continue to die every year as bycatch in US fisheries, but the situation is still an improvement over historic levels and better than in many other parts of the world. The improvement is most marked in the Pacific tuna fishery – although dolphins are still harassed by the millions each year and some hundreds killed, the mortality rate has declined by 95 per cent (Gosliner, 1999).

Yet despite these bright spots, the US no longer leads the battle against commercial and scientific whaling. Because of an unreasonable fear of appearing (once again) to have trampled Native Americans' rights, the US has turned from a world leader in saving whales to a world laggard. Some (including the authors of this chapter) would say that this was dramatically confirmed when the US Commissioner to the IWC, in his capacity as chair of the Commission, began in 2007 to negotiate a compromise to allow a limited return to commercial whaling.

US Cetacean Conservation Policy in the 21st Century

For the US to proactively initiate and negotiate a compromise between the pro- and anti-whaling factions at the IWC flies in the face of overwhelming support from the American public for an end to commercial whaling (Kellert, 1999). Indeed, a recent study of 230 young Americans (aged 18–30) found that a majority believed the US government opposes *all* whaling, while a third thought that aboriginal whaling is supported while other types are opposed (Parsons et al, 2009). A disconnect appears to exist between public perception of US policy and reality.

Rather than leading an effort within the IWC to build its mandate to mitigate the many threats facing whales, including climate change, chemical and noise pollution, and entanglement in fishing gear, the US has chosen instead to try to meet the whalers halfway, all in the name of safeguarding aboriginal subsistence whaling. (In this the US policy can be contrasted with that of Australia, which in 2009 injected AU$1.5 million into whale conservation work in this forum.)

A diplomatic solution, one outside the IWC, may be the only way to end commercial whaling. The US may be the only country capable of achieving such a solution because of its international political strength and influence. This would also be in its best interest, as aboriginal quotas would no longer be hostage to commercial whaling interests.

The intrinsic value of cetaceans and their importance to ecosystems is recognized and protected by the MMPA. Largely due to this seminal piece of legislation, the US has one of the strongest, most precautionary and transparent management regimes to control the 'take' of marine mammals in the world. The law explicitly recognizes that marine mammals should be protected regardless of their conservation status, given the inherent uncertainties arising from a life lived largely below the sea's surface. This statute has provisions to accommodate those who can argue, legitimately and persuasively, that they have a need to kill or cannot avoid killing some number of marine mammals, as long as that number is sustainable and/or negligible.

The US has the potential – a potential it sometimes achieves – to be a world leader in cetacean protection. As human impacts on the cetacean environment progress (see for example, Stachowitsch et al, 2009), the need for this leadership will only become more urgent. Despite an uneven history, the US now has the necessary public support and legislative tools to secure the best possible future for cetaceans under its jurisdiction and even in international waters, but only if short-term political interests do not supersede conservation needs.

Note

1 This shift towards compromise actually began in 1994, when the US and Norway agreed that the former would work quietly to assist the latter's return to commercial whaling (under its objection to the moratorium) using the IWC's newly adopted method of calculating quotas, in exchange for Norway's agreement not to trade whale products internationally.

References

Anonymous (1973) 'Help for whales', *Time Magazine*, 16 July, www.time.com/time/magazine/article/0,9171,907581-1,00.html

Forestell, P. (2009) 'Public attitudes and popular culture', in W. F. Perrin, B. Würsig and J. G. M. Thewissen (eds) *The Encyclopedia of Marine Mammals*, 2nd edition, Academic Press, New York City, pp898–913

Gambell, R. (1991) 'International management of aboriginal subsistence whaling', presented at the Second Annual Conference of the International Association for the Study of Common Property, Winnipeg, Manitoba, 26–30 September, http://dlc.dlib.indiana.edu/archive/00003372/

Gosliner, M. L. (1999) 'The tuna-dolphin controversy', in J. R. Twiss Jr and R. R. Reeves (eds) *Conservation and Management of Marine Mammals*, Smithsonian Institution Press, Washington, DC, pp120–155

House of Representatives (1990) 'H. Concurrent Resolution 287: A concurrent resolution for a United States policy of promoting the continuation, for a minimum of an additional 10 years, of the International Whaling Commission's moratorium on the commercial killing of whales, and otherwise expressing the sense of the Congress with respect to conserving and protecting the world's whale populations', 101st Congress, House of Representatives, US Congress, 29 June, Washington, DC

House of Representatives (2001) 'H. Concurrent Resolution 180: Expressing the sense of the Congress that the United States should reaffirm its opposition to any commercial and lethal scientific whaling and take significant and demonstrable actions, including at the International Whaling Commission and meetings of the Convention on International Trade in Endangered Species, to provide protection for and conservation of the world's whale populations to prevent trade in whale meat', 107th Congress, House of Representatives, US Congress, 26 June, Washington, DC

House of Representatives (2008) 'H. Concurrent Resolution 350: Expressing the sense of the Congress that the United States, through the International Whaling Commission, should use all appropriate measures to end commercial whaling in all of its forms, including scientific and other special permit whaling, coastal whaling, and community-based whaling, and seek to strengthen the conservation and management measures to facilitate the conservation of whale species, and for other purposes', 110th Congress, House of Representatives, US Congress, 14 May, Washington, DC

Hoyt, E. (2002) 'Whale watching', in W. F. Perrin, B. Würsig and J. G. M. Thewissen (eds) *The Encyclopedia of Marine Mammals*, Academic Press, New York City, pp1305–1310

Kellert, S. R. (1999) *American Perceptions of Marine Mammals and Their Management*, The Humane Society of the United States, Washington DC

Lavigne, D. M., Scheffer, V. B. and Kellert, S. R. (1999) 'The evolution of North American attitudes toward marine mammals', in J. R. Twiss Jr and R. R. Reeves (eds) *Conservation and Management of Marine Mammals*, Smithsonian Institution Press, Washington, DC, pp10–47

Lilly, J. C. (1967) *The Mind of the Dolphin: A Nonhuman Intelligence*, Doubleday, New York City

McIntyre, J. (1974) *Mind in the Waters*, Scribner's, New York City

Miller, R. J. (2002) 'Exercising cultural self-determination: The Makah Indian Tribe goes whaling', *American Indian Law Review*, vol 25, pp165–273

Norris, K. S. (1974) *The Porpoise Watcher: A Naturalist's Experience with Porpoises and Whales*, W. W. Norton, New York City

Nowak, R. M. (1999) *Walker's Mammals of the World*, 6th edition, The John Hopkins University Press, Baltimore

Parsons, E. C. M., Rice, J. P. and Sadeghi, L. (2009) 'Awareness of whale conservation status and whaling policy in the US – A preliminary study on American youth', *Anthrozoös*, vol 23, pp119–127

Payne, R. S. and McVay, S. (1971) 'Songs of humpback whales', *Science*, vol 173, pp585–597

Read, A. J., Drinker, P. and Northridge, S. (2006) 'Bycatch of marine mammals in US and global fisheries', *Conservation Biology*, vol 20, pp163–169

Rose, N. A., Parsons, E. C. M. and Farinato, R. (2009) *The Case Against Marine Mammals in Captivity*, The Humane Society of the United States and the World Society for the Protection of Animals, Washington, DC

Senate (1998) 'S.R. 226: Expressing the sense of the Senate regarding the policy of the United States at the 50th Annual Meeting of the International Whaling Commission', 105th Congress, Senate, US Congress, 9 May, Washington, DC

Senate (2005) 'S.R. 33: Expressing the sense of the Congress regarding the policy of the United States at the 57th annual meeting of the IWC', 109th Congress, Senate, US Congress, 16 May, Washington, DC

Stachowitsch, M., Parsons, E. C. M. and Rose, N. A. (eds) (2009) 'State of the cetacean environment report 2008', *Journal of Cetacean Research and Management*, vol 11 (Suppl.), pp288–302

UN (1982) 'LOS – United Nations Convention on the Law of the Sea of 10 December', United Nations, www.globelaw.com/LawSea/lsconts.htm

Woolley, J. T. and Peters, G. (2009) 'The American Presidency Project', online, University of California, Santa Barbara, California, www.presidency.ucsb.edu/ws/?pid=3387

6

Whales in the Balance: To Touch or to Kill? A View of Caribbean Attitudes towards Whales

Nathalie Ward

Two conflicting images are imprinted on my mind. In one, two boys in a colourful, double-ended dingy lean over its edge, arms outstretched with hopeful hands, eager to touch the humpback calf. In the other, two men rest their hands on the gunwale of the pitch-tarred hull, as a humpback whale surfaces, turning with a trusting eye to investigate them, just as the harpoon strikes her head. Two hands – one reaches to touch, the other to kill.

There is an interlude between feeling and action, when the hand is poised to touch or kill. Different island communities place different values on their shared marine mammal resources, depending on societal and cultural beliefs. Therefore, conflicts between and within island nations (as well as with developed nations) over how to equitably use, distribute and protect these resources are inevitable.

The iconic humpback whale symbolizes the disparity between how some Caribbean island states *and* the European and North America countries perceive 'the whale'. Humpback whales migrate between developed North America and the Caribbean island nations so they serve as a fitting contrast of cultural values. To each, the whale represents a 'provider', yet each culture has a different interpretation of the symbol. While elevated to near-deified status in the European and American anti-whaling culture, the whale represents a consumptive resource in the Caribbean pro-whaling culture of the Eastern Caribbean.

To most Europeans and North Americans, the whale is a titanic figure that has acquired a special status as a tangible symbol for environmental conservation – a proverbial 'sacred cow'. This image of the whale is an emblem of humanity's emotions – the whale provides these nations with a global totem symbolizing their heightened concern over the planet's wellbeing as much as the wellbeing of whales

Figure 6.1 *Flensing of a humpback whale on Semple Cay in Bequia, Saint Vincent and the Grenadines*

Source: N. Ward

(Barstow, 1991). These nations nurture a staunch belief in the whale's value as a benign and *non-consumptive* resource.

To many Caribbean nationals, the whale is a resource for sustainable use – a fishery to be utilized. The whale 'provides' them with a subsistence or *consumptive* need, which can include nutritional, economic, cultural, historical and/or spiritual needs. Some Caribbean peoples derive a direct value from whales, as both a consumptive and productive use value, such as in aboriginal and directed fisheries, where meat and oil are used for the household or sold in the marketplace (Ward, 1995).

Whale meat was once a viable source of protein in many Eastern Caribbean islands (see Figure 6.1). While some people are ambivalent about eating whale meat, there is a general consensus (among those who influence whaling policy) that if whale hunting doesn't threaten extinction, there is no reason the meat can't be consumed by those who want it. This feeling merges with nationalism in West Indians (Caribbean people) who want to preserve whaling as part of their culture and resent being told what to do by 'outsiders'. Anti-whaling rhetoric about 'barbaric' and 'blood-thirsty' Caribbean peoples prompts increased antagonism.

Whaling proponents in the region assert that it is one thing to postulate that whales are special and have rights, but it is an infringement of human rights to condemn the cultural practices of people relative to their use of animals, and therein, to attempt to force them to give up these practices. They state that we must clearly recognize that there is a difference between imposing a moral or ethical standard on Euro-Americans and imposing such standards on the Caribbean community.

Caribbean nationals (i.e. primarily government officials and politicians in the Eastern Caribbean) contend that they should not be victimized by the previous exploitative whaling practices of developed nations, and nor should they be held accountable for the developed world's lamentable whaling history. They believe they have been backed into a cultural corner by the aggressive tactics of anti-whaling aficionados who have launched tourist boycotts and unleashed insulting attacks on whaling nations. Some Caribbean states are of the opinion that whaling nations are vilified (by conservationists) for continuing to tap what has been a traditional resource for them. As a consequence, what whales 'provide' results, they claim, in an uncompromising dichotomy between conservation-minded European and American nationals and pro-whaling Caribbean nationals.

The marine mammal issue is one of the most intractable disputes involving marine resources in the Caribbean. The primary tension in developing nations is reconciling protectionist objectives – endangered species preservation, biological diversity and ecosystem maintenance – with the utilitarian values of often impoverished, subsistence-oriented populations.

Achieving effective marine mammal conservation in the Caribbean is a challenging task and requires a long-term commitment. There is tremendous political and economic pressure in the region to maintain and expand fisheries and coastal development to address the poor socioeconomic conditions (Afrol News, 2010; Caribbean Net News, 2010a, 2010b; IPS News, 2010.) How does one reconcile the need to protect species and maintain the integrity of the habitat with the need of local people to make a livelihood and maintain cultural integrity?

Whale Watching: An Alternative to Whaling

The potential economic benefits to island nations from whales are now twofold. There is tourism based on ecological ideals such as whale watching and there is the economical gain from supporting whaling, but the two do not mix.

Over the past 20 years, Japan has provided financial support for the fisheries ministries of various Caribbean nations in the Eastern Caribbean (Jamaica Observer, 2010; MOFA, 2010). This economic incentive has strengthened the Caribbean nations' support for Japanese whaling and has influenced the 'votes' of the governments of Antigua and Barbuda, Grenada, St Kitts and Nevis, St Lucia and St Vincent and the Grenadines. Each of these countries voiced their support for Japanese whaling aspirations as recently as 2009 at the IWC annual meeting held in Portugal. This leaves Dominica as the only Eastern Caribbean member country of the IWC that does not support Japan's drive to resume commercial whaling. Further, Dominica has taken a lead in saying 'no' to whaling ventures in Caribbean waters and instead favours whale watching.

Fortunately, whale watching has proven to be a lucrative financial alternative to whaling and is gaining momentum. Whale watching is a growing industry in the Caribbean. In fact, it is the fastest-growing sector of the region's tourism industry and brings much-needed financial revenue to the region. By 2007, the industry was generating an estimated $22 million annually (O'Connor et al, 2009).

Visitors can now go whale watching in a number of Caribbean countries including Antigua and Barbuda, Dominica, the Dominican Republic, Grenada, Guadeloupe, Martinique, St Lucia and St Vincent and the Grenadines. Dominica leads the Eastern Caribbean islands in developing both the whale-watching industry and operational guidelines in the region. St Lucia, Grenada and St Vincent and the Grenadines now realize the huge potential that exists and top the list of countries in the region developing their whale-watching operations. Of these island nations, St Lucia is most active in seeking to grow its whale-watching industry, and recently estimated its earnings from whale watching to be around $2.9 million annually (O'Connor et al, 2009).

The Economists at Large (O'Connor et al, 2009) report, *Whale Watching Worldwide*, explores whale watching on a country-by-country basis and provides figures (such as number of operators and annual revenues) for Antigua and Barbuda, Dominica, Grenada, St Kitts and Nevis, St Lucia and St Vincent and the Grenadines. Whale-watch operators realize the positive impact that whale watching brings to the Caribbean and have joined hands with tour and hotel operators in the region to form their own association, CARIBwhale. The association hopes to encourage and promote responsible whale watching, in particular helping small operations to develop.

Other non-Caribbean island regions (such as the Azores) have successfully gone through the transformation from whaling to whale watching. The Azores is considered one of the world's best places to encounter whales in their natural habitat. Similarly, in the Caribbean, the Dominican Republic has protected its whales since 1986, establishing a humpback whale sanctuary and encouraging responsible whale-watching encounters in Silver Bank, Samana Bay and other protected areas.

These are success stories that deserve our admiration as well as our patronage. We should promote these 'blue tourism' destinations that are protecting their marine mammal resources. Whale watching offers a productive and financially viable alternative to hunting whales. It is an industry that can remain in the hands and voices of the community.

ENGAGING CONSERVATION CITIZENSHIP

Policy decisions in the Caribbean are often made without fully informing or involving the local populations that are most affected. Whaling is a prime example. Policy-makers present the whaling issue to Caribbean citizens as a colonial European or American attitude 'telling' developing nations 'what to do'. In general, the public is largely uninformed or misinformed because advocacy for whaling rests primarily in the voices of politicians. For example, most local citizens are unaware of the 'barter' between Japan and pro-whaling nations for votes in the IWC. The ordinary citizen is not cognizant of the long-range implications of Japanese economic development and various fishery assistance packages, which may translate to depleting or 'fishing out' their local waters by Japanese interests (Ward, 2001).

Endangered species conservation efforts are critical in developing nations, where the resources may be declining at a faster rate than protection standards can be implemented. Species in the Caribbean continue to be lost through incremental decision making by political institutions (Horrocks et al, in press). Attempts to address these issues solely in scientific and management terms are likely to fail.

In the Caribbean, the development of a compelling rationale and effective strategy for conserving marine mammals first requires the recognition that contemporary conservation problems are primarily the result of political and socioeconomic forces. Successful conservation management will only occur if both local perceptions and attitudes towards marine mammal issues are considered.

As long as human perceptions and attitudes on environmental issues set people apart from one another, conservation citizenship[1] to protect marine mammals can only be attained by finding a means to ameliorate these tensions. A profound shift in understanding and changing human attitudes is essential.

TO TOUCH OR TO KILL?

The question remains: can one shape the direction that the hand takes – to touch or to kill? This is the moment where educators, like myself, need to intervene to focus our attention on how to protect an endangered and migratory species – the iconic humpback whale.

What are the building blocks to create a conservation ethic? It seems that the foundation, upon which all else rests, is an assessment and understanding of cultural identity with the resource. We must ask: what are local people's attitudes and values about whales? What is/was their cultural identity with the whale? What are their perceptions of European nations and America, with whom they share the species? These questions suggest a fundamental need for educators to reassess the relationship between culture and the environment.

Educators must immerse themselves in the textures and nuances of the culture in order to gain insights regarding the similarities and differences between cultures (for example in endangered species conflicts). Efforts to understand cultural distinctions in knowledge and attitudinal orientations may reveal the role of culture-specific environmental concerns. If educators identify a priori attitudes that can act to promote conservation (i.e. 'whales provide more money to an entire community through whale watching rather than whaling'), they can provide a window of opportunity wherein the learner's experience is altered to consider new perspectives.

It is crucially important for the educator to find multiple 'windows' for teaching in order to develop education programmes that vigorously challenge misconceptions of animals and their environment. An educator who has gained these insights is more likely to create the *cognitive dissonance* for the learner that is necessary for emotive and transformative thinking. Understanding how a person thinks, feels and behaves towards the resource or opposing culture (i.e. conservation psychology) is the link needed to promote relevant and proactive conservation education. We are remiss if we do not consider the enormous complexity of the issue as inextricably woven into the fabric of West Indian society.

Citizen Science: An Investment Model

In order to combat misconceptions, our goal must be to revolutionize community initiatives and to engage the public in science's valuable role in increasing our understanding about the world around us. In the case of whales, it is important to present an informed debate so that citizens can make decisions about which developments to pursue and which to reject. Traditionally considered distinct from science, education and outreach can effectively be twinned with research through institutional collaborations.

Since there is a lack of information about marine mammals in the wider Caribbean region, citizen science partnerships can help provide more data than scientists can collect alone. Citizen science programmes allow for large-scale and simultaneous data collection by having individuals from the public observe and record data on natural events. They then report their data to a central repository for analysis (Cohn, 2008). These programmes have been particularly useful in providing information on aspects of breeding biology and migration (in birds and butterflies) (Jasonoff, 2003; Brossard et al, 2005).

To effectively promote marine mammal conservation in the Caribbean, we need to identify a harmonized Caribbean strategy to make the Caribbean a 'safe zone' for whales and dolphins (see Figure 6.2). A useful model would be to develop a citizen science programme using the humpback whale as a flagship species to promote trans-boundary protection of marine resources.[2] Such a programme could recruit local citizens to share their knowledge of the species and their cultural insights. Local citizens can also aid in the collection of biological or behavioural information about humpbacks that can be incorporated into research studies. In

Figure 6.2 *'Let dolphins swim free for all generations to see' – educational poster sponsored by WDCS, UNEP, the Caribbean Environment Programme, Caribbean Conservation Association and the Marine Education and Research Centre*

Source: N. Ward

effect, an educator/researcher can more effectively promote and elicit conservation action by harnessing cultural insights and local involvement in science.

In summary, most information and scientific understanding generated by researchers remain in the hands of scientists, academics and policy-makers geographically and conceptually distant from the region of study. What better investment is there than designing research programmes that incorporate the resource management needs of local groups and put results in a form that communities can employ when making resource management decisions?

Notes

1 Concepts of environmental problems are perceived differently in developing countries with subsistence economies. Local people are less likely to mobilize around environmental issues unless issues are characterized in compelling themes relevant to the utilitarian values of their culture (for example, human health).
2 The humpback whale is an international citizen – a migratory animal with no need of passport to travel some 1500 miles from its northern feeding grounds in US Stellwagen Bank National Marine Sanctuary waters to its tropical mating and calving grounds off the Dominican Republic Humpback Sanctuary (and other Caribbean islands).

References

Afrol News (2010) 'Cameroon, Togo, Gambia "bought by whaling nations"', 23 July, www.afrol.com/articles/16642

Barstow, R. (1991) 'Whales are uniquely special', in N. Davies, A. M. Smith, S. R. Whyte, and V. Williams (eds) *Why Whales?*, Whale and Dolphin Conservation Society, Bath, UK, pp4–7

Brossard, D., Lewenstein, B. and Bonney, R. (2005) 'Scientific knowledge and attitude change: The impact of a citizen science project', *International Journal of Science Education*, July, vol 27, no 9, pp1099–1121

Caribbean Net News (2010a) 'Commentary: Caribbean credibility at stake in IWC vote', Sir Ronald Sanders, 4 June, http://upgraded.antiguaobserver.com/?p=34682

Caribbean Net News (2010b) 'Japan pays St Kitts-Nevis and Grenada for support on whaling, says British newspaper', www.caribbeannewsnow.com/caribnet/archivelist.php?news_id=23577&pageaction=showdetail&news_id=23577&arcyear=2010&arcmonth=6&arcday=14=&ty=

Cohn, J. P. (2008) 'Citizen science: Can volunteers do real research?', *BioScience*, vol 58, no3, pp192–197

Horrocks, J. A., Ward, N. and Haynes-Sutton, A. M. (in press) 'An ecosystem-approach to fisheries: Linkages with sea turtles, marine mammals and seabirds', in L. Fanning, R. Mahon and P. McConney (eds) *Marine Ecosystem Based Management in the Caribbean*, Amsterdam University Press, Amsterdam, The Netherlands

IPS News (2010) 'Caribbean under fire for pro-whaling stance', 20 June, http://ipsnews.net/news.asp?idnews=51887

Jamaica Observer (2010) 'Regional states could determine future of whaling', 22 June, www.jamaicaobserver.com/news/Regional-states-could-determine-future-of-whaling_7732100

Jasonoff, S. (2003) 'Technologies of humility: Citizen participation in governing science', Special Issue: Governance of and Through Science and Numbers, *Minerva Journal*, vol 41, no 3, pp223–244

MOFA (2010) 'Japan–Saint Vincent and the Grenadines relations', Ministry of Foreign Affairs of Japan, www.mofa.go.jp/announce/press/2006/11/1124.html#4

O'Connor, S., Campbell, R., Cortez, H. and Knowles, T. (2009) 'Whale watching worldwide: Tourism numbers, expenditures and expanding economic benefits', a special report prepared by Economists at Large for the International Fund for Animal Welfare, Yarmouth, MA

Ward, N. (1995) *Blows, Mon, Blows! A History of Bequia Whaling*, Gecko Productions, Inc. ,Woods Hole, MA

Ward, N. (2001) 'Beliefs and attitudes of Caribbean girls about whales: An approach to understanding cultural identity with implications for conservation education', postdoctoral dissertation, Antioch University, Keene, NH

7

The British and the Whales

Mark Peter Simmonds

INTRODUCTION

The UK could be accused, on historic grounds, of inconsistency in its attitude towards whale-kind. In the closing decades of the 20th century, its approach switched from rampant exploitation to benign conservation and welfare, rapidly moving from one end of the spectrum of behaviour towards animals to the other. The UK was a major player in commercial whaling right up until the 1960s. Whaling and whale products helped to underpin Britain's Industrial Revolution and, long before this, provided food and a livelihood for many. A few people grew very wealthy on what eventually became a relentless, cruel and unsustainable exploitation of these animals. Then, in just the space of a few decades, the UK emerged as a major opponent of commercial whaling, reflecting strong domestic anti-whaling opinion. Culturally there is now a strong positive attitude to whales and dolphins in the UK, as evidenced by their use in many logos and much advertising. The UK is also, almost uniquely, four countries where there are no dolphinaria, although many British tourists make special journeys to visit captive dolphins elsewhere. Perhaps it is the fascination that we have with these animals that drives our ongoing desire to meet them at close quarters.

When a large whale rather unexpectedly visited London by way of the River Thames in the midwinter of 2006, the whole of the UK seemed to stop and watch. Hundreds of people turned out along the banks of the river and millions more watched on TV. In fact real-time media coverage extended across the world, including a live feed that dominated the BBC and Sky digital news channels for much of 21 January, the day of the main rescue attempt (see Figure 7.1 and front cover, lower photograph). A practical and compassionate response to the stray northern bottlenose whale, *Hyperoodon ampullatus*, came from a coalition of rescue organizations led by British Divers Marine Life Rescue (BDMLR), and

Figure 7.1 *Northern bottlenose whale in the Thames, 2006*

Source: Rex Features/Brendan Beirne

supported by a small group of veterinary experts. Most of the people involved in this rescue, and other similar attempts to help stricken marine mammals around the British Isles, are volunteers; and such rescue efforts are supported exclusively by a number of animal charities and coordinated through a national forum, the Marine Animal Rescue Coalition (MARC). MARC has no legal standing and no income but despite this has facilitated the development of sophisticated protocols for rescue. In some ways, this situation typifies the strong feelings many of the British population have nowadays for whales. Many of us are fascinated by these animals and willing to give generously of time and other resources to help them.

Ancient Whaling

Not long ago, a large oil-rich whale, such as the northern bottlenose whale, swimming up-river into London might have received a very different reception because it would have been swimming straight into a major whaling port. Britain's history of whale exploitation begins long ago. The Ancient Britons, like many other early maritime cultures, certainly welcomed the arrival of stranded cetaceans. For them the freshly dead or dying whales found on the shore would have been a helpful, if unpredictable, source of muscle and blubber for food, and bones (and probably baleen) from which tools and utensils were fashioned. Ancient remains of whales in Scotland suggest this opportunistic use of whales may have occurred as early as the Mesolithic or middle Stone Age (8500–4000BC) (Martin, 1995; Mulville, 2002). On the Western Isles, to the northwest of mainland Scotland, large quantities of cetacean bones can be found dating from the Bronze Age (1000BC) until the Norse period (up to the 12th century). However, it remains difficult to say when hunting of cetaceans actually began in the British Isles (Mulville, 2002). Organized and even commercial capture of cetaceans probably arrived with the seafaring Vikings, and while the Basques of France are better known as early whalers, the Vikings probably started whaling some centuries before them (Roman, 2006).

The value of stranded or hunted whales was formally recognized in the 14th century when, in 1324, a statute was passed that gave the English sovereign qualified rights to cetaceans stranded or caught in the waters of England and Wales (BNHM, 2009).The Scottish crown quickly claimed similar rights. It is clear that the precise nature of the animals concerned was not well understood at the time and they were famously described as the 'Fishes Royal'. Cetaceans are known to have been consumed by Henry III, Edward III, Henry V, Henry VII, Henry VIII and Elizabeth I. More recent consumption in Britain occurred during the food shortages caused during and after World War II, when whale meat was marketed from fishmongers (and whale-derived margarine was also important). The royal right to stranded cetaceans was passed to the British Museum of Natural History in 1913 and this has facilitated an important programme of research ever since, with thousands of animals recorded to date.

Early cetacean hunting in Britain was opportunistic and basic. When whales were sighted offshore, people would set out in small boats to try to drive them aground. This is similar to the hunting method still practised in the Faroe Islands (a kingdom under Denmark some 200 miles to the northwest of Scotland), although these days the hunt in the Faroe Islands includes such modern elements as motorized vessels and mobile phones.

Sometimes, whales would also 'helpfully' strand themselves on British shores without any apparent encouragement. In the late 17th century, for example, on remote Tiree, the most westerly of the islands in the Inner Hebrides, about 160 'little whales' (most probably pilot whales) were stranded without any human involvement. Food was scarce and the locals apparently ate them all, commenting that the 'sea-pork is both wholesome and very nourishing meat' (Martin, 1995). Sometimes remarkably large schools were also driven ashore. In 1741, in the Bay of Hillswick, Shetland, some 360 whales were recorded as taken in a drive hunt and, in 1841, 287 whales met the same fate on the west side of Orkney. The oil from this last assisted stranding fetched GB£398, a small fortune at the time for the local community. These whales were again most likely to be pilot whales or, as the Scots called them *caain* whales (from the Scottish word *ca*, meaning to drive).

Meanwhile, the Basque whalers were far more organized and efficient than the British whalers, and capable of taking far larger 'booty'. They removed an estimated

Figure 7.2 *An unknown whaler in the mouth of a whale about to be flensed at a land station*

Source: Aberdeen Art Gallery and Museums Collections

40,000 right whales from the Atlantic between 1530 and 1610 (BMNH, 2009) from a population that never recovered on this side of the Atlantic. (The efficiency of the Basques almost certainly helps to explain one of the greatest of all whale mysteries, why the Atlantic gray whales went extinct 300–400 years ago; long before modern whaling started).

Modern Whaling

British industrial whaling began in Scottish waters in 1903 supported by strong Norwegian investment (Tønnessen and Johnsen, 1982) (see Figure 7.2). There were three whaling land stations in Shetland and one on the Isle of Harris. They all closed during World War I and only partially reopened afterwards, with the last of the stations (Olna Firth in Shetland) closing for the final time in 1929 (Martin, 1995). The Harris vessels hunted mainly around St Kilda, while those from the Shetland Islands sought their quarry to the north and northwest of the islands, an area regarded as the most productive in the North Atlantic. During this period, some 10,000 whales were taken. Products included whalebone, fertilizer, bone meal and meat. The main prize by now though was whale oil and the main species taken was the fin whale.

In the 18th century it was whale oil that lit the factories (allowing them to work 24 hours a day) and fired the street lamps (making the streets safer for the developing middle classes) (Archibald, 2004). Whalebone was also valuable, and culturally highly significant, as whalebone stays and corsets created the slim-line look favoured by those women of the day who could afford it. The capital, London, was the major whaling port for many years but whaling was also important for many northern ports as well, including Dundee, Hull, Montrose, Leith, Aberdeen and Peterhead, all of which hosted sizeable fleets (Archibald, 2004). Many smaller ports also hosted whaling vessels and, by 1861, some 6000 men were involved in the British whaling industry and many thousands more involved in ship building, maintenance and the many ancillary trades that supported the fleets.

Between 1906 and 1912, the three Shetland stations jointly produced some 12,000 barrels of oil, which was something of a record at the time (Tønnessen and Johnsen, 1982). Each barrel fetched about £3 4s 1d. In 1950, the Bunaveader station in Harris reopened after World War II, and operated for two seasons with just one catcher boat that was crewed mainly by Norwegians. In June 1952, the station went into liquidation having killed 53 whales, and that marked the end of shore-based whaling in Britain.

There was an overlap between shore-based whaling and British distant water whaling, which began in the mid-18th century as the government of the day avidly sought whale oil, which was regarded as the 'perfect lubricant' for the machines manufacturing woollen textiles (Archibald, 2004). The first subsidized whaling trip was probably from Leith in 1750, and Aberdeen, Dundee and Peterhead soon became important ports involved in this expeditionary whaling (Martin, 1995). The hunting method employed at this time was to travel to the whale grounds in large vessels and then close upon the whales using rowing boats. Capture was affected with a hand-held harpoon. The exhausted whale, after it had been 'played' by several crews using the rope attached to the harpoon securing it, was usually finally dispatched by lances. Blubber was stored on board the large vessels in barrels and rendered into oil back at port. The target species at this time was the bowhead whale and once this slow-moving whale was hunted-out in one Arctic whaling ground the whalers would simply move on to the next. By the mid-1880s, whaling boats were steam powered, with a bow-mounted harpoon gun that fired a grenade harpoon, allowing the faster rorqual whales, including the blue whale, to be targeted.

Once the Arctic grounds had been heavily exploited, interest moved to the other pole. In 1908, two companies, Christian Salvesen and Sons of Leith, jointly established a whaling station in South Georgia. Operations continued from there until Antarctic whaling was halted in 1963 (Martin, 1995). Valuable whaling in Antarctica soon led to a dispute over which countries owned which territories.

Britain was so keen at one point to drive home its claims that it requested help from the International Court of Justice in The Hague (Tønnessen and Johnsen, 1982). Argentina and Chile opposed this and Norway, a great whaling rival of Britain, also entered the dispute. In Royal Letters Patent of 21 July 1908, it was formally established that South Georgia, South Shetland, South Sandwich and Graham Land were 'a part of the British dominions as dependencies of our colony of the Falkland Islands'. Norway later acquiesced to this decree. The primary reason for Britain's historical interest in these remote locations is very well illustrated by the valuable 1910–1911 catch from South Georgia of 6529 whales.

At this time, interest in which fleet was killing animals from which whale 'stock' (and with whose permission) necessitated greater understanding of the animals, their discrete populations, movements and their migrations. Logbooks to detail which species were being killed, and where, were placed on whaling vessels. The London Natural History Museum intervened as it became apparent that a largely uncontrolled and unsustainable slaughter was taking place. In 1911, the Museum first suggested that extinction was likely if the situation continued. These concerns resulted in Major G. E. H. Barret-Hamilton being dispatched to South Georgia to study the 1913–1914 whaling season (Tønnessen and Johnsen, 1982). Sadly, he did not complete his work and died in South Georgia in January 1914, but his reports were the first scientific investigations into the whales being exploited in the area. Further studies and growing concerns about dwindling whale 'stocks' eventually led to agreements to limit whaling, to the establishment of the IWC and, finally, to the global moratorium on commercial whaling agreed by the IWC in 1982 and enacted in 1986–1987.

The 1961–1962 Antarctic whaling season is regarded as a turning point in British whaling history (Tønnessen and Johnsen, 1982). Economic returns were particularly poor, petrochemicals had now replaced whale oil in all its previous usages and the Japanese and Russian whaling fleets were expanding their activities. The last whale to be hunted by Britain was in 1963 (Archibald, 2004), and it is reasonable to assume that the cessation of whaling at this point resulted from a combination of both economic factors as well as conservation-related considerations.

Throughout much of British whaling history, the true nature of whales as large warm-blooded mammals that suckle their young was not widely appreciated. The first real insights into their biology did not come until the 1800s. At this time, cetaceans started to be properly described (Würsig et al, 2009). Among those unravelling the truth about whales was the British zoologist, John Edward Gray, who described all the marine mammals in the British Museum's collection; a collection that he also helped to build. His treatise on the collection was published in 1866. In 1870 he published again, this time on the 'geographical distribution of the Cetacea'. In fact, he authored well over a thousand papers, many of which were about aquatic species, and 16 of the marine mammal species that he identified and described still stand today.

Figure 7.3 *Francis Buckland applies 'medicine' to a porpoise – an illustration from his book*

Source: Buckland (1868)

Whales and Dolphins in Tanks

Like most other European nations, and as many other countries around the world still do, the UK at one time hosted dolphinaria. Indeed efforts aimed at publicly displaying cetaceans can be dated back to Victorian times. The popular writer and gentleman naturalist, Francis Buckland, recalled his attempts in the early 1860s to gain a live harbour porpoise, *Phocoena phocoena*, for the London Zoological Gardens in his book *Curiosities of Natural History* (Buckland, 1868). The zoo was first sent a porpoise in November 1862. The poor creature had been caught off Brighton and was transported to London's Bond Street where it spent 'several hours panting on a fishmonger's slab'. Not surprisingly it arrived at the zoo in poor condition. Buckland tried to revive it, giving it first 'a dose of ammonia' and later 'a good glass of stiff brandy and water' (see Figure 7.3). He noted that both administrations caused the porpoise's respiration to rise and he took this as a good sign. However, it died shortly after this.

Buckland conducted a post mortem, recording that the fisherman that caught it 'had kindly given their victim a gentle reminder by means of a tap on the side of the head, which had left a considerable bruise; and they had also … acted very cruelly towards their poor prisoner' by blinding it. Buckland commented 'I have since learnt that it is the custom of fishermen in some parts of England, when they catch a shark, dog-fish, grampus, porpoise or any other enemy who is apt to damage their nets, to put the poor beast's eyes out. I trust, however this custom is not universally prevalent among our fishermen. It is an un-English action' (Buckland, 1868). Buckland also felt that he had learnt much from this first effort to add to the London menagerie, including the fact that the porpoise was able to survive for some time out of water. Buoyed up by his new knowledge, he placed adverts in public journals to try to obtain other 'sea-pigs' for the zoo. Several were acquired but all died after a few days.

Is this Victorian mixture of concern for the porpoise's wellbeing, combined with a desire to exploit it by exhibiting it, still evident in the British attitude? Certainly dolphin shows were popular in the UK in the 1960s and 1970s, when over 30 shows ran involving more than 300 dolphins (Cooper, 2003). The reason that such animals are no longer kept and exhibited in Britain seems to be the result of a significant public campaign to end this practice, combined with a number of other factors, including possibly the enactment of the US MMPA, which made sourcing dolphins from that side of the Atlantic far more difficult and expensive. A growing awareness of the issues surrounding the dolphin shows (including the low survival rates of captives and the impacts of capture and trade of dolphins as, for example, detailed in *The Rose Tinted Menagerie* (Johnson, 1994)) and a grassroots campaign generated much media interest and public support. The campaign included regular protests outside the establishments holding the animals, the lobbying of members of parliament, and at least one attempt to rescue an animal from a dolphin show and release it into the Irish Sea (which failed). There was also an associated 1991 film, *Into the Blue – Dolphin Rescue*, narrated by British actor Michael Caine. A government-commissioned report on British dolphinaria (combined with the public concern), then led to a new law, passed in August 1993, setting very high standards for cetacean accommodation that are arguably prohibitively expensive to meet.

The public campaign had many facets. At the Brighton dolphinarium and aquarium, for example, potential visitors were encouraged not to enter by protestors and outside Windsor Safari Park, where dolphins and orca were being kept, symbolic crosses were placed on the grass verges by the gates (Alan Knight, personal communication). Each cross was intended to commemorate a dolphin that had died there. Campaigners on at least one occasion also invaded the Windsor dolphinarium stage during a show. In response, Winnie, the Park's orca, was directed to spit water at them (which the audience apparently enjoyed) but such high-profile and media-worthy high-jinks also contributed to the demise of the British dolphin shows.

Several dolphins held in the UK were reintroduced to the wild via a rehabilitation facility in the Caribbean Islands of the Turks and Caicos. This was no small undertaking. The transfer of a bottlenose dolphin, *Tusiops truncatus*, known as 'Rocky' from Morecambe Marine Land to the Caribbean, and his eventual release back into the wild, became the subject of a children's book by Virginia McKenna (McKenna, 1997). His journey began when a petition signed by 4000 people was handed to the local council in August 1989. The Born Free Foundation and other NGOs then took up Rocky's case and with support from 'dozens of celebrities' launched their 'Into the Blue' project. The *Mail on Sunday* newspaper ran an appeal and £100,000 was raised in support of the dolphin's relocation. British Airways Cargo helped by offering free space in a 747 aircraft for Rocky and his carers. Two other dolphins from UK facilities joined Rocky in the Turks and Caicos rehabilitation facility and, in September 1991, they were all released together. The release seems to have been a success and McKenna (1997) reports that 'our dolphins' were seen over 60 times either alone, in pairs or all together. She adds, 'Even if we never see them again, we believe that freedom to race and play and live life as a wild dolphin, even if only for a short time, is surely preferable to years spent in a pool "scented" with chlorine' (McKenna, 1997).

The last British dolphinarium, which was in Yorkshire, closed in 1993. More recently there have been rumours of attempts to re-establish dolphinaria in the UK, including a possible revision of the regulations on the standards required to keep such animals. However, given that public opinion seems to have changed little in this regard, and that many of the organizations and people who oppose cetacean captivity still stand ready to strongly denounce dolphinaria, it seems an unlikely future development. However, as described below, many British people still go to dolphin shows elsewhere in the world and there is clearly a significant financial reward for those that do display dolphins and whales. Arguably it is a quintessentially British situation that despite the campaign and undoubted strength of public opposition, there is no actual ban on the keeping of these animals in the UK, just some very stiff regulations that might yet be revised.

The Ongoing Perils of Popularity

The WDCS sponsored an IPSOS MORI poll of the UK public in 2005. One of its more surprising findings was that 45 per cent of the general public had seen whales or dolphins in a zoo, aquarium or dolphinarium within the last ten years. This contrasted with the statistic that three-quarters of the general public had *never* been whale or dolphin watching in the wild. At least some of the people who reported seeing cetaceans in captivity may have been thinking back to UK facilities even though, as described above, they were closed more than ten years before the survey. Nonetheless, the evidence that British people are visiting dolphinaria overseas is also supported by prominent advertising campaigns in the UK encouraging 'dolphin tourism' to various parts of the world. Oddly, the MORI poll confirmed that the majority of the general public still disagreed that it is acceptable for such animals to be on display.

Interestingly the IPSOS MORI poll revealed something else rather curious about those who *had* seen these animals in the wild: around one quarter over the course of the preceding ten years had also actually swum with dolphins. In fact, the ambition of the British to get into the water with dolphins is well established. For example, 'swimming with dolphins' emerged as the number one 'dream activity' in a BBC poll of 20,000 people in 2003.

Despite the fact that the majority have yet to actually see them, the UK still hosts populations of bottlenose dolphins around its shores that are relatively accessible. There is one such resident population in Cardigan Bay, and some of these animals can be regularly seen from the seafront in New Quay, Ceredigion. Another population is found along the coasts of the Moray Firth, extending south to St Andrews and beyond. Again there are vantage points and boat trips along this coast that make these animals relatively easy to view.

Occasionally, in recent decades, the UK has also hosted several of what have become known as 'solitary sociable' dolphins. These are animals that, for unknown reasons, have become separated from their schools and instead live alone and over time, through a process of habituation, come to socialize with people. This same process often seems to put them increasingly in jeopardy. The most famous solitary sociable dolphin was 'Luna', the young orca (or killer whale, *Orcinus orca*, the largest species of the dolphin family) who lived apart from others of her kind in Nootka Sound on the west coast of Vancouver Island, Canada. Her story is now the subject of an award winning documentary film. In fact the 'solitary sociable' phenomenon affects a number of the more highly social cetacean species and is being increasingly reported around the world. The majority are bottlenose dolphins and, in 1987, there were four solitary bottlenoses at various points around the UK coast and keeping them safe became a significant issue. Such animals become increasingly interested in socializing with people, so they often live in shallow waters where there are many risks, some coming directly from the people themselves. On one level it is people interacting with these animals, and thereby gradually habituating them to human contact, that cause the dolphins to lose any natural wariness and seek human company in waters that are often busy, polluted and frequented by fishermen. There is also a risk to those that choose to enter the water with such animals.

Take the case of 'Dave' (Simmonds and Stansfield, 2007; Eisfeld et al, 2010): in April 2006, a solitary bottlenose dolphin appeared off the Kent coast of south England at the popular seaside resort of Folkestone. Nicknamed 'Dave' by locals, she was later found to be a juvenile female. She established a relatively small range (as many solitaries do) and was consistently located in a zone very close to shore that was just a few kilometres long and a few hundred metres wide; most of the time she could be easily seen from shore. This same small area was used by many local people and tourists for swimming, boating, kayaking and fishing. Bottlenose dolphins are rarely seen on the Kent coast these days and Dave's very exposed situation gained lots of attention in local and national media. There were immediate concerns about her welfare and the MARC forum set up a working group to consider her situation and offer advice on her welfare. This resulted in various public meetings and educational events. Volunteers formed regular patrols on the adjacent shore giving out leaflets and other advice. Posters warning of the dangers of interacting with a wild solitary dolphin were positioned around her haunts. Despite all this effort, people flocked to see and swim with her.

Those who closely observed Dave noticed that her behaviour changed over a period of a year, in close accordance with the 'solitary dolphin stages' of habituation previously reported (Wilke et al, 2005). When she first arrived off the coast near Folkestone, she had been reluctant to associate with swimmers or boaters and kept her distance. However, by June 2007, interactions with Dave had become commonplace and she was regularly soliciting human contact. Up to 20 to 30 people, some hanging off her dorsal fin trying to hitch rides, could be seen in the water with her in the late summer of 2007 (see Figure 7.4). Additionally, the high number of boats, some of whom were purposely searching for and following Dave (although she could be seen perfectly well from shore) was a major concern. By the end of September, she was reported to have started becoming much more boisterous with people in the water (again in line with the behaviour of other solitary sociable dolphins). Her reported behaviours now included at least a couple of occasions when she had prevented swimmers from getting out of the water. She was also seen spinning kayaks around. This all sounds like fun, until the relative sizes and competencies of a dolphin versus a human in the water are considered. Some people were clearly terrified by the dolphin when she first appeared next to them in the water and at least twice people had to be helped out of the water after she had tried to stop them leaving.

By September and October 2007, Dave was reportedly travelling a little further from what had become her usual home range (Eisfeld et al, 2010). Sadly, she was also getting more and more injuries, including nicks and scratches. Deeper wounds resembled propeller marks. On 15 October she was seen to have been severely

Figure 7.4 *'Dave' the bottlenose dolphin and friends off Folkestone in 2007*

Source: Terry Whittaker

wounded and lost about half of one tail fluke. She most likely sustained this injury from entanglement in fishing gear or possibly a propeller strike. This wound was clearly life threatening and the voluntary community monitoring her called in a veterinary expert, who, due to her 'tameness', was able to administer antibiotics, attempting to mitigate the risk of infection. He was also able to remove some angling line and hooks embedded in her dorsal fin.

For a few days after she received this major wound, her movements seemed limited, then she started swimming strongly again. (Meanwhile, experts had opined that other dolphins have survived even having lost such a large portion of their tail flukes.) However, she was not sighted after 9 November and given her wound, her friendly nature, her previous faithfulness to a well-known small inshore range, and the fact there has never been another record of her (despite some efforts on both sides of the English Channel to find her), it seems most likely that she died, either as a result of the wound or some other incident.

Her story arguably tells us a few things about modern British society and its attitudes towards such animals:

- The enthusiasm of many to interact with dolphins seems to heavily outweigh any warnings that this may be dangerous for the dolphin and the person involved.

- A similar enthusiasm for this single remarkable animal motivated a team of people to give up their time to try to protect her.
- There are laws in the UK intended to protect such animals but they are rarely used and difficult to apply, although, in fact, in Dave's case, two men were prosecuted under the 1981 Wildlife and Countryside Act for disturbing her.

This last point may highlight a further difference between the situations in the UK and US. The explosion of interest in marine mammals led to highly significant new laws for cetaceans in the US (as described by Rose et al, in this volume). In the UK we saw something similar in the Wildlife and Countryside Act, which has been amended several times but remains far less comprehensive.

All four of the solitary dolphins present in the UK in 1987 appeared to have short lives because of their particular behaviour. The dolphin that based itself in busy Portsmouth Harbour was killed by a tug boat propeller (a similar fate to that which befell Luna). A dolphin living around Maryport in Cumbria washed ashore dead having succumbed to an infection, almost certainly facilitated by the accidental wounds she received in near-shore pathogen-contaminated waters. A third dolphin known in the Thames Estuary has disappeared. A sad litany of events. For those that spent time with Dave, hers is a particularly sad story and their concerns will have been deepened because they knew her as an individual. (Many unknown dolphins also die in UK waters each year in fishing nets.)

In 2000, dolphins topped a poll conducted by BBC *Wildlife* magazine of the nation's favourite animals. The seemingly rapid change of heart in the UK about cetaceans – moving swiftly from accepting their lethal exploitation to firmly opposing it in just a few short years – deserves some further consideration. This change has not been unique to the UK. It also happened in the 1960s and 1970s in North America (as considered by Rose et al in this volume), across the rest of Europe and in Australia and New Zealand (Lavigne et al, 1999; Forestell, 2009). It has been suggested that four key factors registered on the collective consciousnesses of people to bring about the change:

- Dolphins (and whales) were found to have cognitive capacities rivalling human capacities.
- Watching dolphins and whales is an exciting and entertaining activity.
- The worldwide slaughter of whales is both unethical and biologically a disaster.
- Marine mammals demonstrate a number of adaptations to the marine environment that may have important application to human technology (Forestell, 2009).

Another theory is that this change resulted from a rapid urbanization of the human population, increased knowledge of marine mammals through new research and extensive media exposure, and a genuine shift in ethical values (Lavigne et al, 1999).

This author sides with the notion that there has been a genuine shift in ethical attitudes in the UK and that this is a part of a growing awareness of the welfare needs of animals. Discussion with those campaigning against the captivity of cetaceans in the 1960s and 1970s reveals how they were indeed inspired by new knowledge about these animals. The advent of domestic television, complete with wildlife documentaries and strong messages about animal rights from the 1950s onwards, simply changed the way that the world was viewed and continues to do so. However, there has been, and to some extent still is, a backlash against the supposed special status of cetaceans. More recently, we can see the pro-wildlife use movement (led by those countries still championing whaling) is also using the increasingly sophisticated tools of 21st-century communication, including websites, blogging and tweeting to get across their arguments in favour of whaling and other lethal uses of marine mammals. (Future readers may laugh at the notion these techniques were actually sophisticated, but they are for this time and represent as much a revolution in how people receive attitude-shaping knowledge as radio and television did before them.)

New Points of Conflict

At the 61st meeting of the IWC on the Portuguese island of Madeira in May 2009, the UK delegation made the following intervention:

> *Chairman, the people of Britain both at home, here in Madeira and in many other places around the world enjoy their whales alive. Well-managed whale watching can be regarded as a sustainable use of these animals and as they are animals of the global commons we claim our rights to this peaceful and low impact enjoyment. And before others step in (as they have in the past) to state that whaling and whale watching can co-exist, we do not agree. You can watch a whale many many times – you kill it and eat it just the once.*
>
> *Whales are not only a global resource; they are creatures of great splendour and awe – generators of inspiration and respect for nature, charismatic symbols of our global environment; let us follow the example set here in Madeira – once the centre of a very major whaling industry – and move to a more harmonious relationship with them.*

Does this rhetoric from the UK delegation in the whaling forum match the British government's own treatment of these animals back home? The website of the relevant ministry (Department of the Environment, Food and Rural Affairs) proudly states: 'UK policy is to maintain the IWC's international moratorium on commercial whaling that was introduced in 1985/86. All whales are protected by law in UK waters' (DEFRA, 2009). In fact, a slew of international treaties that the UK is party to relate to these animals and require action on the UK's part (Parsons et al, 2010a, 2010b). These have caused various national laws to be generated and

Box 7.1 The main international agreements and domestic laws affecting cetaceans in UK waters

- The Bern Convention on the Conservation of European Wildlife and Natural Habitats 1979
- Bonn Convention (Convention on the Conservation of Migratory Species of Wild Animals)
- United Nations Convention on the Law of the Sea (UNCLOS)
- Convention on Biological Diversity
- Washington Convention (1948)
- Convention on International Trade in Endangered Species of Wild Fauna and Flora (CITES)
- Agreement on the Conservation of Small Cetaceans of the Baltic, North East Atlantic, Irish and North Seas (ASCOBANS)
- Oslo/Paris Convention (for the Protection of the Marine Environment of the North-East Atlantic) (OSPAR)
- Habitats (and Species) Directive Council Directive 92/43/EEC
- Domestically:
  - the Whaling Industry (Regulations) Act 1934, as amended by the Fishing Limits Act, 1981
  - Wildlife and Countryside Act (as variously amended)
  - The Conservation of Habitats and Species Regulations, 2010 in respect of England and Wales (which are intended to transpose the Habitats and Species Directive into law in the UK) and the regulations' equivalent in Scotland
  - Various regulations relating to offshore developments
  - Local and national biodiversity action plans
  - Seismic guidelines (relating to the use of seismic exploration looking at submarine oil and gas deposits

Note: For more detail see Parsons et al (2010a, 2010b)

European law has also required actions including revisions to this body of law. All in all, there is a complex web of various commitments and somewhat confusing legislation that relates to these animals (see Box 7.1).

These animals are intended to be highly protected and the net effect of UK and EU laws is that vessels cannot be used to take whales in British waters and it is forbidden to kill, harm or disturb cetaceans, although various caveats apply to this. In addition, two species (there are over two dozen recorded in UK waters) qualify for the development of special areas of conservation (SACs): the bottlenose dolphin and the harbour porpoise.

The UK is also a party to ASCOBANS – which aims to restore and/or maintain biological or management stocks of small cetaceans at the level they would reach when there is the lowest possible anthropogenic influence. ASCOBANS proposes to reach these aims through coordinating and implementing conservation measures for small cetaceans (DEFRA, 2009).

However, the UK's record of actual practical conservation of whales, dolphins and porpoises has been criticized (Parsons et al, 2010a, 2010b). For example, despite increasing reports of fast-moving vessels chasing dolphins, only two prosecutions using the Wildlife and Countryside Act have been successful to date. Furthermore, the legal protection of cetaceans has now seemingly largely been transferred to the Conservation of Habitats and Species Regulations (and their equivalents in Scotland) producing what appears to be a rather confusing situation, including differences between Scottish and other UK law.

Similarly, there has been a significant problem with cetacean bycatch (the incidental capture of animals in fishing gear) to the south of the UK. Hundreds of cetacean corpses in the 1990s suggested a high mortality. European nations, including the UK, have not adequately addressed this matter, or indeed bycatch more generally, across the region.

Certainly both of these examples present some very difficult challenges. In the first instance proving that a dolphin has been disturbed (or indeed even can be disturbed) by water craft requires a magistrate or a jury to understand quite a lot about cetacean biology. Providing solid evidence of an offence out at sea is also difficult, although the new generation of portable electronic devices, including mobile phones with video capability, provide new opportunities to capture such incidents.

Perhaps the best evidence that the UK is still neglecting the welfare of its own cetaceans comes from the failure to assign SACs for harbour porpoises anywhere within its waters (Parsons et al, 2010a, 2010b) and also the recent changes to British law that, rather surprisingly, have downgraded the protection given to this species. I will not explore the reasons why this might have happened further here, except to note that there is an increasing movement of industry into UK waters – including renewable energy generating devices and this may mean that the interests of such industries (supported at the highest levels in government) may now be in growing conflict with the requirements of cetaceans for a quiet and healthy environment.

THE 'MONSTER' WHALES OF WINTER

So the monster whale did sport and play
Amongst the innocent little fishes in the beautiful Tay,
Until he was seen by some men one day
And they resolved to catch him without delay. (McGonagall, 1884)

I have described the enormous public interest in the 'Thames Whale' in January 2008 and noted that its positive reception would not necessarily have been replicated even 50 years earlier. During a recent visit to the Aberdeen Maritime museum, I learnt of a comparable visit to a city by a large whale, the 'Tay Whale', which visited Dundee via the River Tay in the winter of 1883–1884. In this case, the ocean giant was a humpback whale, *Megaptera novaeangliae*, and the town it rather unwisely visited was, at that time, Britain's premier whaling port. The story of the whale was first immortalized by McGonagall (1884) in a famed narrative poem, and most recently by Crumley in a fine fictionalized account (Crumley, 2008).

The rare opportunity to see the whale caused thousands of people to turn out and enjoy the spectacle and, at first, they cheered at its surfacing and acrobatics. Beyond the whaling community, overwintering in dock at that season, it is unlikely that anyone would have seen such an animal before. Eventually, however, there were calls for it to be dispatched so it was hunted and killed, and the crowds applauded this too.

The demise of the Tay Whale was particularly cruel. On 31 December it was struck by several harpoons but swam on towing two six-oared rowing boats, a steam launch and a steam tug around the Tay estuary and out to sea. This went on all night, with further projectiles being fired into the copiously bleeding animal. Then, at dawn, a freshening wind put strain on the harpoon lines and they broke (Williams, 1996). Seven days later the whale's body was retrieved floating six miles off Inverbervie. The hunters auctioned the carcass (which is almost certainly the animal shown in Figure 7.5) and it was purchased for £226 by John Woods, an oil merchant from Dundee. The 'monster' was placed on public display. Admission cost six pennies or one shilling (depending on time of day) and 12,000 people came to see it on the first Sunday alone, with special buses being provided to transport them. On 25 January, the famous surgeon, Professor John Struthers, carried out a public dissection accompanied by music from the band of the 1st Forfarshire Rifle Volunteers. Later with a wooden backbone and frame introduced, the remains were stitched back to their original form and the exhibit taken on a tour of Aberdeen, Edinburgh, Liverpool, London and Manchester.

McGonagall (1884) chronicled this in the following verse:

> *Then hurrah! for the mighty monster whale,*
> *Which has got seventeen feet four inches from tip to tip of a tail!*
> *Which can be seen for a sixpence or a shilling,*
> *That is to say, if the people all are willing.*

After some other adventures, its bones are now displayed in the Barrack Street Natural History Museum in Dundee. Incidentally, the bones of the Thames Whale – prepared and stored after a daily newspaper raised the funds – reside somewhere usually out of sight inside the research collection of the Natural History Museum in London.

Are the people who cheered the death of one whale in the 19th century and the people who tried to will the survival of the other in the 21st so very different? Probably not, but the deaths of whales no longer meet any pressing human needs in the British Isles and nor do they provide income to British people. Now we have a better, and widely shared, insight into the whale as something quite different from

Figure 7.5 *Whale on the beach at Stonehaven, 1884*

Note: While the image is labelled 'finner' whale – a term generally used for the fin whale – this is clearly a humpback and Stonehaven was the original landing site for the 'Tay whale'.

Source: Photographed by George Washington Wilson from the Aberdeen Art Gallery and Museums Collections

a monster. It therefore seems unlikely that attitudes in Great Britain will change around again, despite the well-funded campaigns run against them by nations (see for example the chapter in this volume by Martinsen about Norway). The whales also have champions who work on their behalf introducing them to a receptive public in the way that primates have previously been introduced to us – as sentient creatures sharing our planet and imbued with rights. So it is that the work of Hal Whitehead for sperm whales can be compared with that of Jane Goodall on the behalf of chimpanzees.

However, there are problems ahead. In British and adjacent waters we can see a major expansion of industrial activities taking place and the cetaceans may be 'in the way' of this. How the administrations in the UK, Europe and elsewhere will act in response to growing ecological problems and to meet the needs of their human populations may again test the 'special relationship' that we have with the whales and dolphins.

Acknowledgements

This chapter benefited from the assistance of many people and thanks in particular go to Nicky Kemp, Alan Knight, Sue Fisher, Philippa Brakes, Chris Butler-Stroud, Mick Green, Sonja Eisfeld and Cathy Williamson, and also to the Aberdeen Art Gallery and Museums Collections and Terry Whittaker for permission to reproduce the photographs used here.

References

Archibald, M. (2004) *Whalehunters Dundee and the Arctic Whalers*, Mercat Press Ltd, Edinburgh

BNHM (British Natural History Museum) (2009) *The National Whale Stranding Recording Scheme*, British Natural History Museum, website: www.nhm.ac.uk/research-curation/research/projects/strandings

Buckland, F. T. (1868) *Curiosities of Natural History*, Second Series, R. Bentley, London

Cooper, A. (2003) 'Behind the captive smile: A historical perspective of dolphins shows in the UK', unpublished but available at www.prijatelji-zivotinja.hr/index.en.php?id=87

Crumley, J. (2008) *The Winter Whale*, Birlinn Ltd, Edinburgh

DEFRA (Department of the Environment, Food and Rural Affairs) (2009) *Whales and Dolphins*, www.defra.gov.uk/wildlife-pets/wildlife/protect/whales/index.htm

Eisfeld, S. M., Simmonds, M. P. and Stansfield, L. R. (2010) 'Behavior of a solitary sociable female bottlenose dolphin (*Tursiops truncatus*) off the coast of Kent, SE England', *Journal of Applied Animal Welfare Science*, vol 13, pp31–45

Forestell, P. H. (2009) 'Popular culture and literature', in W. F. Perrin, B. Wursig and J. G. M. Thewissen (eds) *Encyclopedia of Marine Mammals*, 2nd edition, Academic Press – Elsevier, San Diego, New York, London, pp898–913

Johnson, W. (1994) *The Rose Tinted Menagerie*, Heretic Books, London

Lavigne, D. M., Scheffer, V. B. and Kellert, S. R. (1999) 'Conservation and management of marine mammals', in Twiss, J. R. Jr and Reeves, R. R. (eds) *Conservation and Management of Marine Mammals*, Smithsonian Institutional Press, Washington DC and London, pp10–47

Martin, A. (1995) *Fishing and Whaling*, National Museums of Scotland, Edinburgh

McGonagall, W. T. (1884) *The Tay Whale*, www.mcgonagall-online.org.uk

McKenna, V. (1997) *Back to the Blue*, Templar Publishing, London

Mulville, J. (2002) 'A whale of a problem? The use of zooarchaeological evidence in modern whaling', in Monks, G. (ed) *The Exploitation and Cultural Importance of Sea Mammals*, Proceedings of the 9th Conference of the International Council of Archaeozoology, Durham, August, pp154–166

Parsons, E. C. M., Clark, J., Warham, J. and Simmonds, M. P. (2010a) 'The conservation of British cetaceans: A review of the threats and protection afforded to whales, dolphins and porpoises in UK Waters, Part 1', *International Journal of Wildlife Law and Policy*, vol 13, no 1, pp1–62

Parsons, E. C. M., Clark, J. and Simmonds, M. P. (2010b) 'The conservation of British cetaceans: A review of the threats and protection afforded to whales, dolphins, and porpoises in UK waters, Part 2', *Journal of International Wildlife Law & Policy*, vol 13, no 2, pp1–72

Roman, J. (2006) *Whale*, Reaktion Books, London

Simmonds, M. and Stansfield, L. (2007) 'Solitary sociable dolphins in the UK', *British Wildlife*, vol 19, no 2, pp96–101

Tønnessen, J. N. and Johnsen, A. O. (1982) *The History of Modern Whaling*, C. Hurst and Company, London; Australian National University Press, Canberra

Wilke, M., Bossley, M. and Doak, W. (2005) 'Managing human interactions with solitary dolphins', *Aquatic Mammals*, vol 31, no 4, pp427–433

Williams, M. J. (1996) 'Professor Struthers and the Tay Whale', *Scottish Medical Journal*, vol 41, pp92–94

Würsig, B., Perrin, W. E. and Thewissen J. G. M. (2009) 'History of marine mammal research', in Perrin, W. F., Würsig, B. and Thewissen, J. G. M. (eds) *Encyclopedia of Marine Mammals*, 2nd edition, Academic Press/Elsevier, San Diego, New York, London, pp565–569

8

Whales in Norway

Siri Martinsen

Introduction

When Norway, acting against the moratorium on commercial whaling agreed by the IWC in 1982, decided to restart its commercial whaling activities in 1993, the foreign onlooker could easily be led to believe that the issue of continued whaling was a matter of great importance to most Norwegians. It is, however, the claim of this chapter that the 'pro-whaling Norwegian' is a creation of an effort on the part of Norwegian politicians in the late 1980s and early 1990s, with the aim of making sea mammal hunting an issue of national interest. Pro-whaling attitudes are not necessarily a feature that is closely bonded to Norwegian identity or history in general, neither do they necessarily hold a prominent place in the mind of Norwegians of today.

Undoubtedly, attitudes towards whales in the political and official Norwegian arenas have greatly coincided with the views of the whalers themselves – whales have been seen as a resource, animal welfare concerns were not viewed as relevant (until pressure from abroad became too problematic), and a broad debate about the animal welfare aspects seemed to be highly unwanted. This is not surprising as the Ministry of Fisheries, which has responsibility for sea mammals, has been and indeed still is coloured by being a ministry of industry and a ministry for the interests of the fishermen, whalers and sealers.

Attitudes towards whales in Norway vary both with time and between groups: for example, minke whales have variously been seen as 'helper', 'resource', 'pest' and 'fellow being'. The whaling issue must also be viewed alongside the general awareness in Norway concerning animal welfare issues. It can now be argued that attitudes of care and concern for whales as individual animals should include knowledge about the animals' abilities to perceive pain and suffering. Ignorance about the suffering, and a sense of naivety towards the extraction of animal products, and animal exploitation in general, once prevented concern but this is no longer the case. Attitudes towards whaling are not in a special position in this respect: at the same time as modern minke whaling started in the 1920s, so did another enterprise increasingly criticized as cruel: fur farming. But in the early decades of this industry and up to the 1980s, there was not much debate about this issue either. The products were detached from the animal, and many people concerned (or even involved) in animal welfare would wear fur without a thought of this being a contradiction. It is only in the last 20 years that criticism of legal animal abuse has become mainstream and naivety about the cruelty inherent in animal industries before the 1980s is not, of course, limited to Norway.

It the last 20 years much has changed in Norway. For example, the industry of fur farming has grown unpopular with the majority of the public (Opinion, 2000) and scepticism towards many forms of animal use has become more widespread (Strømsnes et al, 1995). Therefore, it would seem likely that scepticism towards whaling should grow as well. However, the strong governmental campaign steering the attitudes of 'good Norwegians' in a pro-whaling direction did indeed have an influence, and the information gap about sound reasons to oppose whaling would thus be expected to be harder to fill in Norway than it may be in some other countries.

The Minke: A Helper from God?

A common claim from Norwegian officials, as well as the whalers themselves, is that opposition towards whaling is rooted in a religious belief that whales are 'holy' and that the concern for them is a concern for a 'totem-animal' rather than the real biological animals of several different species (Stenseth et al, 1993; Frøvik and Jusnes, 2006; Ministry of Fisheries and Coastal Affairs, 2009). This claim is hard to verify by studying the arguments of animal welfare and environmental organizations, but respect for whales as part of folklore and religion exists in different cultures – notably in Norwegian coastal culture.

Norwegian fishermen used to view the minke whale not as a food resource but as a 'helper' sent from God, and this attitude can be traced back in early literature (i.e. 'Kongespeilet' or 'The King's Mirror' written around 1250) (Stenseth et al, 1993). The minke whale, and other whales, were said to lead the fish towards the shore where they were easily caught. At this time it seems that the fishermen were not well disposed towards the hunting of whales, and a conflict developed between them and whalers from the 1860s (Bogen, 1933; Ministry of Justice, 1994). The coastal fishermen protested against the whale killings and claimed it made it harder for them to catch fish as the natural balance was being broken. The whalers, the 'gold diggers' of the time, responded with allegations of superstition and envy, and claims were made that if the fish did not show up they would just have to switch to other businesses. The government looked to researchers to resolve the conflict and claimed that catching whales did not have anything to do with the poor fishing conditions.

Thus, there was no political will to initiate protection, but regulations did come into place through the 1880 Whale Law to ensure that whaling was not carried out too near the shore. The Norwegian Whale Protection Committee, however, was not satisfied and called whaling 'an enterprise corrupting the country' (Bogen, 1933). The struggle continued and, in 1904, Norway put in place its Whale Protection Law to end coastal whaling in the areas where the protesting fishermen lived. As we now know, this did not, however, protect whales effectively. Norwegian whaling companies expanded into other countries and other waters, and the huge whaling industry of this era was only just beginning to reap its profits.

Contrary to the claim of present-day politicians and the promoters of whaling that it is a 'tradition from ancient times' (see Stenseth et al, 1993; Parliamentary Debate, 2010), early industrialized whaling appears to have been very much a 'one man show' orchestrated by the 'father' of the exploding harpoon, Svend Foyn. It is true that whales were also sporadically hunted and killed in earlier times, as were many other animals that, unlike the whales, are not necessarily still perceived as imperative to kill today. But whaling was not seen as traditionally important in the beginning of the 1900s, though it certainly became *economically* important for the men who were involved in this new 'goldmine' of the sea.

Animal Welfare in Norway During the Growth of Whaling

A position of reverence for a species may entail concern and compassion for the individuals of the species, but one does not necessarily follow the other. Was this the case for the Norwegian fishermen? Did they also consider the welfare of the whales during whaling? If they did, this is not well documented and, indeed, one source claims that they were not concerned about the whales for the animals' own sake (Stenseth et al, 1993). But the question must be asked in a broader sense: in the beginning of industrialized whaling, were there concerns regarding animal welfare in Norway?

Yes, there were. The very first attempts to establish animal welfare organizations were made during the last half of the 19th century, the very same time as whaling developed, and the first national animal welfare organization, the Norwegian Society for Animal Protection, was established in 1859. The organizations were made up of veterinarians, teachers, policemen and the like, and the causes that drew their attention were typically the plight of animals whose suffering was happening in 'the open'. Thus the whipping of horses, physical abuse of companion and working animals, the plight of sick, starving and thirsty tame animals and starving wild birds during winter, i.e. 'visible' suffering, provided the main focus for the first Norwegian animal welfarists (Dyrebeskyttelsen Norge, 2010; Dyrebeskyttelsen

Halden, 2010). Also prominent artists in Norwegian culture raised their voices against mistreatment of animals. Bjørnstjerne Bjørnson (Bjørnson and Anker, 1982), Henrik Wergeland (Wergeland, 1897), Camilla Collett (Knudsen, 2009) and Cora Sandel (Knudsen, 2004) were among those who wrote about their love and concern for animals. The artists' personal experiences were often the reason for their involvement, but who would be able to gain personal experience of whaling, except the whalers themselves?

The suffering of the whales could not be easily seen by the Norwegian public in the 1900s and it is still a challenge to represent to them a scene that unveils itself under water. Those who were close enough to the industry to know what whaling was about were too close. They had gathered to make a living, and possibly a fortune, around the men that made the big money at the time. In these communities thoughts were not of the whales but on the money one could get from hunting. In the whaling communities that developed in the south of Norway the whales were assigned the role of resource only, and the developers of the whaling industry became respected heroes for the wealth they accumulated. Consideration for the whale species or individual whales seems to have been non-existent. Whaling was the way to wealth, and more whaling meant more wealth (and the Norwegians were not unique in this – see, for example, the chapter in this volume about British whaling by Simmonds). As with many other enterprises at the time, sons of whalers were expected to be whalers themselves, and this loyalty towards a family occupation may have helped to prevent criticism of the industry. Furthermore, some people still saw whales more as fish – a low status animal, which was, until very recently, thought not to feel pain – despite the fact that zoologists had agreed that whales are mammals back in the 16th century (Mielche, 1951; Stenseth et al, 1993).

However, reports exist from writers onboard whaling vessels who reacted to the cruelty when seeing whaling first-hand. Krarup Nielsen (1921), a doctor on a whaling vessel that set out from Sandefjord, describes the whales dragging the ship behind them and, at the end of the description, remarked: 'it cannot be denied that in these circumstances the poor animal is subjected to great suffering'. Mielche (1951) reports from the season of 1949 that he had 'big tears in his eyes' when the boat chased a blue whale and her calf: 'In my hometown Tjøme I am [a] member of The Society for Animal Protection, and I also came to think of my own wife and our little daughter. I was as soft at heart as a pudding, and I started to pray to the Lord that the harpoon should get out of place.' However, these admissions are only briefly reported, and the main themes of these books are about the wealth, the possibilities and the adventures of whaling.

Still the words of Norwegian explorer Thor Heyerdahl, born in the whaling city Larvik in 1914, may bear witness that not everyone grew up to see the world through the whalers' eyes. His expeditions brought him into contact with whales from a totally different perspective than the whalers:

> *After hearing the peaceful breathing of a whale beside me both day and night on fleets and reed boats out on the world's oceans, it is not only zoological clichés that the whale is a warm blooded mammal, just as much as horse or dog. It is very clear that a whale who swims around and struggles with a harpoon in his body suffers just as much as a moose or a bear would do in the same situation. If Norway wants to be seen as a civilized model country, we should show the numerous and growing crowd of critics that we are just as humane towards warm blooded fellow beings in the sea, as towards those that move on land ...'* (Heyerdahl, 1982)

Early 1990s: Creating a 'Whaling Nation'

Should a 'true Norwegian' be for or against whaling? It was not until the late 1980s and early 1990s that this question carried any meaning. Whaling was not an enterprise everybody identified with, and with increasing regulation, less people were associated with the business (Stenseth et al, 1993). But the Norwegian government wanted it to be different. Previously there had been whaling communities and whaling towns, but now the Norwegian government wanted to create a 'whaling nation'. The objective being that everyone should identify with the whalers.

Why were Norwegian politicians eager promoters of whaling? A range of different factors may have been at play. Traditionally farmers, fishermen and hunters have been well represented in Norwegian politics. As demonstrated by the roles held by Norway's present minister of fisheries in 2010, it is accepted that politicians move between different farmers' and fishers' associations and parliament. Wealth and politics tend to go together and many of the people involved in whaling certainly accumulated wealth. Whalers and fishermen are now often the same people and speak with one voice. Animal welfare, however, is not an economically strong movement in Norway. The whalers themselves give credit for the political pro-whaling attitudes to their own lobbying and their 'powerful connections' (Storhaug, 1994). Some critical voices suggest that a clear pro-whaling stance was a way for Norwegian politicians to gain the support of the coastal communities, and make them overlook more substantial issues where the government did not act in accordance with the will of these communities, i.e. EU-adaption with consequent loss of rights over fishery resources and general centralization (Aasjord, 1993; Hvoslef, 1993).

Could whaling possibly also be a welcome symbol – both for EU-friendly and EU-hostile politicians – to show Norwegians that Norway would 'hold its ground' against foreign political pressure? Regardless of how the attitudes came about, the fact was that the whalers had friends in strong political positions to fight for their interests, and the whales did not. In order to justify the use of resources and energy spent on the whalers' behalf – and the risk of adverse effects for Norway from whaling opposition abroad – all Norwegians needed to be convinced that whaling was important to them. Part of this political campaign was that whaling was important because whales were seen as a threat. The minke whale was thus assigned the role of a pest. And the whale protectors had to be depicted as not worthy of notice: they were described as 'ignorant', 'unscientific' and 'misguided' to help prevent full consideration of their arguments. Their message also needed to be seen as a threat in order to emphasize the importance of not letting them win.

Was there really a strategy as clear and crude as this to influence the Norwegian public's opinion? I argue that there was – there was an open campaign to convince politicians from other countries to accept Norwegian whaling, and a similar campaign was directed towards the public at home. According to accounts from the whalers, Prime Minister Gro Harlem Brundtland applied a rule that every communication about whaling should start with the words 'on scientific ground' (Storhaug, 1994), and the Norwegian scientific establishment has recently been profoundly criticized for its role (Holt, 2010).

A whole range of researchers – anthropologists, philosophers and biologists alike – were recruited to 'educate' the Norwegian public about whaling. Together they crafted a book sponsored by the Ministry of Fisheries and released it to coincide with the first commercial whaling season, 1993: *Vågehvalen – valgets kval* (Stenseth et al, 1993). It was intended to be for 'public education' and included chapters written by the minister of foreign affairs and minister of fisheries. The book is an illustration of the nature of the debate at the time. Chapters included a description of how unintelligent the minke whale is, several claims that the opposition towards whaling is based on belief in the 'super whale', and repeated statements about whales being a considerable threat to fish stocks. However, not everybody would be expected to read a whole book about whaling and so the main effort was channelled through the media.

Politicians were active in promoting the set of arguments in favour of whaling, and newspapers published stories about whales and whaling in which the main sources seemed to be the politicians and the whalers themselves. Whales were to be seen as unworthy of any special consideration. A journalist from one of the biggest Norwegian newspaper was onboard a whaling vessel during the 1993 season and described the whale as 'Seven and a half metres with meat ... a jaw like an old hoover bag ... lots of blood, intestines and other foul-smelling organs' (Ivertorp, 1993).

Was the media naturally receptive and willing to cite the message of politicians and whalers, or had they been cultivated? Perhaps both; in 1990 a public relations organization was set up on the initiative of politicians from Northern counties and the Whalers Association: Survival of the High North (Stenseth et al, 1993). This organization was funded nearly 100 per cent by the Norwegian state (Whalers Association, 2010) and its aim was to promote the interests of sea mammal hunters using grassroots credibility:

> *The goal was to improve the international media coverage, be a voice in international arenas like the IWC, to reduce pressure from the campaigns against Norwegian authorities, and by doing so, increase the probability for more whaling-friendly decisions from authorities ... Norwegian authorities had the same goals, they wanted to improve the media coverage and bring about the development of the hunting industries. Since then, there has been a close political and economic cooperation between Norwegian authorities and the Survival of the High North ... There has been systematic work to reach out to journalists, give them factual information, and correct factual errors when it has been necessary.* (Frøvik and Jusnes, 2006)

What Were the Arguments and Did They Work?

The main argument was as simple as 'whales eat our fish', and this claim was stated over and over again. 'They eat fish. Many fish' was the minimalistic mantra of the Whalers Association (Bastesen, 2003). Politicians described the whales as greedy and 'the rats of the sea' (Stoltenberg, 1992), and a leader of the Fisherboat Owners Association suggested that 'starving children in Africa' should be contrasted with the amount of fish eaten by the whales to show the importance of killing these animals (Pedersen, 1992). The academics contributing to *Vågehvalen – valgets kval* made somewhat more sophisticated comments on the same theme: 'The minke whale exerts an equally large "predator pressure" on fish stocks as humans ... it will cause a considerable loss on the resource store' (Stenseth et al, 1993). Minister of Foreign Affairs Johan Jørgen Holst, stated in the same book that: 'If the whale population grows too big, the fish stocks cannot be managed by humans to a degree that is necessary and desirable to ensure both nutrition and income.'

Did this argument ring true with Norwegians? Personal experience from outreach work in this period suggests that the political campaign did indeed have an effect. Norwegians seemed inclined to accept arguments along the lines that some wild animals are competitors for prey animals of interest to humans, and the whaling issue in some respects mirrored the controversy about wolves and other land predators claimed to be in competition with human interests in both wild animals and livestock (Lislevand, 2007). The 'whales eat fish' argument seemed to make simple sense, especially combined with data showing how numerous the whales were and how much each whale consumed, and the Norwegian public also had little access to research that could question the claims about minke whales as a threat to fish stocks.

In a recent Norwegian White Paper about sea mammal policy, claims about the necessity of whaling are less bold: 'The minke whale population can be expected to stabilize at a slightly higher level than today. With this as a basis the economic viability of other fisheries would not change considerably as a consequence of an end to the hunt' (Ministry of Fisheries and Coastal Affairs, 2004). Despite this acknowledgement, the government continues to assert that 'the responsible taking of sea mammals may also have a beneficial effect for the total use of other marine resources' (Ministry of Fisheries and Coastal Affairs, 2009).

In the campaign in support of whaling, emphasis was also put on how unappealing the minke whale was. It was 'not suited for whale watching', 'not especially social', 'not intelligent', it had 'a comparatively small and simple brain', it was 'not a good learner', had 'only primitive communication abilities' and so forth (Stenseth et al, 1993). Such claims are partly incorrect and partly irrelevant, but the Norwegian public did not have any other information than that from government sources and the occasional voice of animal rights groups trying to oppose this constant influence. For most Norwegians this was, at the time, not enough to spur interest in the minke whales as fellow beings.

Another important, and possibly crucial, argument was that of Norwegian independence: if 'we' want whaling, 'we' should not be told otherwise by the rest of the world. The important thing was that Norway stood up against the rest – this point was very much stressed by Prime Minister Gro Harlem Brundtland (Geelmuyden, 1998). To spark Norwegian nationalism on behalf of the whalers was a clever move. An image was generated of the proud whaler standing alone fighting for his livelihood, but this time it was not against the big beast and the rough sea, but against the beastly bureaucracy and the rough tide of hostile attitudes from Europe and the US. This was an image that some could identify with. The David and Goliath myth about the little Norwegian against the global financial powers conveniently left out important parts of history, such as Norwegian millionaires driving several whale species to the verge of extinction (Krarup Nielsen, 1921; Stenseth et al, 1993), but it was still effective in its simplicity.

In the continuation of this argument, it was also important that the average Norwegian felt that his own interests were threatened. For example, Gro Harlem Brundtland said: 'No, that is not only about the whale! Next time they will protest against us catching fish in nets ...' (Geelmuyden, 1998). 'They' were of course the other countries, but 'they' were also whale protectors in general, more specifically environmentalists and campaigners for animal welfare. This argument took an interesting turn; the main point was to describe animal welfare concerns as 'dangerous', 'illegitimate' and 'emotional' (Stenseth et al, 1993), in contrast to the hard facts of whale counting and stomach content analyses. But the parody went further: the Norwegian pro-whaling campaign invented the 'super whale', claiming that animal welfare groups campaigned for a fictional 'super whale' possessing all assumed positive characteristics of different whale species (Stenseth et al, 1993; Kalland, 1994).

A Glance at the Sealing Issue

The history of Norwegians' attitudes towards whales and whaling would not be complete without some consideration of the closely related sealing issue. In Norway, sealing and whaling have many similarities – both became big industries around the same time and were even initiated by the same man, Svend Foynd (Norsk Polarhistorie, 2010). Overexploitation and killing methods, which appeared cruel to those reporting from the hunt, characterized both industries (Mielche, 1951; Elstad, 2005; Haug, 2008). And yet both enterprises developed a close bond to Norwegian politicians, and still seem to hold a protected position in Norwegian politics today. In modern times, the same arguments are given by politicians to promote the sealing industry as are given for the whaling industry: the sea mammals 'eat all the fish'; Norway should not be told what to do; and concerns of animal cruelty are only due to 'misinformation' (Ministry of Fisheries and Coastal Affairs, 2009). But, moreover, a government-driven pro-sealing campaign just before the whaling issue reached its most intense period, made the climate even more hostile to an open debate about animal welfare and sea mammals (Gardar, 2006).

The Role of Norwegian Environmental Protection Organizations in the Whaling Debate

Norway's most prominent environmental voice, professor of philosophy and founder of 'deep-ecology', Arne Næss, was an outspoken opponent of whaling (Næss, 1992). However during the early 1990s, the Norwegian environmental organizations were to take a different position on the whaling issue. Philosopher Sigmund Kvaløy Sætreng described Norwegian environmental organizations as 'different' from those in other countries, because they 'protect man in nature' (Sørensen, 1993). Some organizations were apparently proud to promote this anthropocentric view. But whatever their initial stance, Norwegian environmental organizations also seemed to be a particular target for the pro-whaling campaign; Survival of the High North had as a specific aim to collaborate with these organizations (Frøvik and Jusnes, 2006), and so did the whalers (Storhaug, 1994). The environmental organizations were important if whaling was to be given the image of a 'green' enterprise (Norwegian Government, 2010), instead of being sullied by the shameful history of overexploitation.

In *Vågehval – valgets kval*, three environmental organizations were invited to write about their attitudes towards whaling: Friends of the Earth, the World Wide Fund for Nature (WWF) Norway, and Nature and Youth (the youth organization of Friends of the Earth) (see Aasjord, 1993; Hvoslef, 1993). WWF Norway provided a moderate critique of whaling at the time and the two other organizations were in favour of the practice, with Nature and Youth's contribution using arguments similar to those presented by the Norwegian government (Sørensen, 1993).

A survey was conducted in 1995 (Strømsnes et al, 1995) where all the environmental protection organizations and NOAH (the only animal welfare organization in the survey) were thoroughly studied. One of the questions asked of the members of the organizations was whether 'animal protection was an important part of environmental protection'. Interestingly, over 90 per cent of the members in all environmental organizations agreed with this statement. There was also a clear majority in the general public in agreement with this statement. This is arguably an indication that the government's attempt to discredit 'animal welfare' arguments failed.

ATTITUDES TO WHALE WELFARE TODAY

As a contrast to the environmental organizations, the national Norwegian animal welfare and animal rights organizations have never been positive towards whaling. The Norwegian Society for Animal Protection prosecuted the Norwegian government for the use of the cold harpoon in 1982. However, the case was dropped when the government stopped using it in 1984 (Hjort, 1985). During the early 1980s, a group called 'Aksjon Vern Hvalen' (Action Save the Whales) existed and, in 1989, NOAH was founded and it promptly launched an anti-whaling campaign as one of its main focuses.

When NOAH and the Norwegian Society for Animal Protection delivered 300 signatures against whaling to the Fisheries Committee of Parliament in 1993, the vice leader Peter Angelsen (*Bladet Tromsø*, 1993) said: 'I cannot on behalf of Parliament receive such nonsense ... But I do it with strong doubt. I also think it is shameful that you attack peaceful fishermen and whalers in coastal Norway like you do.' The attacks he referred to were verbal and written opposition to whaling in media. The two organizations again joined forces in April 2010 to deliver the first 4000 signatures in an ongoing signature campaign against Norwegian whaling. Again the signatures were delivered to the committee of parliament, and the leader Terje (personal communication to NOAH and Norwegian Society for Animal Protection, 2010) came out to receive them. His remarks were: 'Whaling is a difficult issue, there are people who want to conduct this business, but we also must take into consideration animal welfare ... We appreciate your input.' Even though the Parliamentary Debate (2010) on sea mammal policy just a month before had shown that the whalers and sealers still have a strong influence on Norwegian policy in these matters, this response is a sign that animal welfare attitudes have strengthened their position since the early 1990s.

In 1992, a survey showed that 21 per cent of Norwegians were 'against whaling under any circumstances' (Nagasaki, 1994). This number is hard to compare to the 18 per cent who thought it was 'important to ban Norwegian whaling' in 1995 (Strømsnes et al, 1995), as the questions are quite different. In 2009 another survey (Opinion, 2009) was conducted that found that 34 per cent of Norwegians thought that 'Norway should start phasing out commercial whaling with regards to animal welfare concerns in the hunt'; 26 per cent had no opinion on the matter. Only 21 per cent thought it was 'acceptable that it may take several minutes up to over one hour for the whales to die', whereas 50 per cent thought this was not acceptable. Only 7 per cent regularly ate whale meat, and 19 per cent had never tried it. This survey shows that animal welfare concerns regarding whaling are growing, but it also shows that there is a considerable proportion of the population (a third) who still do not have any opinion on the matter.

The newest Norwegian White Paper on sea mammal policy from the Ministry of Fisheries and Coastal Affairs (2009) takes no notice of this development. Stuck in the past, it echoes the refrain from the 1990s: increased economic output from whaling and sealing and strategies to counter the international opposition are the main aims. The killing methods are said to be 'internationally accepted' and existing campaigns against them are 'based on misinformation'. The recent survey data indicate that this repetition of old tunes is not entirely representative of the Norwegian public of today. As is the case for all animal welfare issues, public opinion has to lead the way towards political change for more humane policies.

Although the general awareness of animal welfare issues in whaling is growing in Norway, there is still a knowledge gap to fill about whales and whaling with the Norwegian public. Having been sold the myths of whales being one-dimensional 'vermin of the sea' and 'fish-eating meat-mountains', the interest in and respect for these animals need time to develop further.

Experience from many countries shows that contact with wild animals, combined with knowledge of them, forms a basis for respect. Do Norwegians want to learn more about the whales that visit our waters? Yes, it seems so – 53 per cent of Norwegians answered a recent poll that they would like to go whale watching (Opinion, 2010). Whales may see a better future in Norwegian waters if the government decides to listen to the 53 per cent who want to learn more about whales without killing them, instead of the 7 per cent who still eat a whale steak once in a while.

The journey of the minke whale in Norwegians' minds has moved over time from culturally important 'helper animal' to economically desirable 'resource', further down to despised 'pest'. It might now have the possibility of taking a climb towards its rightful place as 'fellow being' worthy of consideration.

References

Aasjord, B. (1993) 'Hvaldebatten og Naturvernforbundets rolle', in N. C. Stenseth, A. H. Hoel and I. B. Lid (eds) *Vågehvalen – valgets kval*, Ad Notam Gyldendal, Norway, pp167–172

Bastesen, S. (2003) 'Spekk, løgn & hvalfangst', www.folkevett.no/tema/biomangfold-og-naturvern/spekk,-loegn-&-hvalfangst

Bjørnson, B. and Anker, A. (1982) *De gode gjerninger redder verden*, Gyldendal, Norway
Bladet Tromsø (1993) 'Quarrel about whaling', *Bladet Tromsø*, 29 January
Bogen, H. (1933) *Linjer i Den norske hvalfangst historie*, Aschehoug & Co, Norway
Dyrebeskyttelsen Halden (2010) 'Historikk', www.dyrebeskyttelsenhalden.no/default.asp?ArtID=111
Dyrebeskyttelsen Norge (2010) 'Dyrebeskyttelsen Norge – på dyrenes side i 150 år', www.dyrebeskyttelsen.no/nyheter/dyrebeskyttelsen-norge-pa-dyrenes-side-i-150-ar
Elstad, Å (2005) 'Ishavsfangsten 1905', P2, 18 April, www.nrk.no/programmer/radio/p2-akademiet/1.1656268
Frøvik, R. and Jusnes, L. (2006) *Risikoanalyse sel og hval*, Survival of the High North, Norway
Gardar, J. H. (2006) 'En folkefiende vender hjem', *Ny Tid*, 10 February
Geelmuyden, N. C. (1998) '*Grepet i ord*, portrait interview with Gro Harlem Brundtland', first published in edited version in *Det Nye*, 1993, published in full length in Geelmuyden, N. C. (1998) *Ærlighetens komedie*, C. Huitfelds Forlag, Norway
Haug, T. (2008) 'The resources', in D. Pike, T. Hansen and T. Haug (eds) *Prospects for Further Sealing in the North Atlantic*, Proceedings of the 13th Norwegian-Russian Symposium, Tromsø, 25–26 August 2008, IMR/PINRO
Heyerdahl, T. (1982) 'Letter to Norges Dyrebeskyttelsesforbund', *Norges Dyrebeskyttelsesforbund*, 27 November
Hjort, J (1985) 'Letter concerning the case Norwegian Society for Animal Protection against the Norwegian government', Advokatfirma Hjort, Eriksrud, Myhre and Bugge Fougner, 5 February
Holt, S. (2010) 'Norwegian minke whaling in the North Atlantic: A memoir', www.wdcs.org/submission_bin/norwegianminkewhaling,amemoirsh2010.doc
Hvoslef, S. (1993) 'Miljøvernorganisasjonene og hvalfangsten – WWF', in N. C. Stenseth, A. H. Hoel and I. B. Lid (eds) *Vågehvalen – valgets kval*, Ad Notam Gyldendal, Norway, pp173–182
Ivertorp, L. (1993) 'Hvalfangststart med et smell', *Aftenposten*, 25 June
Kalland, A. (1994) 'Super whale: The use of myths and symbols in environmentalism', in *11 Essays on Whales and Man*, Survival of the High North, Norway, www.highnorth.no/library/myths/su-wh-th.htm
Knudsen, K. (2004) *Dyrevenner. Forut for sin tid*, Vesle-Brunen Forlag, Norway
Knudsen, K. (2009) *Camilla Collett. Jeg elsker dyrene*, Vesle-Brunen Forlag, Norway
Krarup Nielsen, A. (1921) *En hvalfangerfærd*, H. Aschehoug and Co, Copenhagen
Lislevand, T. (2007) 'Selsom fangst', *Aftenposten*, 12 April, www.aftenposten.no/meninger/debatt/article1733348.ece
Mielche, H. (1951) *Hval i sikte*, AS John Griegs Forlag, Bergen
Ministry of Fisheries and Coastal Affairs (2004) 'Stortingsmelding nr 27 – norsk sjøpattedyrpolitikk', Ministry of Fisheries and Coastal Affairs, Oslo, Norway
Ministry of Fisheries and Coastal Affairs (2009) 'Stortingsmelding nr 46 – norsk sjøpattedyrpolitikk', Ministry of Fisheries and Coastal Affairs, Oslo, Norway
Ministry of Justice (1994) 'Bruk av land og vann i Finnmark i historisk perspektiv', NOU-report, chapter 7.17, www.regjeringen.no/nb/dep/jd/dok/nouer/1994/nou-1994-21/9.html?id=374583

Næss, A (1992) 'Om høsting av hval', presentation for seminar on whales at Senter for Utvikling og Miljø, Oslo, published in *Natur og Miljø*, Naturvernforbundet, 29 July 1992
Nagasaki, F. (1994) *Public Perception of Whaling*, Institute of Cetacean Research, Japan, http://luna.pos.to/whale/icr_pub_pall.html
Norsk Polarhistorie (2010) 'Svend Foyn, 1809–1894', www.polarhistorie.no/personer/1159343938.61
Norwegian Government (2010) 'Hvalfangst', www.norge.se/Om-Norge/okonomi/trade/marine/whaling/
Opinion (2000) *Holdninger til pelsoppdrett*, Opinion AS, Norway
Opinion (2009) *Holdninger til hvalfangst*, Opinion AS, Norway
Opinion (2010) *Holdninger til hvalfangst*, Opinion AS, Norway
Parliamentary Debate (2010) 'Stortinget – Møte tirsdag den 9 mars 2010 kl. 10', www.stortinget.no/no/Saker-og-publikasjoner/Publikasjoner/Referater/Stortinget/2009-2010/100309
Pedersen, J. (1992) *Vesterålen*, 2 December
Sørensen, H. (1993) 'Miljøbevegelsen og vågehvalfangst – Natur og ungdom', in N. C. Stenseth, A. H. Hoel and I. B. Lid (eds) *Vågehvalen – valgets kval*, Ad Notam Gyldendal, Norway, pp183–186
Stenseth, N. C., Hoel, A. H. and Lid, I. B. (eds) (1993) *Vågehvalen – valgets kval*, Ad Notam Gyldendal, Norway
Stoltenberg, T. (1992) 'Spekk, løgn & hvalfangst', www.folkevett.no/tema/biomangfold-og-naturvern/spekk,-loegn-&-hvalfangst/ and http://vassdragsvern.no/cgi-bin/naturvern/imaker?id=13204
Storhaug, J. (1994) *Bastesen – en kysthistorie*, Gyldendal Norsk Forlag, Oslo
Strømsnes, K., Grenstad, G. and Selle, P. (1995) 'Miljøvernundersøkelsen 1995 – Dokumentasjonsrapport', Rapport 9616, LOS-Senteret, Bergen
Wergeland, H. (1897) *Skrifter i Udvalg*, Olaf Husebys Forlag, Norway
Whalers Association (2010) 'Stortingsmelding nr 46 – norsk sjøpattedyrpolitikk', presentation in Committee Consultation, 12 January, unpublished

9

Of Whales, Whaling and Whale Watching in Japan: A Conversation

Jun Morikawa and Erich Hoyt

In 2009–2010 Jun Morikawa and Erich Hoyt embarked on a correspondence and then met to talk at length about the past, present and the future of whales, whaling and whale watching in Japan. Professor Morikawa, a specialist in Japan's foreign aid programmes and author of *Whaling in Japan: Power, Politics and Diplomacy* (2009), has attempted to unravel the background, logic and future of whaling. Erich Hoyt, author of 18 books mainly on whales, whale watching and marine protected areas (e.g. Hoyt, 2011), has travelled around Japan for nearly 20 years, largely as an invited guest for whale-watching communities speaking at conferences and workshops. The following represents extracts from their discussions.

Introduction

It has been many years since the aftermath of World War II when Japan depended on whales for food. The Americans encouraged Japan to return to whaling after the war. But by the 1980s, whale meat was eaten by fewer and fewer people. In recent studies, young people – children and young adults – no longer choose to eat whale. Some restaurants serve it and sometimes people buy it for use at home, but it is mainly older people who have a nostalgic taste for it. Given this loss of interest in whale meat as a food, is whaling commercially viable? The Japanese government perpetuates the argument that whaling and eating whale meat is part of the country's culture and this still seems to be a key part of Japan's domestic and international agenda. Will it succeed? Are 21st-century Japanese youth interested in pursuing the whaling agenda indefinitely? At the same time, whale watching has started up all around Japan and millions of Japanese seek out whale-watching opportunities on foreign trips to the west coast of North America and to New Zealand and Australia. Is this a significant development? What are the implications of Japan's national policies on marine conservation in Japan? These and other matters are explored here and we use Jun's recent book *Whaling in Japan: Power, Politics and Diplomacy* as an additional resource on these matters.

Erich Hoyt (EH): *Outside of Japan, whaling is often viewed in simplistic terms without realizing the complexity of the industry in Japan. Can you help explain to people the structure of the industry?*

Jun Morikawa (JM): There are three types of whaling being practised in Japan:

1 large-scale de facto state-run and quasi-commercial whaling in Antarctica and the Northwest Pacific;
2 small-scale coastal whaling centred on small-type species; and
3 large-scale dolphin hunting, which constitutes a part of coastal whaling.

When we hear that whaling in Japan is structured in this multi-layered way it gives the impression that Japan's commercial whaling as a whole is running on a commercially viable basis. However, the answer to the question of whether Japan's commercial whaling is viable is 'no', because the conditions that have supported Japan's commercial whaling seem no longer to exist.

The first problem for the whalers is, somewhat ironically, that there is no dependable actor in the Japanese private sector that is willing and able to take on the task of running whaling in the deep seas. As early as the mid-1970s, major Japanese whaling companies such as Maruha Co. and Nissui Co. expressed their intention to withdraw from the whaling business. Their reasons included the depletion of the whales, an emergence of an international climate that did not favour the continuation of commercial whaling, and high fuel prices after the oil shocks of 1973 and 1979, combined with the shrinking domestic market for whale meat. In other words, market forces by then had made whaling companies, in general, realize that large-scale commercial whaling is no longer economically viable.

Nissui Co. abandoned whaling in the seas around Japan in 1976 and Maruha Co. in 1977. Another major whaling company, Kyokuyo, had already done so in 1964. However, the Japanese Fishery Agency used their 'administrative guidance' to resist the desire of the major fishing companies to withdraw. As a result, three major and three medium whaling companies were merged and a new flagship company, the Nihon Kyodo Hogei, was created in 1976 to continue commercial whaling in the Antarctic ocean.

EH: *Why then is there still whaling in Japan? Is it only a limited market to be satisfied? Is the continuation of whaling an economic matter in Japan? Or is it a political matter, or more of a cultural matter?*

JM: The issue has been highly politicized since 1985 when the Parliamentary League for the Promotion of Whaling was created and it would be helpful to understand a little about whaling history and the 'actors' involved.

Japanese people who eat whale meat can be mainly found in the generations over the age of 50 who were born during or just after World War II, or who grew up in the reconstruction period during the early stage of rapid economic growth (from the later part of 1950s to the beginning of the 1960s). During that period there was a food shortage in general and a protein deficiency in particular. Many people from these generations, together with their parents, ate what was then cheap whale meat both at home and through the school lunch programme, implemented to remedy protein deficiency. For the majority of Japanese people, once they had re-established their economic and social bases, they soon returned to their 'traditional' national food culture and forgot about this substitute meat. Hence it was natural that from the late 1960s onwards that consumption of whale meat began to drop sharply.

Thus Japan's so-called national whale meat-eating culture was created under special historical conditions and lasted about 20–25 years. (We might also note that the Fishery Agency was created in 1948, during a time when large-scale whaling was vigorously promoted as a national policy.)

Returning to the present day and the current 'actors': the Japanese Fishery Agency's high-ranking officials do not want to lose their whaling-related budgets, posts or their power. Meanwhile the diminished whaling industry wants financial, administrative and political assistance from the Fishery Agency. Then there are the politicians whose constituencies happen to be located in coastline areas who want to obtain as much support as possible from local electors. They try not to upset the fishermen and their associates by taking an anti-whaling posture. Many of these politicians show a pro-whaling stance not only in their own constituency but also inside of the Diet (Japan's two-chambered national legislature), combining to form a pro-whaling parliamentarian group. In fact within the Diet there is a sort of consensus on whaling that transcends various party lines.

Such politicians try to portray themselves as true nationalists who defend Japan's unique and traditional whaling culture and fight back against what they portray as the emotional, illogical, racist and cultural imperialist forces of the West.

This is an ideal situation for the Fishery Agency since they can count on these politicians' political backing within the Japanese bureaucratic world and the IWC conference room. In return the politicians can count on support from the Fishery Agency to help materialize various local public works projects and so forth.

In the fishing industry in general, as well as in various industrial organizations, there are many retired Fishery Agency officials occupying important jobs. Therefore, one can see very cosy relationships between these power elites. Their linked overarching political aims are firm and clear: maintain and re-expand commercial whaling by all means. And indeed they have a lot of resources to achieve their intended goal.

EH: *Do you think commercial companies will be interested in reinvesting in the future if circumstances change? The IWC discussed a limited lifting of the commercial whaling ban at its 2010 meeting. This was not passed but it gained a surprising amount of support.*

JM: I do not think Japan's former major whaling companies will return to their old business even if the conditions for commercial whaling in the deep sea are officially sanctioned and the Japanese Fishery Agency offers its support.

We can also look at this from a different perspective. Big Japanese fisheries companies now depend very much upon foreign business counterparts for securing, and sustaining in the longer term, large amounts of the various marine products they require.

Foreign countries have also become important export markets for their fish-related products. In another words, Japanese fishing companies must now be careful about how they are being perceived by the consumers of these countries.

From the management strategy point of view too, these Japanese multinational fishery companies have realized the importance and usefulness of cultivating, maintaining and strengthening a good relationship with resource-rich countries and their business counterparts as well as respected local communities. So, they are very conscious of their social responsibilities, especially in the human rights and environmental spheres today.

The US, Canada, Australia and New Zealand are marine resources-rich countries and many of the citizens and NGOs of these democratic countries have a strong focus on nature conservation issues in general, and the whaling issue in particular.

EH: *So there are good commercial reasons for keeping away from the whaling issue for many would-be investors. But what about Japan's own international interests? The annual IWC meeting brings Japan's whaling activities into view each and every year, complete with highly critical comments from many quarters.*

JM: From the commercial sector, the only possible remaining actor will be the Kyodo Senpaku Co., which was established in November 1987 (its predecessor was Nihon Kyodo Hogei Co.). This company has been supplied with various vital resources, including a whaling factory ship and, on 24 March 2006, the company announced that its shares would be transferred to several public-interest corporations including the Japanese Institute of Cetacean Research (ICR).

So, here we have a situation where there continues to be considerable state support and if Tokyo should decide to resume re-legalized large-scale deep-sea and sea-around-Japan (within its own Exclusive Economic Zone) commercial whaling, we can anticipate that a considerable injection of taxpayers' money will follow. However, the views of the Japanese public will have a profound political effect in such matters; hence, in recent years, the Japanese government, especially the Fishery Agency, has intensified its pro-whaling public relations activities internally as well as externally.

EH: *What about the economic viability of coastal whaling?*

JM: According to my understanding there are two kinds of coastal whaling being practised in Japan: type A targets the so-called 'small' whales such as Baird's beaked whales and pilot whales in what is referred to as 'small-type coastal whaling'; type B specifically targets dolphins and porpoises, including in the drive hunts.

Type A coastal whaling is currently being carried out mostly by a handful of small local whaling companies located at four small regional whaling towns, Taiji, Wada, Ayukawa and Abashiri, and uses smaller fishing vessels (less than 48 tons). For example, these whalers were allowed to catch around 182 whales per year in 2001. However, their economic situation has been rather tough and to minimize costs they have tied up half their vessels. The running costs for coastal whaling are high and their earnings rather low. Their economic hardship can also be related to the inflow of considerable amounts of whale meat from the state-sponsored research whaling operations into the already saturated domestic market (noting also that meat from whales bycaught in fishing nets has now been allowed into the domestic distribution markets by the Fishery Agency). Therefore, currently type A coastal whaling has apparently lost its economic viability.

EH: *However, we hear a lot about the traditional nature of this form of whaling and its cultural value. Japan, at the IWC, often seems to seek to place coastal whaling alongside the 'indigenous' or 'aboriginal' whaling claims of other nations, such as the Alaskan Inuit.*

JM: Coastal whalers and the Japanese Fishery Agency have been developing the notion that this whaling has a strong historical tradition and its social and cultural value should be respected and preserved. They claim that Japanese traditional coastal whaling is fundamentally different from that of so-called commercial whaling in general, and therefore Japan should be allowed to maintain this form of whaling. But this notion has many questionable aspects.

For example, is it true that this form of whaling and related social, economic and cultural structures have not experienced any historical transformations since the introduction of modern whaling at the beginning of the 20th century? My observations suggest that the historical traditions related to whaling and relevant communities' social, economic and cultural structures have been modernized; these are 'modernized' and highly commercialized traditions, although the degree and extent of modernization varies between the towns. For example, Abashiri's commercial whaling activities only began during World War I, in 1915, initiated by a newly emerging whaling industry from the mainland. From time to time, some coastal Ainu people had engaged in traditional and sustainable whaling activities; however, the Japanese government has tended to ignore this important traditional activity of the Ainu people.

EH: *There is a debate between countries about whether the smaller species are covered by the provisions of the IWC and Japan claims that they are not. Their large-scale killing has continued in spite of the moratorium declared in 1986. What of the type B whaling, as you call it?*

JM: A massive amount of dolphin meat has been distributed to supermarkets and bars labelled as whale. It has been claimed that the production of dolphin meat rose to 3736 tons in 1988 in order to make up for the lack of whale meat due to the IWC ban on commercial whaling. However, this resource has been managed since 1989 by the imposition of fishing industry regulations and fishing boundaries and this has kept the catch hovering roughly between 1200 and 1700 tons since 1991. Some dolphin meat has been consumed in traditional localized dolphin-eating regions, with some sold in the cities of the Kansai region as 'whale meat'.

According to Professor Toshio Kasuya's comment on the issue,'The species most commonly caught is the Dall's porpoise and the Japanese Ministry of Agriculture, Forestry and Fishery allows over 20,000 dolphins and porpoises to be taken annually' (Kasuya, 2007). A recent estimate is that 23,000 small cetaceans are taken annually, making Japan the top porpoise- and dolphin-eating country in the world. In addition, significant numbers of dolphins and whales are caught in drift and fixed nets and allowed to be distributed into the domestic market. This is discussed further in my book.

EH: *How do Japanese people view these hunts?*

JM: Most know nothing about them. The town of Taiji – as featured in the film *The Cove* – is just one place where coastal dolphin hunting occurs but this hunting is not widespread around the country. Only 7 of the 47 Japanese prefectures were involved in dolphin hunting in 2007: Iwate stands out, 10,218 dolphins were killed there in that year. In Wakayama, where Taiji is located, 1623 were taken, in Hokkaido 806, Miyagi 126, Okinawa 87, and Niigata 2 and, finally, just 1 in Kanagawa.

Furthermore, the price of dolphin meat has dropped from the boom time for it in 1994, when it reached ¥600 per kilo, then down to ¥300 per kilo and, as of spring 2010, the price of dolphin meat dropped further to ¥150 per kilo.

EH: *What do you think about the future of coastal whaling activities and could it again become fashionable to eat whale meat?*

JM: There are a number of reasons why the re-establishment of viable coastal commercial whaling in Japan would not be easy:

1 As I pointed out earlier, the former major and medium-sized whaling companies moved on from the whaling operations a long time ago, and they are unlikely to have any enthusiasm or capability to return to such activities.
2 Then there is the problem of pollution in dolphin and whale meat and blubber. While there is some debate about the significance of this, certain substances accumulate in these animals and concerns have been expressed to the government by citizens' groups investigating food contamination issues. In addition, there have been a number of studies, notably by Professors Tetsuya Endo and Koichi Haraguchi (of the Health Sciences University of Hokkaido, Hokkaido, Japan and Daiichi College of Pharmaceutical Sciences, Fukuoka, Japan, respectively), revealing heavy mercury and other contamination in some cetacean samples marketed for consumption in Japan (Endo et al, 2002, 2004; Simmonds et al, 2002). The results of such studies have not been widely reported inside Japan and domestic opposition to the

government's policy of reintroducing whale meat into school lunches has not yet reached a significant level. And coming back to your question, I suspect that the majority of Japanese consumers who are not used to eating whale meat in their daily lives are unlikely to change their dietary behaviour and even the 'whale meat lovers' might think twice about their diet if they were informed of a dietary risk.

3 The third problem for any renaissance of the industry is the contraction of the whale meat markets. The demand for whale meat in Japan is now negligible. A *Sankei Shimbun* article of 6 March 2006 said that in 2005, consumption of whale meat was roughly 7300 tonnes. This is no more than 0.2 per cent of overall meat consumption, which stood at 3.76 million tonnes. Even when whale meat consumption was compared to tuna, it was only around 1 per cent of tuna consumption and if there were no officially sponsored, nationwide campaigns to encourage people to eat whale meat, consumption would probably be even less.

EH: *Do you sense, as I do, that attitudes are changing in Japan towards whales and dolphins? I have seen some of this on my travels around Japan and have noticed changes even over the last two decades. Where do you think this change is coming from?*

JM: First of all we should remember the existence of other forgotten traditions, as Hiroyuki Watanabe notes in his excellent study *A Historical Sociology of the Whaling Issue* (Watanabe, 2006; published in English translation as Watanabe, 2009). In many coastal fishing villages local traditions quite distinct from the tradition of whaling and eating whale meat still existed and were being faithfully preserved at the beginning of the 20th century. In other words, whales were not eaten and were referred to by the title 'Ebisu-sama', which refers to one of the Seven Gods of Good Fortune (Shichfukujin). These traditional positions and perceptions seem to have somehow survived and, like a groundwater network, are likely to affect the current trend of 'rediscovering nature' within Japanese society.

Indeed, new social 'currents' are emerging rather gradually and spontaneously from the bottom, meaning they have grassroots sympathy and support within Japanese society. One, for example, is the view that whales and dolphins are admirable wildlife and that we should let them live as they like in Mother Nature. There is also a developing preference for the idea of 'non-lethal sustainable use', for example, support for whale and dolphin watching.

These interesting phenomena can be seen, for example, from the Japanese public's more active response and involvement in stranded whale and dolphin rescue operations, and from the steady development of whale and dolphin watching in Japan since the late 1980s.

So in the longer run we shall see a new political and social situation. However, for the moment this is being kept hidden and weakened by the Japanese Government's skilful pro-whaling public relations activities.

Now I must ask you about whale watching. What have you found in your travels around Japan with whale watching? How does it fit in with the whales that

Figure 9.1 *A whale-watching lookout on Chichi-jima in Ogasawara, Japan*

Note: This is one of the best lookouts in the world, providing great views of humpback whales on the breeding grounds from January to March, with passing groups of dolphins and other sightings year round.

Source: E. Hoyt

Japanese like to see on holiday when they visit Hawaii, Australia or New Zealand? Do you think whale watching will catch on in a big way in Japan?

EH: *I suppose that most Japanese don't spend time thinking about whales and dolphins but I have observed that those who do are very passionate about them. Among the 10 per cent of Japanese who travel frequently, many of them have met whales on holiday trips. No one has ever made an estimate of the number of Japanese people watching whales on foreign holidays each year, but it must be several million, based on the percentages of Japanese tourists on whale watches in some of the more popular spots in Hawaii, Alaska, British Columbia, California, Mexico, New Zealand and around Australia. It could be some of the same people taking multiple trips – we don't know – but either way substantial numbers of Japanese have met whales abroad.*

But it's a different matter seeing whales in your own country. I have been on at least 40 whale watches over the years in eight different communities around Japan and I can tell you that the Japanese whale watchers are even more enthusiastic when

Figure 9.2 *Whale watching on Chichi-jima in Ogasawara*

Note: The Ogasawara Whale Watch Association collaborates with local scientists and teachers to present regular talks at this whale-watching lookout. Some operators bring their customers here to orientate them to the islands and the surrounding waters, and to give them their first glimpses of whales.

Source: E. Hoyt

watching whales in their own country. Whale watching started in the southernmost Ogasawara Islands in 1988 when Japanese cartoonist and artist Kyusoku Iwamoto decided to take his bird watching buddies to see humpback whales and various dolphins (see Figures 9.1 and 9.2). The numbers grew from there and it is a healthy industry now in Ogasawara, providing substantial income as part of the local tourism. At the same time whale watching has spread to a dozen other communities all around Japan (see Figure 9.3); it attracts perhaps 98 per cent Japanese people; foreigners are not coming to Japan to go whale watching – as you might imagine there is a bit of an image problem. And in terms of Japanese people watching, I think the numbers are smaller than abroad – we're talking about 100,000 to nearly 200,000 people a year over the past decade going whale watching all over Japan. Still, over a decade, that's well over a million whale watchers, or whale watch trips, *around Japan – some of them are repeat trips by a growing number of dedicated Japanese whale watchers.*

JM: Is whale watching compatible with whaling?

Figure 9.3 *Whale watching on local fishing boats in Tosa Bay, Shikoku Island*

Note: Whale watching of Bryde's whales and various dolphins takes place in Kochi-Prefecture from several ports at Kuroshio-cho, formerly called Ogata, on Shikoku Island.

Source: E. Hoyt

EH: *I know there have been some conflicts in Japan, and other whaling countries, when whales or dolphins are hunted near the tourist sites. For the most part, whalers and whale watchers have not crossed paths. But the larger question of course is whether a country can continue with such diametrically opposed 'uses' of whales and hope to be successful. I think there are some tourists who visit whaling countries to go whale watching out of curiosity or to support the local whale watch industry in the face of whaling. So there may be some added business that way. But I think more business is lost by the incompatible image that is presented to the outside world. I think the economic studies of whale watching over the past decade have clearly shown the substantial value of whale watching, far beyond whaling, even without considering the heavy subsidies of whaling. But more than that, how we treat whales is a reflection of our growing awareness of the loss of wild nature and the contamination of the ocean and I think people everywhere are thinking about these things, and the implications.*

When I go to Japan, one thing that I say when Japanese people ask my views about whaling is that I think that we simply know too much about whales and dolphins to kill them. I think in human history until the past couple of decades, we knew very little about them. In the 1960s we started to get some glimpses of dolphin intelligence in aquariums and in the early 1970s the first studies of these animals in the wild began in the US, Canada, Argentina and Mexico, with humpback, gray, killer and southern right whales. Before that, the study of whales was classical zoology and anatomy, the study of dead animals. But today, through the patient study of these animals in the wild, we know that they are sentient, sharing with humans many of the higher social mammalian traits. Beyond this, we know they sing complex songs that echo across entire ocean basins, that they have powers of sensory discrimination in the sonic realm far beyond our man-made sonar, that they can be very long-lived, even longer than humans, that they use tools and that they pass on their 'culture'. All of this, with their distinctive social and behavioural adaptations, leaves us full of wonder. We know too much.

References

Endo, T., Haraguchi, K. and Sakata, M. (2002) 'Mercury and selenium concentrations in the internal organs of toothed whales and dolphins marketed for human consumption in Japan', *The Science of the Total Environment*, vol 300, pp15–22

Endo, T., Haraguchi, K, Cipriano, F., Simmonds, M. P., Hotta, Y. and Sakata, M. (2004) 'Contamination by mercury and cadmium in the cetacean products from Japanese market', *Chemosphere*, vol 54, pp1653–1662

Hoyt, E. (2009) 'Whale watching', in W. F. Perrin, B. Würsig and J. G. M. Thewissen (eds) *Encyclopedia of Marine Mammals*, 2nd Edition, Academic Press, San Diego, CA, pp1219–1223

Kasuya, T. (2007) 'Japanese whaling and other cetacean fisheries', *Environmental Science and Pollution Research*, vol 10, pp39–48

Morikawa, J. (2009) *Whaling in Japan: Power, Politics, and Diplomacy*, Columbia University Press, New York, NY

Simmonds, M. P., Haraguchi, K., Endo, T., Cipriano, F., Palumbi, S. and Troisi, G. M. (2002) 'Human health significance of organochlorine and mercury contaminants in Japanese whale meat', *Journal of Toxicology and Environmental Health*, vol 65, pp1211–1235

Watanabe, H. (2006) *Hogei Mondai No Rekishi Shakaigaku: Kin-gendai Nihon Ni Okeru Kujira to Ningen* [*A Historical Sociology of the Whaling Issue: Relationships Between Whales and Human Beings in Modern Japan*], Tōshindō, Tokyo, Japan

Watanabe, H. (2009) *Japan's Whaling: The Politics of Culture in Historical Perspective*, Trans Pacific Press, Melbourne, Australia

10

A Contemporary View of the International Whaling Commission

Richard Cowan

For much of the last decade it has become fashionable to decry the activities of the IWC – or rather what is seen as its lack of activity. The IWC is often described as 'dysfunctional' and 'moribund'. Both pro- and anti-whaling nations claim that the Commission is failing to address their concerns. These lines of argument seem to me to ignore the very real achievements of the IWC and to deny the realities of international negotiation.

The adoption by the IWC in 1982 of the moratorium on commercial whaling was, I believe, the boldest step towards environmental conservation taken in any forum in the 20th century. Despite the weaknesses inherent in the IWC's structure – the right of parties to object to, and therefore circumvent, measures adopted by the majority; the unfettered right of contracting governments to issue 'special permits' for whatever lethal scientific research they deem appropriate – the moratorium has resulted in a massive reduction in the number of whales being killed either directly or indirectly for the purposes of harvesting whale products.

At the same time, the IWC has begun to recognize the need to address wider environmental threats to whale populations, threats that arise from climate change, ship strikes, pollution (both by noise and by toxic discharges) and from habitat destruction and degradation. Clearly more work needs to be done in these areas and the IWC's Scientific Committee might usefully devote more of its resources to their examination, at the expense of work currently being done to determine how commercial catch limits might be set. Some of those engaged in whaling operations have also recognized the need to address animal welfare concerns arising from whale hunting, as well as the safety of the hunters. More work needs to be done here too, but the first steps have been taken.

On the other side of the great divide, the whaling nations can and do take sufficient whale meat to satisfy local demand. While their activities certainly attract criticism from the non-whaling countries of the IWC and from wider public opinion, parties to the ICRW have generally interpreted the Convention as providing a degree of legality to these activities. Thus, while many countries make formal diplomatic protests about whaling, none has, in recent years, seen whaling as a cause to introduce trade or other sanctions against the whaling nations, nor to act in ways that would undermine generally good relations with them on other fronts. It seems unlikely that the same degree of international legitimacy would attach to whaling activities carried out outside the ICRW, even within the terms of a new Convention whose members were exclusively actual or potential whaling nations.

We have had some interesting discussions within the IWC and outside it about the possible role that conflict resolution procedures might play in enabling the deadlock within the IWC to be broken. But for conflict resolution procedures to work effectively, it is surely necessary that all parties to the conflict can see that common ground exists between parties and that occupation of that ground is possible without having to sacrifice key principles. If, for both sides, the gain from moving to the common ground is worth the pain involved in doing so, then resolution is possible. Equally, if for one side or the other, the pain involved in moving to the common ground is less than that which might result from their opponents' taking unilateral action, then resolution is still possible, but negotiations are more difficult and more unequal because of the element of duress involved.

For some countries in the anti-whaling group, there is probably no price at which they would willingly endorse a resumption of commercial whaling. For others, there may be such a price, but it is likely to be a high one and appears on the basis of discussions to date to be too high for those who are asked to pay it. At the very least, it would seem to involve:

- the adoption of a very precautionary approach to the setting of catch limits (the Revised Management Procedure (RMP) – the scientific basis for the setting of catch limits – as approved by the Commission in 1992, or stronger);
- significant curtailment (and eventually total cessation) of lethal whaling operations carried out in the name of science;
- the adoption by all parties without objection (either immediately or at any future date) of a set of rules for the conduct of commercial whaling operations; these would have to be sufficiently transparent and robust to ensure that catch limits, once set, would never be exceeded; moreover, they would have to involve full international oversight and be capable of being enforced by the Commission in case of any infractions by parties;
- there would also have to be international oversight of a market-sampling regime to ensure that all cetacean products exposed or offered for sale were derived from legally hunted whales;

- costs incurred in regulating commercial whaling would have to be borne primarily by the whaling industry, or by the governments of countries issuing licences to take whales;
- recognition of existing whale sanctuaries and a willingness to support the creation of new ones;
- deployment of resources within the IWC's Scientific Committee such that a greater proportion of these is used to investigate and better understand the impact on cetacean populations of environmental threats to their survival.

Some countries would go further, insisting that those taking whales should collect (or else provide for the collection of) welfare data on whales struck or killed, and that conditions should be set with respect to visibility and sea state, so as to minimize the stress and suffering caused to hunted whales.

It is perhaps not surprising that these terms have proved unacceptable to the principal whaling nations and their allies. The recognition of the Southern Ocean Whale Sanctuary and the cessation of lethal whaling for the purpose of scientific research would deprive Japan of the greater part of the whale products to which it now has access. The catch limits likely to flow from the operation of the RMP to whale stocks in the North Pacific, far from compensating for this loss, would be likely to increase it by setting catch limits at levels substantially lower than current scientific takes.

If the terms on which a majority of the anti-whaling bloc might be induced to agree to a resumption of commercial whaling are unacceptable to the whaling nations, what opportunities arise for the whaling nations should they decide to leave the IWC and set up an organization more amenable to their wishes? Theoretically, the possibilities are limited only to the extent that non-whaling countries would presumably bar whaling vessels from their Exclusive Economic Zones. But I am convinced that the reality of the situation would be rather different. As I noted earlier, the fact that most parties to the ICRW see the Convention as giving current whaling operations some degree of legitimacy constrains those countries opposed to whaling from taking significant economic or political action against the whalers, as many of the general public in anti-whaling countries would demand. If economic action by anti-whaling countries were to make itself felt, then the public in the whaling countries themselves might seriously question the wisdom of pursuing a policy that is of importance principally to a minority of their populations and that attracts significant adverse international criticism. Would the balance between gain and pain necessarily be weighted in favour of the continuance of whaling operations? I doubt it.

Over the last three years or so, some in the IWC have made strenuous efforts to put together a deal under which virtually all current whaling activities would be legitimized, but with catch limits significantly lower than those that the whaling nations currently award themselves. These efforts produced no definitive result, primarily because the reductions offered by the whaling nations were considered scientifically unsound and inadequate to justify the Commission undermining both the moratorium and its long-standing opposition to 'scientific' whaling, particularly in the Southern Ocean.

It seems to me highly unlikely that this proposed deal, if indeed there ever was a deal on offer, can be resurrected in the foreseeable future, so the IWC is effectively obliged to accept the status quo for the present. On the basis of this brief analysis I am obliged to conclude that there are no obvious routes to breaking the current deadlock that do not involve for either side or both sides an excess of pain over possible gain.

There are times, as every chess player knows, that if you see you cannot win a game, then playing for the draw is the honourable and courageous thing to do. Stalemate need not be seen as negative, it avoids either side feeling that they have lost completely and it preserves, in the instance of the IWC at least, the very considerable benefits that both sides derive from the current situation. More intellectually attractive solutions may be possible in the longer term, if attitudes on either side change. But for the present, both sides have too much to lose and too little to gain to make meaningful compromise possible.

If participants on both sides of the whaling debate were to recognize this, then some of the tension and mistrust that has characterized meetings of the IWC in recent years would disappear. The IWC, relieved of the burden of trying to resolve matters whose resolution has eluded it for at least the last 15 years, could afford to meet less often and in a less combative atmosphere in which it might reasonably address wider issues of whale and dolphin conservation.

Part II

The Nature of Whales and Dolphins

Figure II *Humpback whale spyhops to get a better view above the surface*

Source: Rob Lott

11

The Nature of Whales and Dolphins

Liz Slooten

Humans have in the past defined intelligence, culture and society in ways that focus on human abilities and characteristics, assuming that these characteristics are unique to humans. Research over the last few decades has become less anthropocentric, focusing on the evolution of behaviour in human and nonhuman animals. It is now becoming clear that the behaviour and abilities of humans result from the combination and modification of properties found in other animals (including primates) rather than from 'unique' properties (see for example, Roth and Dicke, 2005).

Research on whales and dolphins shows that they are capable of complex behaviour and cognition, including self-awareness (Reiss and Marino, 2001), behavioural flexibility, innovation and creativity (for example, dolphins creating bubble rings – Reiss, 1998; McCowan et al, 2000). Strong mother–calf bonds and long periods of lactation in many cetaceans facilitate social learning, for example learning about catching prey, spatial orientation, social relationships and codes of behaviour (see, for example, Whitehead, 2003; Sargeant and Mann, 2009a).

At least some of the increase in odontocete (toothed cetacean) brain size was matched by the evolution of a novel nasofacial muscle complex related to echolocation (Fordyce, 2002, 2003). In this volume, Lori Marino explains that cetacean brains, like human brains, have undergone immense increases in size and complexity, but along a different neuroanatomical trajectory and therefore constitute an alternative evolutionary route to complex intelligence on Earth. Different parts of the cetacean brain have increased in size. For example, the paralimbic cortex, unique in mammals, may enable cetaceans to think in ways that we are incapable of. Likewise, the neocortical surface area to total brain weight is much larger in cetaceans than humans. Visual and auditory processing areas are closer and more highly integrated than in most mammals, which appears to allow for highly developed cross-modal sensory processing abilities. This is corroborated

Figure 11.1 *Highly social group of sperm whales*

Note: In the foreground the asymmetrical blowhole characteristic of this species is visible and in the background the large head of another sperm whale is clearly visible above the surface.

Source: Steve Dawson, Otago University

by experimental results showing that dolphins can relate visual and echolocation information about the same object (Pack and Herman, 1995; Herman et al, 1998). Other aspects of brain structure indicate that cetaceans have particularly well-developed cognitive abilities that play a role in high-level cognitive functions such as attention, judgement, adaptive intelligent behaviour and social awareness (see the chapter by Marino in this volume) (see Figure 11.1).

The fact that cetaceans have taken a distinctly different route in their brain development may well contribute to our difficulty in understanding and interpreting cetacean intelligence. Intelligence is notoriously difficult to define and measure. It is certainly impossible to rank animal species by intelligence. It makes more sense to explore the many different forms of intelligence and complexity (behavioural and social) seen in different species (de Waal and Tyack, 2003).

Evidence of cognition, self-awareness and culture (for example, self-recognition, reasoning, transmission of behaviour through social learning) has now been observed in cetaceans as well as other nonhuman animals. That it took so long for this to become clear is due in part to the difficulties of studying the behaviour of cetaceans. The marine environment is relatively inhospitable for humans. In most parts of the world, humans would only survive for a matter of hours in the water without a boat, submarine or at least a wetsuit or drysuit. Even with a boat, it is difficult to remain at sea for long periods and the weather often makes it difficult to find, let alone study whales and dolphins. Added to this, most cetaceans spend more time out of sight – underwater – than they do clearly visible at the surface or at sufficiently close range to be observed underwater (for example, from a glass bottomed boat, or while snorkelling or using scuba gear).

While studies of cetaceans in captivity have historically been useful in assessing the basic cognitive abilities of cetaceans in controlled experiments, more complex behaviour and social interactions can only be studied in free-living animals, for example, a group of bottlenose dolphins feeding through cooperative behaviour, where one individual creates a wall of stirred up mud around a school of fish, while other dolphins catch those fish as they try to escape (Torres and Read, 2009).

Long-term field studies of several species paint a picture of complex social interaction and complex behaviour, some of which is clearly learned. For example, young dolphins learn behaviours such as pushing a sponge around on their nose from their mothers, as well as many other behaviours including foraging strategies and social rules (Connor, 2001; Sargeant and Mann, 2009a). Some cetaceans clearly understand the concept of imitation (see for example, Herman, 2002) and innovation (see Weinrich et al, 1992). Social systems in several species are matriarchal or matrilineal, with strong social bonds between mothers and offspring. In some species, social groups consist of females and their offspring with very little movement between groups. Other species have a more fission–fusion social structure but still have strong mother–offspring bonds and there is evidence for social learning from mothers and other individuals (see for example, Mann et al, 2007; Gibson and Mann, 2008; Sargeant and Mann, 2009a, 2009b).

It is difficult to observe and interpret behaviour, even for species that are visible at the surface a good proportion of the time. An important decision for researchers studying the behaviour of whales and dolphins is whether to observe from close quarters or from further away. One advantage of getting close to the animals in a research boat, or even getting in the water, is that this allows the researcher to see the detail and to use above and underwater cameras. Close observations make it possible to see even subtle behaviour events, and to determine which individual exhibited which behaviour and how other individuals reacted. However, by being close enough to properly observe behaviour, one may well have an effect on the behaviour seen. Keeping your distance (for example, using a theodolite and spotting scope from land) means little if any disturbance. In most cases one can be confident that the animals are not aware of the presence of the observer, and therefore that the observed behaviour is undisturbed. However, from a distance it is generally not possible to make detailed behavioural observations. Group size, movements of individuals and relatively obvious behaviours such as leaps and lobtails can be readily observed from land. However, observation of subtle behaviours such as body contact and identifying which individual used which behaviour are usually not possible from a distance. Studies that use several approaches (for example, land observations, boat-based research and sound recordings) are likely to provide the most reliable results (assuming that boat approaches are made in such a manner as to minimize disturbance).

Individual behaviour events observed in the field can be analysed separately or grouped into behaviour categories such as 'feeding', 'resting' or 'socializing'. Sequence analysis is a good way to categorize behaviour, and arguably the closest we can get to categorizing the behaviours in the way the animals do themselves.

The sequences in which the individual behaviour events are used provide information about the 'meaning' of each behaviour. For example if behaviour events A to E tend to be used close together in time and one of these behaviours has a clear meaning or intention, such as catching a fish, then it would seem reasonable to categorize these as 'feeding' or 'foraging' behaviours. Likewise, if behaviours F to K tend to occur close together in time and these behaviours include mating and other forms of body contact then it seems reasonable to call this category 'social' or 'sexual' behaviour. In a long-term study of Hector's dolphins (see Figure 11.2 and 11.3), sequence analysis categorized most behaviours into such groups of behaviours with a reasonably obvious meaning, except for leaps and lobtails (Slooten, 1994). These 'aerial' behaviours were sometimes associated with social behaviours and sometimes with aggression. They appear to indicate a high level of motivation or excitement, perhaps analogous to humans shouting, jumping up and down or waving their arms. Detailed observations are needed before, during and after the occurrence of leaps and lobtails, in order to determine whether a particular leap or lobtail was used in a social, sexual or aggressive context.

An advantage of categorizing behaviour this way is that it avoids the researcher making subjective interpretations of the 'meaning' of each behaviour (for example, placing behaviours that look similar to human aggressive gestures into the 'aggressive' behaviour category). Another potential use of behaviour sequence analysis is to investigate which behaviours solicit a similar response (for example, all behaviours that precede being bitten by another dolphin) or follow the same behaviour (for example, a bite might be followed by the other dolphin biting back or moving away at speed).

Despite the challenges of studying whales and dolphins in the wild, it is becoming clear that culture is vital to many of these animals. Complex acoustic repertoires, including song in some species, are shared among individuals in the same group or area. For example, an expert in humpback whale vocalizations can tell whether a particular song type is from Tonga, Australia or the Atlantic and even in which year the recording was made. Humpback whale song evolves over time, with novel sounds being incorporated into songs and these innovations being adopted by other whales in the same population, and over time by adjacent populations.

These issues are explored in detail in Part II of the book. Hal Whitehead discusses the importance of culture for whales and dolphins. For example, female sperm whales, pilot whales and orca form permanent, or nearly permanent, social

Figure 11.2 *Hector's dolphins playing with seaweed*

Source: Steve Dawson, Otago University

Figure 11.3 *Another Hector's dolphin investigating seaweed*

Source: Steve Dawson, Otago University

Figure 11.4 *Female Hector's dolphin jumping out of the water close to the photographer's boat, providing the dolphin with an excellent view of the humans on board*

Source: Steve Dawson, Otago University

units. Sperm whale groups containing about a dozen females and their offspring have distinct dialects of codas (Rendell and Whitehead, 2003). They will associate with other social units, but only from the same 'clan'. Likewise, orca groups have their own dialects and distinct behaviours (for example, feeding, beach rubbing and greeting behaviour), even if they coexist in the same area (Bigg et al, 1990; Ford et al, 2000). As Toshio Kasuya pointed out at the Biennial Marine Mammal Conference in Cape Town (Kasuya, 2008) this means that whaling may not only impact population units but also cultural units. Hal Whitehead's chapter further develops this argument. Clearly, the goal of conservation management should be to maintain cultural diversity as well as genetic diversity and sustainable populations. In other words, it is no longer enough to ensure that human activities do not impact cetaceans by ensuring population size and structure are maintained and species are protected throughout their natural range. Knowing that these animals possess social and cultural traits similar to our own brings with it a greater responsibility for ensuring their wellbeing as well as simply their survival (this theme is discussed in the chapter by Brakes and Bass). For example, behavioural changes caused by tourism are no longer a concern only if they affect critical aspects of population dynamics (for example, survival, reproduction, feeding success) but such behavioural changes may be a cause for concern in and of themselves.

The above provides cause for concern over the vulnerability of whales and dolphins when placed in stressful situations such as capture, drive hunts, invasive research and disruption of social networks. As Marino points out in her chapter, 'Beyond our scientific understanding our current knowledge of cetacean brains and cognitive abilities demands we develop a new ethic of respect and coexistence with them.'

References

Bigg, M. A., Olesiuk, P. F., Ellis, G. M., Ford, J. K. B. and Balcomb, K. C. (1990) 'Social organization and genealogy of resident killer whales *(Orcinus orca)* in the coastal waters of British Columbia and Washington State', *Reports of the International Whaling Commission*, Special Issue 12, pp383–405

Connor, R. C. (2001) 'Individual foraging specializations in marine mammals: Culture and ecology', *Behavioral and Brain Sciences*, vol 24, pp329–330

de Waal, F. B. M. and Tyack, P. L. (2003) *Animal Social Complexity: Intelligence, Culture and Individualized Societies*, Harvard University Press, Harvard

Ford, J. K. B., Ellis, G. M. and Balcomb, K. C. (2000) *Killer Whales*, UBC Press, Vancouver, British Columbia

Fordyce, R. E. (2002) '*Simocetus rayi* (Odontoceti: Simocetidae) (new species, new genus, new family), a bizarre new archaic Oligocene dolphin from the eastern North Pacific', *Smithsonian Contributions to Paleobiology*, vol 93, pp185–222

Fordyce, R. E. (2003) 'Cetacean evolution and Eocene-Oligocene oceans revisited', in D. R. Prothero, L. C. Ivany and E. Nesbitt (eds) *From Greenhouse to Icehouse: The Marine Eocene-Oligocene Transition*, Columbia University Press, New York, pp154–170

Gibson, Q. A. and Mann, J. (2008) 'Early social development in wild bottlenose dolphins: Sex differences, individual variation and maternal influence', *Animal Behaviour*, vol 76, pp375–387

Herman, L. H. (2002) 'Vocal, social, and self-imitation by bottlenosed dolphins', in K. Dautenhahn and C. L. Nehaniv (eds) *Imitation in Animals and Artifacts*, MIT Press, Cambridge, MA, pp63–108

Herman, L. M., Pack, A. A. and Hoffman-Kuhnt, M. (1998) 'Seeing through sound: Dolphins perceive the spatial structure of objects through echolocation', *Journal of Comparative Psychology*, vol 112, pp292–305

Kasuya, T. (2008) 'The Kenneth S. Norris lifetime achievement award lecture: Presented on 29 November 2007 Cape Town, South Africa', *Marine Mammal Science*, vol 24, pp749–773

Mann, J., Sargeant, B. L. and Minor, M. (2007) 'Calf inspection of fish catches: Opportunities for oblique social learning?', *Marine Mammal Science*, vol 23, pp197–202

McCowan, B., Marino, L., Vance, E., Walke, L. and Reiss, D. (2000) 'Bubble ring play of bottlenose dolphins (*Tursiops truncatus*): Implications for cognition', *Journal of Comparative Psychology*, vol 114, pp98–106

Pack, A. A. and Herman, L. M. (1995) 'Sensory integration in the bottlenosed dolphin: Immediate recognition of complex shapes across the senses of echolocation and vision', *Journal of the Acoustical Society of America*, vol 98, pp722–733

Reiss, D. (1998) 'Cognition and communication in dolphins: A question of consciousness', in S. R. Hameroff, A. W. Kaszniak and A. C. Scott (eds) *Toward a Science of Consciousness II: The Second Tucson Discussions and Debates*, MIT Press, Cambridge, MA, pp551–560

Reiss, D. and Marino, L. (2001) 'Mirror self-recognition in the bottlenose dolphin: A case of cognitive convergence', *Proceedings of the National Academy of Sciences USA*, vol 98, pp5937–5942

Rendell, L. and Whitehead, H. (2003) 'Vocal clans in sperm whales (*Physeter macrocephalus*)', *Proceedings of the Royal Society of London, B,* vol 270, pp225–231

Roth, G. and Dicke, U. (2005) 'Evolution of the brain and intelligence', *TRENDS in Cognitive Sciences*, vol 9, pp250–257

Sargeant, B. L. and Mann, J. (2009a) 'From social learning to culture: Intrapopulation variation in bottlenose dolphins', in K. N. Laland and B. G. Galef (eds) *The Question of Animal Culture*, Harvard University Press, Cambridge, MA, pp152–173

Sargeant, B. L. and Mann, J. (2009b) 'Developmental evidence for foraging traditions in wild bottlenose dolphins', *Animal Behaviour*, vol 78, pp715–721

Slooten, E. (1994) 'Behavior of Hector's dolphin: Classifying behavior by sequence analysis', *Journal of Mammalogy*, vol 75, pp956–964

Torres, L. G. and Read, A. J. (2009) 'Where to catch a fish? The influence of foraging tactics on the ecology of bottlenose dolphins (*Tursiops truncatus*) in Florida Bay, Florida', *Marine Mammal Science*, vol 25, pp797–815

Weinrich, M. T., Schilling, M. R. and Belt, C. R. (1992) 'Evidence for acquisition of a novel feeding behaviour: Lobtail feeding in humpback whales, *Megaptera novaeangliae*', *Animal Behaviour*, vol 44, pp1059–1072

Whitehead, H. (2003) *Sperm Whales: Social Evolution in the Ocean*, University of Chicago Press, Chicago

12

Brain Structure and Intelligence in Cetaceans

Lori Marino

Introduction

Cetaceans are among the most intelligent of mammals and their adaptation to a fully aquatic lifestyle represents one of the most dramatic transformations in mammalian evolutionary history. Significant modifications occurred at all levels of physiology, morphology and the nervous system. Among the most dramatic but less overt evolutionary changes are those that occurred in the brain. Many cetacean species possess the largest brains of all animals. The most massive brain on Earth belongs to the modern sperm whale (*Physeter macrocephalus*) with an average adult brain size of about 8000 grams (Marino, 2002) as compared with an average adult human brain of approximately 1300g. And, in addition to absolute size, modern cetacean brains are among the largest in relation to body size.

In the past three decades new research has shed light on the complexity of cetacean brains and has begun to lay bare the neurobiological basis for their considerable behavioural abilities. But the story of cetacean brain evolution is one that goes beyond an evolutionary tale about increased brain size. It is also a fascinating example of the way that brain structure–function relationships can follow a complicated pattern of evolutionary divergence and convergence. Our study of cetacean brains has revealed that the human brain is not the only brain that has undergone immense increases in size and complexity. Cetacean brains have as well but along a different neuroanatomical trajectory, providing an example of an alternative evolutionary route to complex intelligence on Earth.

The Origin and Evolution of Large Brains in Cetaceans

Cetacean brains come from relatively modest beginnings. The last terrestrial ancestor of cetaceans, *Pakicetus*, lived about 55 million years ago, looked somewhat like a medium-sized dog and stood on hooved feet that attest to the ungulate ancestry of cetaceans (Gingerich and Uhen, 1998). *Pakicetus* possessed a rather average brain in absolute and relative size (Marino, 2002). The brains of the earliest aquatic cetacean suborder, the archaeocetes ('ancient whales'), were similar in relative size to that of *Pakicetus*. One of the most unexpected findings in recent studies of cetacean brain evolution is that archaeocete brains changed very little over the course of their 13-million-year period of adaptation to a fully aquatic lifestyle. Their relative brain size remained similar to that of *Pakicetus* and there is no evidence for significant changes in brain structure (Marino et al, 2004a). Even the most recent archaeocetes, who were fully aquatic for at least 10 million years, showed no evidence of significant changes in brain size. This evidence refutes a long-standing hypothesis that cetacean brains became large and complex as a direct result of adaptation to an aquatic existence. All of the evidence taken together suggests that the cognitive abilities of archaeocetes were probably the aquatic version of a pakicetid and not nearly as cognitively or behaviourally complex as in modern cetaceans.

The first and most substantial change in cetacean brain size and structure came about approximately 35 million years ago and coincided with the emergence of early versions of the Neoceti ('new whales'), which probably included both odontocetes and mysticetes (Marino et al, 2004a) (although there remains considerable controversy over the origin of mysticetes and their phylogenetic relationship to odontocetes). Many of the initial changes to the cetacean brain, i.e. enlarged cerebral hemispheres, are shared by both modern suborders. But because most of the data on cetacean brain evolution come from odontocetes, extrapolations to mysticetes should be made conservatively, bearing in mind that both suborders adapted early on to two very different behavioural niches, which then led to differences in modern brain structure (see Figure 12.1).

The first major change in odontocete relative brain size was due mainly to a significant decrease in body size along with a rather more moderate increase in absolute brain size. This combination of changes drove the Encephalization Quotient (EQ) levels of these early groups significantly beyond the range of the archaeocetes. Because brain and body size are positively correlated, brain size is often expressed as an EQ (Jerison, 1973), which is a value that represents how large or small the average brain of a given species is compared with other species of the same average body weight. The average EQ for archaeocetes was 0.5, below average, while that of the early odontocetes soared to 2.0, above average, with, importantly, no overlap in EQ across the two groups (Marino et al, 2004a). One subset of odontocetes, the superfamily Delphinoidea (porpoises, oceanic dolphins and toothed whales), underwent a second significant average increase in absolute brain size by 15 million years ago and today represents the most highly

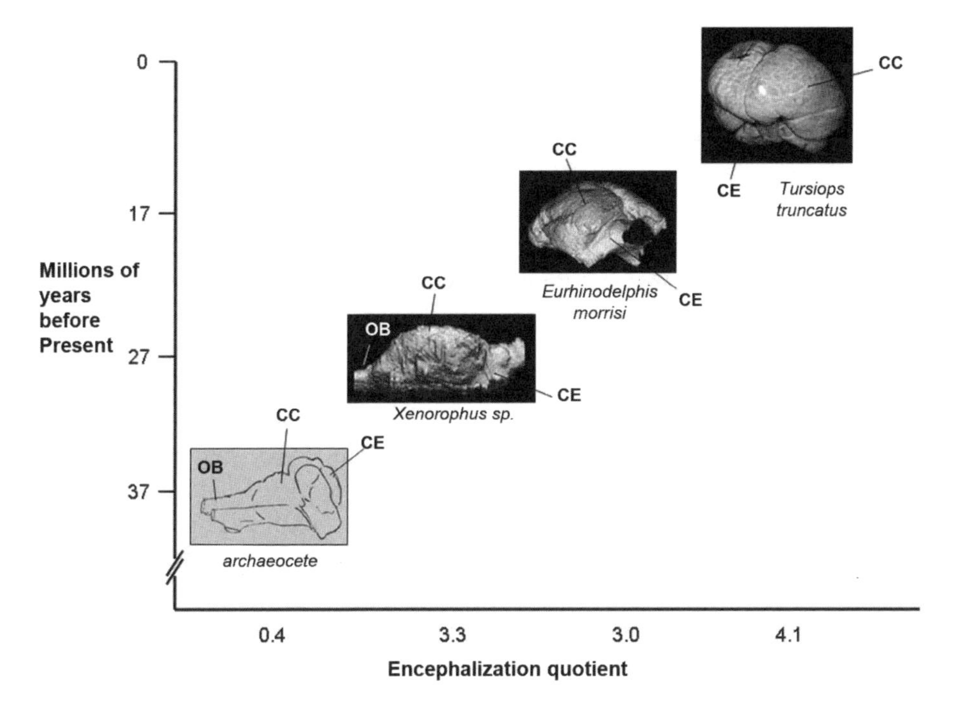

Figure 12.1 *Sagittal (sideways) images of cetacean brains at four stages of evolution along a timeline showing approximate Encephalization Quotient and some major structures*

Note: Images are: line drawing of the brain of a 37-million-year-old archaeocete from an endocast, three-dimensional computed tomography reconstruction of the brain of a 27-million-year-old extinct odontocete, *Xenorophus sp.* (Marino et al, 2003), three-dimensional computed tomography reconstruction of the brain of a 14-million-year-old extinct odontocete, *Eurhinodelphis morrisi* (Marino et al, 2003), and three-dimensional magnetic resonance imaging reconstruction of a modern bottlenose dolphin, *Tursiops truncatus*, brain (Marino et al, 2001a). Images are not to scale. Species do not represent direct ancestor–descendent phylogenetic lineages. OB = olfactory bulb, CC = cerebral cortex, CE = cerebellum. Structures not labelled are absent.

encephalized group of species next to our own, with EQs ranging from 1.76 to 4.95 (modern great apes and some monkeys possess EQs of 1.5 to 3.0 (Marino, 1998)). There is no evidence for further evolutionary changes in odontocete brains in the past 15 million years and essentially no information on recent brain evolution in mysticetes.

Changes in Brain, Behaviour and Ecology

Along with their modest relative size, archaeocete brains were characterized by small elongated cerebral hemispheres ending rostrally in large olfactory peduncles and bulbs (Edinger, 1955). Their brain morphology was almost indistinguishable from that of the pakicetids and many other predatory terrestrial species of the time. However, after the initial change in relative size, Neoceti cetacean brains underwent a number of structural and organizational evolutionary modifications with important functional consequences. The signature changes occurring in odontocetes were the substantial hyperproliferation and reorganization of the cerebral hemispheres, enlargement of auditory processing areas, development of a unique neocortical architecture, and reduction of olfactory structures with a reproportioning of associated limbic regions. These changes, along with many others, placed odontocetes on an unusual evolutionary trajectory that would result in brains that are among the most massive, complex and sophisticated in the animal kingdom.

Discussion and debate about the ultimate causes of the initial major shift in cetacean brain evolution have been ongoing for decades. No definitive conclusions have been drawn to date but palaeoclimatological and palaeobiological findings offer compelling clues. Archaeocetes underwent several changes in body structure and physiology during their adaptation to a fully aquatic lifestyle. In addition to the obvious significant changes in post-cranial skeleton, there were modifications of skull structure that led to a change in chewing style in archaeocetes from their terrestrial counterparts. Skull elongation and the migration of the nares to the top of the head, called 'telescoping', produced significant changes in the architecture of a number of structures including the cranial vault and, hence, the brain. Archaeocetes also began the pattern of change in dentition from heterodonty that culminated in homodonty in odontocetes and baleen in mysticetes. In addition, changes in jaw and ear bone structure indicated they developed increased underwater directional hearing abilities. But there is no evidence for echolocation in archaeocetes (see Uhen, 2007, for a review). Finally, as mentioned, fossil cranial morphology indicates they still possessed olfactory abilities.

The emergence of the Neoceti from the archaeocetes around 35 million years ago signalled a host of significant transformations that were accompanied by dramatic changes in behavioural ecology. The initial major increase in relative brain size in early odontocetes was due to a substantial decrease in body size along with a moderate increase in brain size. The most plausible explanation for this change in brain–body allometry is that cooling in temperate to polar latitudes triggered changes in diversity and productivity in oceanic food chains (Salamy and Zachos, 1999) that led, in turn, to changes in behavioural ecology (foraging opportunities and predation risk, for instance) (Lindberg and Pyenson, 2007). The moderate increase in odontocete brain size was matched by the evolution of a novel echolocation-related nasofacial muscle complex (Fordyce, 2002, 2003) and concurrent changes in ear structures associated with echolocation (Fleischer, 1976). Heterodonty was replaced entirely by long rows of prong-like uniform teeth adapted for grabbing and gulping whole prey items. Taken together, the decreased body size, homodonty, changes in skull and ear bone morphology, and brain enlargement all lead to evidence for a new echolocating niche for the early odontocetes. These early odontocetes were fast and sleek with new and enhanced sensory-perceptual abilities.

The environmental changes that occurred about 35 million years ago spurred a different set of physical and behavioural adaptations in mysticetes that led to the development of baleen, filter-feeding and large body size. Explanations for why two different suborders emerged so early on in cetacean evolution await further data and analysis.

The Massive Modern Cetacean Brain

Modern cetacean brains are among the largest of all mammals in both absolute mass and in relation to body size. The largest brain on Earth today, belonging to the sperm whale with an average weight of 8000g for adults (Marino, 2002), is about 60 per cent larger than the elephant brain and six times larger than the human brain. In terms of EQ, the EQ for modern humans is 7.0. Our brains are seven times the size one would expect for a species with our body size. Almost all odontocetes (the toothed whales, dolphins and porpoises) possess above-average encephalization levels compared with other mammals. Numerous odontocete species possess EQs in the range of 4 to 5, that is, they possess brains 4 to 5 times larger than one would expect for their body weights. Many of these values are second only to those of modern humans and significantly higher than any of the nonhuman anthropoid primates (Marino, 1998).

EQs of mysticetes are all below 1 (Marino, 2002) because of an uncoupling of brain size and body size in very large aquatic animals. It turns out that brains do not correlate well with body size at the two extremes. That is, very large animals, such as whales, tend to have smaller brains compared to their weight (and smaller animals, such as mice, tend to have larger brains for their body size). One explanation is that bodies may become very large by increasing components that do not require the same degree of neural control as the basic body plan. In cetaceans this decoupling of brain and body size may be exaggerated because in an environment of 'aquatic weightlessness' whale bodies can become massive without a concomitant increase in innervations associated with gravity-loaded tissues such as muscles. This has allowed mysticete bodies to become massive without a correlated increase in brain size. However, mysticete brains are large in absolute size and exhibit high degrees of cortical complexity confirming that mysticete brains have, in addition to odontocete brains, undergone substantial elaboration during the course of their evolution (Oelschlager and Oelschlager, 2002). Many of the features of odontocete brain anatomy indicative of progressive elaboration are shared with mysticetes.

Modern Cetacean Brain Structure and Organization

Modern cetacean brains are vastly different from those of their ancestors and also highly unusual compared with other mammals. These changes were first apparent in the fossil record with the emergence of the Neoceti and continued to be refined over the next 20 million years as gradual changes in brain shape are documented in the fossil record (Marino et al, 2004a). Interestingly, there is no evidence for change in brain size or structure in modern cetaceans after the second significant increase in brain mass observed in *Delphinoidea* by 15 million years ago. Therefore, cetaceans have possessed their large complex brains for at least 15 million years without the necessity of further modifications, at least at the gross neuroanatomical level. This is an interesting point when comparing this fact to the history of human brain evolution. Human brains enlarged rapidly over a couple of million years. Therefore, with respect to time (and also the ability to avoid ecosystem destruction), it could be argued that cetaceans are much more successful as large-brained mammals than humans.

To gain proper perspective on the changes that occurred in the cetacean brain it is important to consider that some of those changes were probably specific functional adaptations while others were neuroanatomical accommodations to those adaptations and to general increases in brain size. Specific functional adaptations are those that developed in response to a specific need or evolutionary demand. Examples include the increase in size of auditory brain structures to allow for an increase in processing of sound. Even highly specific functional changes in the brain cannot occur in isolation. There are changes throughout the entire brain that accommodate the emergence of any new function. Additionally, when brains enlarge over evolutionary time they do not do so uniformly. That is, they do not become large versions of the smaller form. When brains enlarge they do so by changing their general architecture to maintain connectivity over longer distances. These allometrically driven modifications typically result in increased modularization of functional areas in the brain. Therefore, brains cannot enlarge without also reorganizing, thus creating new areas and new features simply to maintain the same level of connectivity as in a small brain. And it is important to keep in mind that the modern cetacean brain, like any other brain, is a product of all of these processes.

Enlargement of auditory structures

Cetaceans are highly reliant on audition, and all brain structures associated with hearing, including the auditory nerve, auditory midbrain regions and cortical projection zones, are massive in cetaceans and particularly odontocetes. Moreover, it is a common misconception that cetacean vision diminished substantially because of enhanced audition. With the exception of freshwater dolphins (who are almost blind) cetaceans have acute vision in and out of water. Generally, auditory structures are relatively larger in odontocetes than mysticetes because of the odontocete ability to echolocate (Oelschlager and Oelschlager, 2002). Likewise, the primary and secondary auditory projection zones on the cerebral surface are extensive (Supin et al, 1978).

One of the most intriguing and informative aspects of enhanced auditory processing in cetaceans is the evidence that they have integrated this perceptual system into the general cognitive domain in a way that may be unprecedented in the animal kingdom. For instance, echolocation in odontocetes may have evolved initially as a sensory device to provide information about the environment. The large brains of odontocetes were initially used, at least partly, for processing this entirely new sensory mode. Experimental evidence and behavioural field observations suggest echolocation plays a role in more highly integrated cognitive processing and communication in modern odontocetes. For instance, echolocation sounds are often produced simultaneously with other sounds that are communicative in nature and many authors have noted that sonar provides an efficient way to keep track of a moving social group over distances (Wood and Evans, 1980; Jerison, 1986; Ridgway, 1986). Therefore, the initial function may have evolved into an increasingly complex cognitive-behavioural ecological lifestyle over time.

Reduction of olfaction and reproportioning of limbic structures

As in other aquatic mammals, olfaction (sense of smell) in cetaceans has been severely reduced in mysticetes and entirely lost in odontocetes. Foetal odontocetes possess small olfactory structures (Buhl and Oelschlager, 1988; Marino et al, 2001b) that regress completely by birth. Infrequently, a short olfactory peduncle is observed in adult sperm whales (*Physeter macrocephalus*) and northern bottlenose whales (*Hyperoodon ampullatus*) (Oelschlager and Oelschlager, 2002). Olfaction in adult mysticetes is vastly reduced but they have maintained small olfactory bulbs, a thin olfactory peduncle and an olfactory tubercle (Oelschlager and Oelschlager, 2002).

The limbic system is an evolutionarily old set of structures that supports a variety of functions, including memory, emotional processing and olfaction. It includes the hippocampus, amygdala, mammilary bodies and the limbic cortex. Limbic structures most intimately connected with olfaction, the hippocampus, fornix and mammillary bodies have been greatly reduced in cetaceans (Jacobs et al, 1979; Morgane et al, 1980).

An interesting juxtaposition to the small limbic system is the large amygdala (Schwerdtfeger et al, 1984) and extremely well-developed cortical limbic lobe and entorhinal region above the corpus callosum in cetaceans (Oelschlager and Oelschlager, 2002; Marino et al, 2004b). This combination of a vastly reduced hippocampus and a highly elaborated adjacent limbic cortical zone leads to intriguing questions about whether there was a transfer of learning and memory functions from the olfactory-based hippocampal domain to these other cortical regions. The hippocampus plays important roles in short- and long-term memory and spatial navigation in all vertebrates. Yet cetacean memory, learning and spatial abilities are uncompromised and, quite to the contrary, sophisticated and robust. An extensive body of experimental work on short- and long-term memory, as well as spatial memory in bottlenose dolphins, demonstrates that these capacities and the learning abilities they make possible are highly developed (see Marino et al, 2008, for a review). We have learned from field studies of bottlenose dolphins and killer whales (*Orcinus orca*) that they live in large highly complex dynamic societies that include long-term bonds, higher-order fluid alliances and cooperative networks that rely directly upon memory and learning (Baird, 2000; Connor et al, 2000; Lusseau, 2007). Furthermore, the impressive navigatory abilities during their migrations are hallmarks of mysticete behaviour that attest to their long-term spatial memory (Stern, 2009). Therefore, there is ample indirect evidence to suggest that memory-based systems in cetaceans have been transferred from the hippocampus primarily to adjacent highly elaborated regions of the brain.

Enlargement and rearrangement of cerebral hemispheres

The cetacean forebrain is organized around three concentric tiers of tissue that include the limbic and supralimbic regions but also the entirely unique paralimbic cortex. The function of the paralimbic lobe in cetaceans is largely unknown and is another signpost of the radical departure of the cetacean brain from the general mammalian pattern. The paralimbic lobe is also a provocative reminder that cetaceans may be able to think in ways that we are entirely incapable of.

In addition to its unusual organization, the cetacean forebrain is among the most highly convoluted of all mammals, revealing that there was a substantial increase in neocortical surface area and volume in cetacean evolutionary history. One measure of convolution is the 'gyrification index', which compares neocortical surface area to total brain weight. The index for modern humans is approximately 1.75. Known gyrification indices for odontocetes range from 2.4 to 2.7, substantially exceeding that of modern humans (Ridgway and Brownson, 1984). The gyrification index is positively correlated with brain mass across the mammals, and cetacean (as well as human) brains appear to be consistent with this pattern.

The pattern of elaboration of the neocortex (the evolutionarily newest region of the forebrain) in cetaceans has resulted in a highly unusual configuration as well. The map of sensory projection regions (the cortical regions that receive sensory information) in the cetacean brain stands in striking contrast to that of other large-brained mammals. In primates, for instance, the visual and auditory projection regions are located in the occipital and temporal lobes, respectively. This means that visual information is first processed in the cortex in the back of the brain

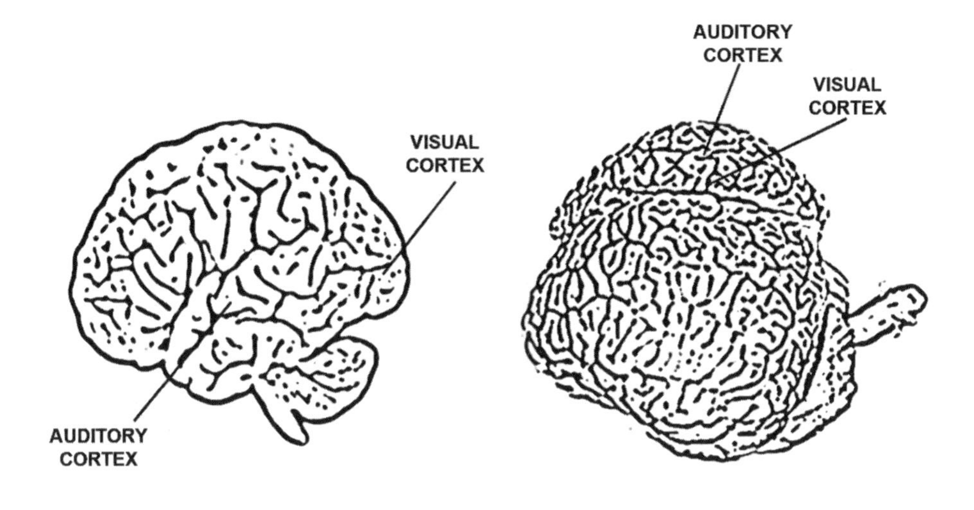

Figure 12.2 *Illustrations of the modern human brain and bottlenose dolphin brain showing the location of the primary visual and auditory cortex in each*

Source: based on photographs from the University of Wisconsin and Michigan State Comparative Mammalian Brain Collections, as well as from those at the National Museum of Health and Medicine (www.brainmuseum.org)

(occipital region) and auditory information on the side of the brain (temporal region). An expanse of non-projection or association cortex intervenes between these two regions. Therefore, visual and auditory information must be sent to this intervening cortex from the projection zones if they are to be integrated.

In cetaceans, by contrast, the visual and auditory projection zones are located in the parietal region atop the hemispheres and are immediately adjacent to each other (see Figure 12.2). This arrangement of cortical adjacency is unusual and suggests that not only is the surface map of the cetacean neocortex different from most mammals but the relationship between the visual and auditory processing areas is closer or more highly integrated than in most mammals. Cetaceans possess a vast expanse of non-projection or association cortex for even higher-order cognitive information processing that lies outside of the visual-auditory region in the remaining temporal and occipital regions. This idiosyncratic pattern of visual-auditory adjacency may allow for highly developed cross-modal sensory processing abilities in cetaceans. There is evidence for this idea from experiments on mental representation in bottlenose dolphins. In these studies, the dolphin was required to examine a complex three-dimensional object visually, and then select a match by echolocating on two or more visually opaque underwater boxes, each containing an object. In other trials the dolphin was allowed to echolocate on the object and had to use only vision to choose the same object above water. In either case, the dolphin's matching performance for various objects was close to perfect. These findings strongly suggest that the mental representations of objects perceived through echolocation are integrated with or closely coordinated with those developed through vision (Pack and Herman, 1995; Herman et al, 1998).

The 'new' cetacean cortex

For several decades, the common view has been that cetacean cortical regions are homogeneous and fairly simple in architecture (Gaskin, 1982; Kesarev, 1971). This view engendered much confusion and debate about how a relatively 'primitive' brain could lie behind the considerable cognitive and behavioural complexity in cetaceans for which evidence was accruing (Glezer et al, 1988). But recent studies reveal a different picture, pointing to extensive neocortical complexity and variability in both odontocetes and mysticetes (Hof et al, 2005; Hof and Van der Gucht, 2007). The cellular architecture of various regions of the cetacean neocortex is characterized by a wide variety of organizational features, i.e. columns, modules, layers, that are associated with complex brains. Furthermore, there is substantial differentiation across the various neocortical regions.

Whereas there appears to be a high degree of organizational complexity throughout the cetacean neocortex, there are specific regions that are especially notable in their apparent degree of elaboration. The cingulate and insular cortices (both situated deeper within the forebrain) in odontocetes and mysticetes are extremely well developed (Jacobs et al, 1979; Hof and Van der Gucht, 2007) and the expansion of these areas in cetaceans is consistent with high-level cognitive functions such as attention, judgement and social awareness (Allman et al, 2005). Moreover, recent studies show that the anterior cingulate and insular cortices in larger cetaceans contains a type of projection neuron, known as a spindle cell or Von Economo neuron (Hof and Van der Gucht, 2007). Von Economo neurons are highly specialized projection neurons considered to be involved in neural networks subserving aspects of social cognition (Allman et al, 2005) and have thus far been found in humans and great apes (Allman et al, 2005) and elephants (Hakeem et al, 2009). Spindle cells are thought to play a role in adaptive intelligent behaviour and the presence of these neurons in cetaceans is consistent with the complex cognitive abilities found in this group.

Despite general similarities in level of cellular and organizational complexity, there are striking differences in the specific connectivity patterns of cetacean brains and primate brains. Specifically, cetacean neocortex is characterized by five layers instead of the six typical of primates and many other mammals. Cetacean neocortex possesses a very thick layer I in combination with the absence of a granular layer IV. In primates, a granular layer IV is the primary input layer for fibres ascending from the midbrain to the cortex and this layer is also the source of important connections within the neocortex. However, granular layer IV is absent in cetaceans and this has compelling implications for how information reaches the cetacean cortex and then gets distributed to other areas (Glezer et al, 1988). The current prevailing view is that the thick Layer I is the primary layer receiving incoming fibres. This means that the way information gets to the cortex and is distributed is distinctly different in cetaceans from primates and other mammals. More importantly, the vastly different connectivity patterns of primate and cetacean brains are another compelling example of two distinctly different evolutionary trajectories taken towards neurological and behavioural complexity.

Implications for Intelligence in Dolphins and Whales

There is strong evidence that the initial dramatic change in cetacean brain size and organization was correlated with striking transformations in behavioural ecology and the perceptual-cognitive-emotional faculties that would underlie such a change. Cetacean brains are massively elaborated but in very different ways from the brains of other mammals, including primates. Despite these striking differences in neuroanatomical organization, particularly at the neocortical level, cetaceans and primates (as well as other species) exhibit very similar cognitive and behavioural capacities that imply a high level of convergence in the factors that led to the evolution of brain and intelligence in cetaceans and primates.

One of the most striking examples of cognitive convergence between dolphins and primates is self-recognition. Bottlenose dolphins have convincingly demonstrated that they use a mirror to investigate their own bodies, showing that they have a sense of self (Reiss and Marino, 2001). These findings are consistent with further evidence for self-awareness and self-monitoring in dolphins and related cognitive abilities (see Simmonds, 2006, for a recent review) that underwrite the complex social patterns observed in many cetacean species (see Marino et al, 2008, for a review). In particular, the highly elaborated cingulate and insular cortex in cetacean brains are consistent with the idea that these animals are highly sophisticated and sensitive in the emotional and social domains. It may be that many cetacean species have achieved a level of social-emotional sophistication not achieved by other animals, including humans.

All of the above factors provide cause for concern over the vulnerability of whales and dolphins when placed in stressful situations, including capture, drive hunts and the disruption of social networks. Moreover, conservation measures that do not take into account the psychological complexity of these animals will do little to alleviate suffering on an individual level. Beyond our scientific understanding, our current knowledge of cetacean brains and cognitive abilities demands we develop a new ethic of respect and coexistence with them.

SUMMARY

55 million years ago cetaceans (dolphins, whales and porpoises) embarked on a dramatic shift from a terrestrial to an aquatic lifestyle that resulted in brains that are among the most massive, complex and sophisticated in the animal kingdom. In particular, the cetacean cerebral cortex (the part of the brain involved in higher-order cognition) evolved along a very different trajectory from other mammals, resulting in a highly unusual arrangement of functional areas and an entirely unique structure, the paralimbic lobe. Yet, despite the vast differences in cortical organization, cetaceans and humans (as well as great apes) share a number of complex cognitive abilities, such as self-recognition. Cetaceans and humans, therefore, are a striking example of evolutionary convergence in psychology among mammals. These similarities, importantly, mean that cetaceans, like humans, are vulnerable to emotional and social stresses that can lead to considerable harm. This important point is critical to guiding the ethics of how we interact with and treat cetaceans.

REFERENCES

Allman, J. M., Watson, K. K., Tetrault, N. A. and Hakeen, A. Y. (2005) 'Intuition and autism: A possible role for Von Economo neurons', *Trends in Cognitive Science*, vol 9, pp367–373

Baird, R. (2000) 'The killer whale: Foraging specializations and group hunting', in J. Mann, R. C. Connor, P. Tyack and H. Whitehead (eds) *Cetacean Societies: Field Studies of Dolphins and Whales*, University of Chicago Press, Chicago, pp127–153

Buhl, E. H. and Oelschlager, H. A. (1988) 'Morphogenesis of the brain in the harbour porpoise', *Journal of Comparative Neurology*, vol 277, pp109–125

Connor, R. C., Wells, R., Mann, J. and Read, A. (2000) 'The bottlenosed dolphin: Social relationships in a fission-fusion society', in J. Mann, R. C. Connor, P. Tyack and H. Whitehead (eds) *Cetacean Societies: Field Studies of Dolphins and Whales*, University of Chicago Press, Chicago, pp91–126

Edinger, T. (1955) 'Hearing and smell in cetacean history', *Monatschrift fur Psychiatrie und Neurologie*, vol 129, pp37–58

Fleischer, G. (1976) 'Hearing in extinct cetaceans as determined by cochlear structure', *Journal of Paleontology*, vol 50, pp133–152

Fordyce, R. E. (2002) '*Simocetus rayi* (Odontoceti: Simocetidae) (new species, new genus, new family), a bizarre new archaic Oligocene dolphin from the eastern North Pacific', *Smithsonian Contributions to Paleobiology*, vol 93, pp185–222

Fordyce, R. E. (2003) 'Cetacea evolution and Eocene-Oligocene oceans revisited', in D. R. Prothero, L. C. Ivany and E. Nesbitt (eds) *From Greenhouse to Icehouse: The Marine Eocene-Oligocene Transition*, Columbia University Press, New York, pp154–170

Gaskin, D. E. (1982) *The Ecology of Whales and Dolphins*, Heinemann, London

Gingerich, P. D. and Uhen, M. D. (1998) 'Likelihood estimation of the time of origin of Cetacea and the time of divergence of Cetacea and Artiodactyla', *Paleo-electronica*, vol 2, pp1–47

Glezer, I., Jacobs, M. and Morgane, P. (1988) 'Implications of the "initial brain" concept for brain evolution in Cetacea', *Behavioral and Brain Sciences*, vol 11, pp75–116

Hakeem, A. Y., Sherwood, C. C., Bonar, C. J., Butti, C., Hof, P. R., and Allman, J. M. (2009) 'Von Economo neurons in the elephant brain', *The Anatomical Record*, vol 292, pp242–248

Herman, L. M., Pack, A. A. and Hoffman-Kuhnt, M. (1998) 'Seeing through sound: Dolphins perceive the spatial structure of objects through echolocation', *Journal of Comparative Psychology*, vol 112, pp292–305

Hof, P. R. and Van der Gucht, E. (2007) 'The structure of the cerebral cortex of the humpback whale, *Megaptera novaeangliae* (Cetacea, Mysticeti, Balaenopteridae)', *The Anatomical Record*, vol 290, pp1–31

Hof, P., Chanis, R. and Marino, L. (2005) 'Cortical complexity in cetacean brains', *The Anatomical Record*, vol 287A, pp1142–1152

Jacobs, M. S., McFarland, W. L. and Morgane, P. J. (1979) 'The anatomy of the brain of the bottlenose dolphin (*Tursiops truncatus*). rhinic lobe (rhinencephalon): the archicortex', *Brain Research Bulletin*, vol 4, pp1–108

Jerison, H. J. (1973) *Evolution of the Brain and Intelligence*, Academic Press, New York

Jerison, H. J. (1986) 'The perceptual world of dolphins', in R. J. Schusterman, J. A. Thomas and F. G. Wood (eds) *The Perceptual World of Dolphins*, Lawrence Erlbaum, New Jersey, pp141–166

Kesarev, V. S. (1971) 'The inferior brain of the dolphin', *Soviet Science Review*, vol 2, pp52–58

Lindberg, D. R. and Pyenson, N. D. (2007) 'Things that go bump in the night: Evolutionary interactions between cephalopods and cetaceans in the Tertiary', *Lethaia*, vol 40, pp335–343

Lusseau, D. (2007) 'Evidence for social role in a dolphin social network', *Evolutionary Ecology*, vol 21, pp357–366

Marino, L. (1998) 'A comparison of encephalization between odontocete cetaceans and anthropoid primates', *Brain, Behaviour and Evolution*, vol 51, pp230–238

Marino, L. (2002) 'Brain size evolution', in W. F. Perrin, B. Wursig and H. Thewissen (eds) *Encyclopedia of Marine Mammals*, Academic Press, San Diego, pp158–162

Marino, L., Sudheimer, K., Murphy, T. L., Davis, K. K., Pabst, D. A., McLellan, W., Rilling, J. K. and Johnson, J. I. (2001a) 'Anatomy and three-dimensional reconstructions of the bottlenose dolphin (*Tursiops truncatus*) brain from magnetic resonance images', *The Anatomical Record*, vol 264, pp397–414

Marino, L., Murphy, T. L., Gozal, L. and Johnson, J. I. (2001b) 'Magnetic resonance imaging and three-dimensional reconstructions of the brain of the fetal common dolphin, *Delphinus delphis*', *Anatomy and Embryology*, vol 203, pp393–402

Marino L., Uhen M. D., Pyenson, N. D. and Frohlich, B. F. (2003) 'Reconstructing cetacean brain evolution using computed tomography', *The New Anatomist*, vol 272B, pp107–117

Marino, L., McShea, D. and Uhen, M. D. (2004a) 'The origin and evolution of large brains in toothed whales', *The Anatomical Record*, vol 281A, pp1247–1255

Marino, L., Sherwood, C. C., Tang, C. Y., Delman, B. N., Naidich, T. P., Johnson, J. I. and Hof, P. R. (2004b) 'Neuroanatomy of the killer whale (*Orcinus orca*) from magnetic resonance images', *The Anatomical Record*, vol 281A, pp1256–1263

Marino, L., Butti, C., Connor, R. C., Fordyce, R. E., Herman, L. M., Hof, P. R., Lefebvre, L., Lusseau, D., McCowan, B., Nimchinsky, E. A., Pack, A. A., Reidenberg, J. S., Reiss, D., Rendell, L., Uhen, M. D., Van der Gucht, E. and Whitehead, H. (2008) 'A claim in search of evidence: Reply to Manger's thermogenesis hypothesis of cetacean brain structure', *Biological Reviews of the Cambridge Philosophical Society*, vol 83, pp417–440

Morgane, P. J., Jacobs, M. S. and McFarland, W. L. (1980) 'The anatomy of the brain of the bottlenosed dolphin (*Tursiops truncatus*): Surface configuration of the telencephalon of the bottlenosed dolphin with comparative anatomical observations in four other cetacean species', *Brain Research Bulletin*, vol 5 (Suppl 3), pp1–107

Oelschlager, H. A. and Oelschlager, J. S. (2002) 'Brains', in W. F. Perrin, B. Würsig and H. Thewissen (eds) *Encyclopedia of Marine Mammals*, Academic Press, San Diego, pp133–158

Pack, A. A. and Herman, L. M. (1995) 'Sensory integration in the bottlenosed dolphin: Immediate recognition of complex shapes across the senses of echolocation and vision', *Journal of the Acoustical Society of America*, vol 98, pp722–733

Ridgway, S. H. (1986) 'Physiological observations on dolphin brains', in R. J. Schusterman, J. A. Thomas and F. G. Wood (eds) *Dolphin Cognition and Behavior: A Comparative Approach*, Lawrence Erlbaum, New Jersey, pp31–59

Ridgway, S. H. and Brownson, R. H. (1984) 'Relative brain sizes and cortical surface areas of odontocetes', *Acta Zoologica Fennica*, vol 172, pp149–152

Reiss, D. and Marino, L. (2001) 'Self-recognition in the bottlenose dolphin: A case of cognitive convergence', *Proceedings of the National Academy of Sciences USA*, vol 98, pp5937–5942

Salamy, K. A. and Zachos, J. C. (1999) 'Latest Eocene–Early Oligocene climate change and Southern Ocean fertility: Inferences from sediment accumulation and stable isotope data', *Palaeogeography, Palaeoclimatology and Palaeoecology*, vol 145, no 1–3, pp61– 77

Schwerdtfeger, W. K., Oelschlager, H. A. and Stephan, H. (1984) 'Quantitative neuroanatomy of the brain of the La Plata dolphin, *Pontoporia blainvillei*', *Anatomy and Embryology*, vol 170, pp11–19

Simmonds, M. P. (2006) 'Into the brains of whales', *Applied Animal Behaviour Science*, vol 100, pp103–116

Stern, S. J. (2009) 'Migration and movement patterns', in W. F. Perrin, B. Würsig and J. G. M. Thewissen (eds) *Encyclopedia of Marine Mammals*, Academic Press, San Diego, pp726–729

Supin, A. Y., Mukhametov, L. M., Ladygina, T. F., Popov, V. V., Mass, A. M. and Poliakova, I. G. (1978) 'Electrophysiological studies of the dolphin's brain', *Izdatel'ato Nauka*, Moscow

Uhen, M. (2007) 'Evolution of marine mammals: Back to the sea after 300 million years', *The Anatomical Record*, vol 290, pp514–522

Wood, F. G. and Evans, W. E. (1980) 'Adaptiveness and ecology of echolocation in toothed whales', in R. Busnel and F. Fish (eds) *Animal Sonar Systems*, Plenum, New York, pp381–426

13

Communication

Paul Spong

About the first question visitors ask, when they walk into our lab and observe a recording of orca acoustics in progress, then notice the pile of tapes, growing daily, that represents decades of such effort, is: 'What are they saying?'. Our answer amounts to 'don't have a clue', though this is usually followed by references to what we do know, or think we understand, i.e. that orcas use echolocation when they are hunting, that we can tell when they are excited and when they are resting, and when they are engaged in the ordinary social exchanges that keep them in touch with one another over considerable distances in the ocean. But beyond that, our understanding lies in a murky grey mist, as if we're adrift in fog on a flat ocean, barely seeing beyond our bow, and without even the hint of a wake to tell us in which direction we should go.

Without question, orcas communicate, as do all other species of whales, certainly with their own kind and quite possibly with others, but the precise content of their communications remains elusive. Thanks to John Ford's work on orca acoustics (Ford, 1989, 1991) we've known since the 1970s that orcas use dialects that are unique to their family groups, and that these dialects are passed down through the maternal lines. Orca dialects are no doubt very useful as tools that promote group cohesion and cultural identity (see Figures 13.1 and 13.2). It is also clear that differences in dialects between groups within a community pose no barriers to social communication. Groups having no commonality in their dialects are quite comfortable with one another when they meet, going about their mutual business in agreeable harmony, enunciating whatever it is they are talking about calmly or with great excitement, even in complete silence, as their moods or circumstances take them. The upshot is a mystery, perhaps akin to that Egyptologists faced when they first encountered hieroglyphics, so what we may need is something like a Rosetta Stone for orca – possible one day, perhaps, but at this stage very much out of discernible reach. In the meantime, we are left with

Figure 13.1 *The closest communication among orcas occurs between mothers and their offspring, here A42 (Holly) with her 2008 baby (A88)*

Source: Paul Spong/OrcaLab

Figure 13.2 *Orca A42 (Holly) with her youngest daughter (A88) and eldest son (A66)*

Source: Paul Spong/OrcaLab

speculation; some observers believing that orcas are engaged only in repetitive 'I'm here' kind of statements that enable everyone in a group to know where the others are within an acoustic space; others believing it possible or likely that precise details of what individuals are encountering are being encoded and transmitted, and even that exchanges can include complex thought content. The common element in these views is that communication is shared; the difference is in the level of sophistication ascribed to the whales.

Sperm whales, possessed of our planet's largest brains ever, display social and acoustic traits that are quite similar, though not identical to those of orcas; they also display traits that are similar to those of the largest-brained terrestrial mammals, elephants. Sperm whale populations are organized as societies that occupy loosely defined ocean realms; roles within the society are structured in space and time to provide the essential means of life and living to all members. Some basic facts were known from observations of old-time whalers and data from the modern whaling industry, but it wasn't until the non-intrusive deep ocean studies, patiently conducted from sailboats by Jonathon Gordon, Hal Whitehead and Lindy Weilgart, beginning in the 1980s, that details began to emerge (Whitehead and Gordon, 1986; Whitehead and Weilgart, 1991).

It turns out that sperm whales, like orcas, use sounds that vary between groups and amount to group-specific dialects, and they display behaviour consistent with the notion that they too, possess culture. Communication in sperm whales is accomplished using 'codas', which consist of a series of complex clicks, with non-random spacing between clicks and variability in the number of clicks; it has been speculated that the sonic structure of codas may be amenable to mathematical analysis. What are we to make of this, and what role does communication play in the functioning of sperm whale (and other cetacean) societies? It is easy to believe that big-brained, sentient, acoustic, social beings are capable of profundities that are expressed in their communications, but very difficult to prove. Inescapably, therefore, one's view of communication in and among whales comes down to one's core set of beliefs about them.

Whatever may be sorted out ultimately, sounds made by whales have intrigued people since ancient times. Aristotle, 2300 years ago, used the voice of dolphins to help separate them from fish; in past centuries, whalers in wooden ships drifted off to sleep to the songs of whales; and recent decades have seen remarkable progress in our observational skills and knowledge base, if not exactly in our understanding of whale communication. The quest began in earnest in the 1960s, initially with John Lilly's studies (Lilly, 1961, 1967) of dolphin brains and vocal mechanisms, and his conclusions about their potential role in sophisticated communication; his work, which was not universally embraced by the scientific community (more rejected than accepted), was followed by the collaboration between Roger and Katy Payne and Scott McVay (Payne and McVay, 1971), which led to the discovery that humpback whales create songs that bear remarkable resemblances to works of music created by humans; this knowledge caused such a stir, that the immediate product, an LP disk titled *Songs of Humpback Whales* (1970, CRM Records) became a pop culture icon of the 1970s; as a result, humpback songs were included in the Golden Record[1] carried aboard the Voyager spacecrafts launched in 1977, with the purpose of taking evidence of the diversity and complexity of life on Earth to the far reaches of our solar system and beyond, on the off chance that it might be discovered by beings in some distant place and time, who would be able to decipher the content and know with certainty that our tiny jewel of a planet had nurtured intelligence; the languages of human cultures were there, and so were the voices of whales.

In the five decades that have elapsed since the 1960s, considerable effort has been devoted to further describing the vocal capabilities of dolphins, including their sound production mechanisms; we are gaining a fascinating picture of the precise control over phonation that is exercised by dolphins and other odontocetes via controlled movement of air inside the head; pairs of tiny 'monkey lips' located under the blowhole give dolphins at least two simultaneous and independent sound production sources (Cranford, 2000) – an extraordinary capability that carries implications for potential communication complexity. What functional use is made of this capability is presently unknown, but little pieces of the puzzle are emerging.

Studies of 'signature whistles' in dolphins suggest that these sounds are used as individual identifiers corresponding to 'voices' in large groups, where knowing the location of close companions or relatives is important to group function and cohesion. Peter Tyack's method for determining the individual origin of vocalizations in groups (Tyack, 1985), which he developed with captives, and with colleagues applied to wild dolphins, has produced intriguing results, but the technique has a drawback in requiring physical attachment of monitoring gear; it's use is therefore limited to controlled situations, i.e. captivity or very accessible wild groups. Signature whistles are retained by individuals for at least ten years, lending credence to the notion that communication between individuals in groups of dolphins is structured and that it endures over time. Moreover, complex variations occur within basic signature whistles, making it likely that they contain far more than identifier information (Sayigh et al, 2007). To dolphins, then, the cacophony we hear via our hydrophones probably makes perfect sense – everyone knows who is speaking, where they are and (presumably) what is being said.

The songs of humpbacks have been studied in numerous populations and locations since their discovery in the 1960s (Payne and McVay, 1971). Researchers generally agree that the songs are made exclusively by males (Darling et al, 2006) but their precise purpose remains uncertain, even controversial. Given that the songs are mostly sung during the winter season, when humpbacks gather in warm shallow waters to have babies and find mates, it seems arguable and logical that the songs are a form of male display that serve to attract females; but if this is the case, the precise meaning, and the mechanism, remain elusive. Perhaps the annual humpback song fest, in which a new version of last year's song spreads rapidly across whole ocean basins, is a form of grand harmony, in which voices, though separated by thousands of kilometres, rise in joyous unison, with some being more perfectly in tune than others; but perhaps the answer is more mundane, even that the songs are a precursor to the blood sports some insist male humpbacks engage in as they compete for female access and attention. The 'Gentle Giant' is nowhere to be found in this scenario; yet the evidence is overwhelming that humpbacks routinely display extraordinary restraint in the face of mortal danger, as when they are entangled in fishing gear and being assisted by humans; the aftermath of a successful release is often so poignant as to be tear-provoking, with the whale acknowledging the rescuers with gentle touches and eye-to-eye contact, that, as communication, leave no shadow of doubt as to intent and meaning. It seems ironic that communication between whales and people may be easier to understand than communication between whales themselves. Perhaps this is because we are active participants; also, the questions asked are simpler. If we were to confine ourselves to describing the role that touch plays in interactions between members of cetacean groups – certainly vital – we might find the road to understanding easier. But that is not where we're at – we want to know what whales are saying.

A basic problem in taking this quest forward is in knowing who the voices we are hearing belong to. We know decisions are being taken in simple situations like travelling or foraging, and sometimes we're correct in interpreting what we hear; we can quite often predict when a group of orcas is about to turn, but we don't know if the decision to turn is made by one individual, or if it's a collective decision. When we put a hydrophone into the water and make a recording, we are usually listening to a group comprising numerous voices; sorting them out is a necessary first step to understanding the conversation (if there is one). Otherwise, we're in a situation analogous to having a single microphone picking up the sounds from a big room in which people are partying – there are many voices, some close, some distant, some soft, some loud, and we are listening to all of them at the same time; to further complicate the situation, loud music may be playing as well. Cetacean researchers have adopted various approaches to the problem, one being illustrated by Tyack's work with dolphins, i.e. physically attaching monitoring devices; another is reflected in studies that use a fixed or towed array of hydrophones, which enable the sources of sounds to be localized, and thus the individuals making them to be identified (Pavan and Borsani, 1997). Perhaps, by creating a library of identified sounds and sources, 'conversations' could be mapped out; describing the conversation is probably a long way from understanding it, but it could be a start.

At OrcaLab, we've been recording the voices of orcas every time we've heard them for 40 years, and though we haven't progressed in understanding what orcas are saying, we do know some things, for example what the family groups sound like. Taking small steps from here, initially by creating a digital archive of the 20,000 plus hours of recordings we've accumulated, it seems possible that we will eventually understand some basics, in particular, who is speaking. Our colleagues at the University of Victoria are developing tools and machine learning capabilities that should take us some way towards that first step (Tzanetakis et al, 2007; Orchive Project, 2007).

On a personal note, I have long wanted to know the voice of Nicola, the first orca I came to recognize as an individual, in 1970. Some day, perhaps soon, I may. Nicola has long disappeared from the scene, but her daughter, grandchildren and great grand offspring still visit Johnstone Strait and Blackfish Sound, so her legacy lives on. Thinking about that gives me perspective, and makes me smile.

Acknowledgements

Many thanks to Lindy Weilgart, David Bain and Helena Symonds for helpful consultations.

Note

1 http://voyager.jpl.nasa.gov/spacecraft/goldenrec1.html

References

Cranford, T. W. (2000) 'In search of impulse sound sources in odontocetes', in W. W. L. Au, A. N. Popper and R. R. Fay (eds) *Hearing by Whales and Dolphins*, Springer Handbook of Auditory Research series, Springer-Verlag, New York, pp109–156

Darling, J. D., Jones, M. E. and Nicklin, C. F. (2006) 'Humpback whale songs: Do they organize males on the breeding grounds?', *Behaviour*, vol 143, pp1051–1101

Ford, J. K. B. (1989) 'Acoustic behaviour of resident killer whales (*Orcinus orca*) in British Columbia', *Can. J. Zool.*, vol 67, pp727–745

Ford, J. K. B. (1991) 'Vocal traditions among resident killer whales (*Orcinus orca*) in coastal waters of British Columbia', *Can. J. Zool.*, vol 69, pp1454–1483

Lilly, J. C. (1961) *Man and Dolphin*, Doubleday, New York

Lilly, J. C. (1967) *The Mind of the Dolphin*, Doubleday, New York

Orchive Project (2007) 'Orchive', http://orchive.cs.uvic.ca/main/index

Pavan, G. and Borsani, J. F. (1997) 'Bioacoustics research on cetaceans in the Mediterranean Sea', *Marine and Freshwater Behaviour and Physiology*, vol 30, pp99–123

Payne, R. S. and McVay, S. (1971) 'Songs of humpback whales', *Science*, vol 173, pp585–597

Sayigh, L. S., Esch, C., Wells, R. S. and Janik, V. M. (2007) 'Facts about signature whistles of bottlenose dolphins, *Tursiops truncatus*', *Animal Behaviour*, vol 74, pp1631–1642

Tyack, P. J. (1985) 'An optical telemetry device to identify which dolphin produces a sound', *J. Acoust. Soc. Am.*, vol 78, no 5, pp1892–1895

Tzanetakis, G., Lagrange, M., Spong, P. and Symonds, H. (2007) 'Orchive: Digitizing and analyzing orca vocalization', Proc. RIAO Conf. Large Scale Semantic Access to Content (Text, Image, Video, and Sound), Pittsburgh, Pennsylvania

Whitehead, H. and Gordon, J. (1986) 'Methods of obtaining data for assessing and modelling sperm whale populations which do not depend on catches', *Reports of the International Whaling Commission*, Special Issue 8, pp149–166

Whitehead, H. and Weilgart, L. (1991) 'Patterns of visually observable behaviour and vocalisations in groups of female sperm whales', *Behaviour*, vol 118, pp275–296

14

Lessons from Dolphins

Toni Frohoff

Rather than explore what we've learned *about* dolphins and other cetaceans, in this chapter I explore what we have learned *from* them. For it is in looking at *who* they are, rather than *what* they are, that their most vital individual and collective characteristics can be revealed through science (Frohoff and Peterson, 2003; Bekoff, 2010; Whitehead, this volume).

Primatologists have pioneered more reflective and individually focused scientific approaches and cetologists can learn a great deal from them. Frans de Waal (2009), Jane Goodall (2010), William McGrew (2004) and others have been examining various perspectives of the social, cultural and even political aspects of nonhuman primates. While the study of wild cetaceans (cetology) was still in its infancy, these primate researchers were already referencing how primate qualities such as morality, humanity, culture and empathy are relevant to humans. Now that we have both neurobiological and behavioural documentation of these very same qualities in species such as dolphins and elephants, the standards by which we view our own characteristics relative to other animals have expanded and diversified (see below) (Plotnik et al, 2006; Hakeem et al, 2009; Marino, this volume; Whitehead, this volume).

Despite the logistical challenges of studying dolphins in the wild (as explained by Slooten, this volume), an increasing number of my colleagues prefer to avoid research on captive cetaceans unless it is specifically and genuinely directed towards the wellbeing of the individuals being studied. Captive studies are limited in that they are fraught with problems given the confounding influences of stress and trauma associated with the captive environment (Sweeney, 1990; Frohoff, 2004), as evidenced by higher mortality (Small and DeMaster, 1995; Woodley et al, 1997). In addition, capture of individuals for captivity places burdens on the often already depleted populations from which they are removed (Reeves et al, 2003).

So, bearing all this in mind, what are the main lessons I have learned to date?

1 *Dolphins are unique individuals with distinct personalities.* The more we learn about cetaceans, the more we see that their populations and cultures are comprised of myriad unique personalities, with entire books devoting much attention to the uniqueness of individuals (Mann et al, 2000; Herzing, 2002; Dudzinski and Frohoff, 2008; Lemieux, 2009). Each dolphin's uniqueness strongly supports the importance of individual psychological wellbeing and the need for greater protection of individuals as well as the populations.
2 *Dolphins can be the ultimate socialites.* Certain individuals have demonstrated strong social bonds that can rival the most poignant of human love stories. This bonding includes bestowing frequent physical affection on one another through pectoral fin rubs (Dudzinski et al, 2010); decades (if not lifetimes) of close association (Mann et al, 2000); supporting one another in times of physical distress, including holding stricken individuals at the water's surface (Connor and Norris, 1982); exhibiting signs of grief when companions or young die (Dudzinski et al, 2003); and exhibiting signs of exceptional altruism both within and across species (see category five, below).

   In addition, the synchronization between dolphins is often exquisite in the way that they swim, surface and breath together, elegantly coordinating their movements, and even sometimes doing this during sleep (see for example, Goley, 1999). Yet conversely some dolphins either only very occasionally, or in some special cases seemingly never, interact with their conspecifics, either through choice or due to geographic isolation (see for example, Frohoff et al, 2006). These individuals often initiate close, sociable contact with humans that often leads to their detriment, warranting increased attention to this matter (see for example Simmond's description of 'Dave' in this volume).
3 *Dolphins are not only intelligent but they also appear to be wise.* The quintessential question received by cetologists is 'just *how* smart are dolphins?'. While it is encouraging that dolphin intelligence is often recognized, it seems so limited that the gold standard appears to be how their intelligence relates to our own. Regardless of how many anthropocentrically designed cognitive experimental hoops dolphins jump through, our species has yet to appreciate that there are different types of intelligence in other species. Given the ecological crises precipitated by our species, I've come to believe that human concepts of cognition need a more critical review anyway. We may be intelligent but are we wise?

   Moreover, 'how *wise* are dolphins?'. Post-reproductive matriarchs are presumed 'wise' and older members of orca, sperm whale, and some other cetacean societies are believed to have a prominent role in their societies (e.g. Mann et al, 2000). Interestingly the same is true of elephants which exhibit many behavioural and neurobiological similarities to some cetaceans (Mann et al, 2000; Plotnik et al, 2006)

4 *Dolphins are emotional.* There is ample anatomical and behavioural evidence that dolphins are not only self-aware but also emotionally sensitive and psychologically complex (Frohoff, 2004; Simmonds, 2006). Thanks to the work of neurobiologists, we now have a completely different view of the dolphin brain than we did before (see Marino, this volume). The dolphins' limbic system – sometimes referred to as the seat of emotion – contains an additional and unique extension called the paralimbic lobe. Recent discoveries such as the existence of spindle and, potentially, mirror neurons in the cetacean, ape and elephant brains confirm that (like so many things) these features are not uniquely human as previously thought (Hof et al, 2005; Marino, this volume). They form an important neurobiological basis for empathy. In other words, dolphins, whales, elephants and perhaps other animals are highly wired *not only for intelligence*, but also for *psychological and social complexity*, and for *empathy*.

In addition, mirror recognition studies conducted on dolphins have demonstrated that they have self-awareness (an aspect of sentience) (Reiss and Marino, 2001). Catherine Kinsman first noted behaviour consistent with mirror responses in solitary belugas during interaction with underwater cameras (personal communication, 2002). In the Baja lagoons in Mexico, I am also exploring the potential for conducting mirror response studies with gray whales who approach small boats; a rare opportunity for looking into the minds of myticetes in a minimally intrusive way and one that may yield results supporting their increased protection.

5 *Dolphins can save or take human lives.* The line between fact and fable is blurred in multiple accounts of dolphin–human interactions from classical antiquity to the present. Most beguiling to us are the modern-day, reputable reports of their 'heroic acts' in aiding unknown humans in distress (Connor and Norris, 1982). Dolphins have attempted – and sometimes apparently succeeded – in saving the lives of people who are exhausted, drowning or, in some instances, being harassed by sharks and also guided boaters who were lost at sea. I have personally witnessed wild dolphins coming to the aid of a swimmer in distress (Dudzinski and Frohoff, 2008).

However, before anticipating that dolphins will always come to our aid, we may do well to consider that dolphins are among the most mischievous of animals, and on the other end of this scale, aggressive behaviour towards humans (as well as towards other cetaceans) also occurs. Captive or solitary dolphins have injured or, on a remarkably few occasions, killed people. Labelling these animals as villains often happens rather than considering that they may be victims of stressful or traumatic situations or environments.

6 *Most dolphins can take us or leave us.* Dolphins are unique among other wild animals in that they are sometimes friendly towards humans, even without being provisioned with food. But lest we flatter ourselves into believing that they always want to be near us, we only have to look at the serious short- and

long-term harm that can occur when we don't respect their needs (see Frohoff et al, 2006; Lemieux, 2009).

7 *Dolphins have a lot to teach us ... but are we willing to listen?* A transition from using dolphins as *resources* to be exploited, to a situation in which they are *respected* for who they are, requires a shift in perspective. Contemporary knowledge of neurophysiology and behaviour in cetaceans, combined with increasing opportunities for studying those wild dolphins who initiate sociable interactions with humans, may provide unprecedented scientific opportunities for an era of less-invasive cetacean research described as 'interspecies collaborative research' (discussed further in Frohoff and Marino, 2010; Marino and Frohoff, in press). Although by no means do I advocate that researchers exploit or intentionally habituate free-ranging animals for the sake of research, given the dangers that such misplaced trust in humans can have for cetaceans (Frohoff et al, 2006; Simmonds, this volume).

Could it be that the wisdom of members of other species and cultures – both *other* and *older* than our own – can offer lessons about the expansion of our own humanity; especially as it pertains to our role in our shared environments? Humans, scientists included, may be surprised at what we can learn not only from cetaceans and other animals, but also about ourselves; particularly as we relate to the natural world.

References

Bekoff, M. (2010) *The Animal Manifesto: Six Reasons for Expanding Our Compassion Footprint*, New World Library, Novato, California

Connor, R. C. and Norris, K. S. (1982) 'Are dolphins reciprocal altruists?', *The American Naturalist*, vol 119, no 3, pp358–374

de Waal, F. (2009) *Primates and Philosopher: How Morality Evolved*, Princeton University Press, Princeton and Oxford

Dudzinski, K. M. and Frohoff, T. (2008) *Dolphin Mysteries: Unlocking the Secrets of Communication*, Yale University Press, New Haven, Connecticut and London

Dudzinski, K. M., Sakai, M., Masaki, M., Kogi, K., Hishii, T. and Kurimoto, M. (2003) 'Behavioral observations of adult and sub-adult dolphins towards two dead bottlenose dolphins (one female and one male)', *Aquatic Mammals*, vol 29, no 1, pp108–116

Dudzinski, K. M., Gregg, J. D., Paulos, R. D. and Kuczaj, S. A. (2010) 'A comparison of pectoral fin contact behaviour for three distinct dolphin populations', *Behavioural Processes*, vol 84, pp559–567

Frohoff, T. (2004) 'Stress in dolphins', in M. Bekoff (ed) *Encyclopedia of Animal Behavior*, Greenwood Press, Westport, Connecticut, pp1158–1164

Frohoff, T. and Marino, L. (2010) 'Towards a new ethical research paradigm on cetacean cognition', Program from the Twelfth International Conference of the American Cetacean Society, Monterey, California

Frohoff, T. and Peterson, B. (eds) (2003) *Between Species: Celebrating the Dolphin-Human Bond*, Sierra Club Books, San Francisco

Frohoff, T., Vail, C. S. and Bossley, M. (2006) 'Preliminary proceedings of the workshop on the Research and Management of Solitary, Sociable Odontocetes Convened at the 16th Biennial Conference on the Biology of Marine Mammals, San Diego, California, December 10, 2005', *International Whaling Commission Scientific Committee*, SC/58/13, IWC, Impington, UK

Goley, P. D. (1999) 'Behavioral aspects of sleep in pacific white-sided dolphins (*Lagenorhynchus obliquidens*)', *Marine Mammal Science*, vol 15, pp1054–1064

Goodall, J. (2010) *Through a Window: My Thirty Years with the Chimpanzees of Gombe*, Mariner Books, Boston, New York

Hakeem, A. Y., Sherwood, C. C., Bonar, C. J., Butti, C., Hof, P. R. and Allman, J. M. (2009) 'Von Economo neurons in the elephant brain', *The Anatomical Record*, vol 292, pp242–248

Herzing, D. L. (2002) *The Wild Dolphin Project: Long Term Research of Atlantic Spotted Dolphins in the Bahamas.* Wild Dolphin Project, Jupiter, Florida

Hof, P., Chanis, R. and Marino, L. (2005) 'Cortical complexity in cetacean brains', *The Anatomical Record*, vol 287A, pp1142–1152

Lemieux, L. (2009) *Rekindling the Waters: The Truth about Swimming with Dolphins*, Troubador Press, Leicester, UK

Mann, J., Connor, R. C., Tyack, P. L. and Whitehead, H. (2000) *Cetacean Societies: Field Studies of Dolphins and Whales*, University of Chicago Press, Chicago and London

Marino, L. and Frohoff, T. (in press) 'Towards a new paradigm of non-captive research on cetacean cognition', *PLoS ONE*

McGrew, W. (2004) *The Cultured Chimpanzee: Reflections on Cultural Primatology*, Cambridge University Press, Cambridge

Plotnik, J. M., de Waal, F. B. M. and Reiss, D. (2006) 'Self-recognition in an Asian elephant', *Proceedings of the National Academy of Science of the United States*, vol 103, pp17053–17057

Reeves, R. R., Smith, B. D., Crespo, E. A. and Notarbartolo di Sciara, G. (2003) *Dolphins, Whales and Porpoises: 2002–2010 Conservation Action Plan for the World's Cetaceans*, IUCN/SSC Cetacean Specialist Group, IUCN, Gland and Cambridge

Reiss, D. and Marino, L. (2001) 'Self-recognition in the bottlenose dolphin: A case of cognitive convergence', *Proceedings of the National Academy of Sciences USA*, vol 98, pp5937–5942

Small, R. and DeMaster, D. P. (1995) 'Acclimation to captivity: A quantitative estimate based on survival of bottlenose dolphins and California sea lions', *Marine Mammal Science*, vol 11, pp510–519

Simmonds, M. P. (2006) 'Into the brains of whales', *Applied Animal Behaviour Science*, vol 100, pp103–116

Sweeney, J. C. (1990) 'Marine mammal behavioral diagnostics', in L. A. Dierauf (ed) *CRC Handbook of Marine Mammal Medicine: Health, Disease, and Rehabilitation*, CRC Press, Boston, MA, pp53–72

Woodley, T. H., Hannah, J. L. and Lavigne, D. M. (1997) 'A comparison of survival rates for captive and free-ranging bottlenose dolphins (*Tursiops truncatus*), killer whales (*Orcinus orca*) and beluga whales (*Delphinapterus leucas*)', IMMA Technical Report No. 97-02, IMMA, Ontario, Canada

15

Highly Interactive Behaviour of Inquisitive Dwarf Minke Whales

Alastair Birtles and Arnold Mangott

INTRODUCTION

In the early 1990s, as the Great Barrier Reef (GBR) live-aboard dive industry began to expand along the shelf-edge Ribbon Reefs north of Cairns, reports began to surface of small, conspicuously marked whales approaching scuba divers. The late Dr Peter Arnold of Townsville's Museum of Tropical Queensland published the first account of these encounters (Arnold, 1997) and had previously documented this new form of minke whale (Arnold et al, 1987). As the dive fleet grew and the frequency of these encounters increased, there was growing concern over potential impacts on the whales (Aw, 1996). In 1996, Peter and one of us (Alastair Birtles) joined John Rumney and Andy Dunstan (General Manager and Operations Manager, respectively) on their vessel *Undersea Explorer* for the first of what became 15 consecutive years of field work. Our initial objective was to find out what we could about the biology and behaviour of these poorly known and elusive little whales, but this rapidly evolved into trying to ensure that the growing whale-watching industry would be sustainable.

STUDYING WHALES AND MANAGING PEOPLE

Inquisitive, indeed apparently almost insatiably curious, sometimes staying around for hours and repeatedly swimming past snorkellers and scuba divers, coming closer and closer until occasionally they were within arm's length, these whales were providing extraordinarily powerful experiences for a growing number of divers. We soon realized that interactions between people and the whales were two-way

Figure 15.1 *A headrise and breath for a nearly vertical and very curious dwarf minke whale near swimmers*

Note: The deployment of a rope for the swimmers to hold on to helps to keep them relatively stationary and has been an important step in developing the sustainability of this industry.

Source: Matt Curnock, Minke Whale Project and James Cook University, Townsville, QLD, Australia

and that the human activities would have to be managed in relation to the whales. The divers were drifting around on the surface in the open ocean and sometimes actively swimming towards the whales. How to immobilize them? The heavy ropes normally used for mooring the vessel were thrown over the side and the swimmers held onto these, thus remaining relatively stationary, one of the first steps towards the sustainable management of this fledgling industry (see Figure 15.1).

The patterns of black, grey and white markings, especially in the shoulder area, allowed identification of individual whales, even in real-time while in the water with them (Arnold et al, 2005). The photo-ID study quickly built up to the stage that we knew we were seeing over 100 different whales each year. The first resighting of a whale within a season was in 1997 and the first whale(s) returning from a previous year soon followed in 2000 (Birtles et al, 2002a).

We submitted the first code of practice for swimming with whales to the Whale Watching Sub-committee of the IWC Scientific Committee in 1998 (Arnold and Birtles, 1999). With some Australian Government Cooperative Research Centre funding 1999–2001 (our first minke grant), we began to study not only the whales (Birtles et al, 2001, 2002a), but also visitor experiences through questionnaires and interviews (Birtles et al, 2002b; Valentine et al, 2004). The GBR Marine Park Authority granted special permit endorsements to just nine vessels in 2003, creating the first fully regulated swim-with-whales industry in the world. The permit conditions required the vessels to follow our Code of Practice and to use our whale sightings sheets. The Code was subsequently revised (Birtles et al, 2008) and we analysed nearly 1500 industry minke encounters (2003–2008) (Birtles et al, 2010). In addition, over 20 honours and postgraduate student projects and many volunteer researchers have helped with the Minke Whale Project (MWP).

However, it is important to recall how partial our insight into the whales' lives really is. The earliest GBR minke sightings are April and the latest September, but 90 per cent of encounters are in June and July, and the residence times of individual whales have a mean of around ten days and maximum of a month (Birtles et al, 2010). They are only at the surface for a few seconds every few minutes to breath. A fleeting glimpse of a dorsal fin only the size of a bottlenose dolphin's; a smoothly curving back – and they never 'fluke up' (lift their tail fins in the air) before diving. However, in the water we can see their whole bodies, twisting, turning, looping the loop, barrel rolling, veering, diving – and occasionally coming into a vertical submerged tail stand near us and revolving gently as they bring both eyes to bear on us (see Figure 15.2). Much like us, their vision will help them perceive at most 30 to 40 metres around them and we have very little appreciation of the acoustic world that must dominate their senses beyond such paltry distances (Gedamke et al, 1997).

Clearly their reappearances throughout June and July at exactly the same locations, year after year, mean that their visits to the GBR must be of considerable significance. At least two whales have returned in five consecutive years and several have been resighted multiple times over periods of up to eight or nine years. Therefore, they must have a very good sense of timing, geography and of navigation – but exactly why they come and where they come from are still mostly a mystery.

Interpreting Whale Behaviour

We have used a variety of research tools, including hydrophones and video and have identified over 30 distinct surface and in-water behaviours with two major functions: (1) displays with potential social functions, and (2) behaviours attributed to exploration (Mangott, 2010). Potential social behaviours include *belly presentations*, *courtship behaviours* and *bubble releases*. These behaviours have a distinct visual component and are particularly evident in larger social groups (i.e. more than six animals present). *Belly presentations* of dwarf minke whales, for example, are often directed to conspecifics but also to the vessel and swimmers. A cetacean flashing its bright white belly may signal its position (Norris and Dohl, 1980) and/or indicate its movement intentions (Herman and Tavolga, 1980). *Belly presentations* are also interpreted as acts of courtship or sexual solicitation (Caldwell and Caldwell, 1972; Tyack, 2000). The northern GBR is thought to be a breeding

Figure 15.2 *A close binocular examination for the author (Alastair Birtles) by a young female dwarf minke ('Strumpet') as she performs repeated close approaches (<3m) and occasional very close approaches (<1m), often culminating in a submerged tail stand followed by a headrise and partial pirouette*

Note: In this photograph the whale is moving from a vertical submerged tail stand into a headrise. The body is turned to enable the whale to look forward at the photographer and the end of the rostrum is above the surface as she was beginning a headrise.

Source: Alastair Birtles, Minke Whale Project & James Cook University, Townsville, QLD, Australia

ground for dwarf minke whales (Birtles et al, 2002a) thus it is possible that the observed social displays may be part of the reproductive process.

The fact that individual whales return to the GBR each austral winter (Birtles and Arnold, 2002; Sobtzick, in prep), the presence of all age classes (Dunstan et al, 2007), including cow–calf pairs (Birtles et al, 2002a) and the apparent social aggregations, all indicate that this region is significant to this species.

Exploration or curiosity is widespread in the animal kingdom (for example, insects, fish, birds, rodents, primates, cetaceans: Hinde, 1966; Dahlheim et al, 1981; Herzing, 1999; Morgan and Sanz, 2003; Compagno et al, 2005) and is important ecologically and in an evolutionary sense. A natural exploratory drive enables animals to learn from and respond to their environment, facilitates finding food or potential mates, and helps with avoiding predators (Mason, 1973; Norris and Dohl, 1980; Fairbanks, 1993) – all factors that are crucial for the survival of their species. Many behaviours displayed by dwarf minke whales in interactions with vessels and their swimmers seem to be directly attributed to exploration. Such behaviours include *close* (1–3m) and *very close approaches* (1m or less) to swimmers, *headrises* and *spyhops*, *motorboating* and *pirouettes*. In all these behaviours the animals have been observed to visually inspect swimmers or divers. In some instances the positioning of the animal to the human observer is indicative of binocular vision. Binocularity enhances depth acuity (Fox, 1978) and improves accuracy for localizing objects (Dawson, 1980). During *headrises*, *pirouettes* and *belly presentations* to humans, both eyes of the dwarf minke are protruded and seem to focus on the observer.

These whales repeatedly approach and initiate contact with dive tourism vessels, maintain close contact (within 60m) with them for prolonged periods (mean ± SE = 2.8 ± 0.2 hours) and aggregate in particular around swimmers (Mangott et al, 2011). Most marine and terrestrial wildlife species either avoid or tolerate human observers, and the initiation of contact is rarely made by the animals. Although voluntary approaches are often observed in cetaceans (see for example, Caldwell and Caldwell, 1972; Dahlheim et al, 1981; Herzing, 1999; Roden and Mullin, 2000), most interactions between people and cetaceans are short or infrequent.

The whales' exploratory drive appears to increase with time, perhaps as they desensitize to the disturbance. The already very close individual whale–swimmer passing distance (mean ± SE = 6.80 ± 0.08m; N = 1602) substantially decreases both over the short term (within the first 90 minutes of an interaction) and longer term (over the six weeks of a typical field season). The closeness of the approaches by dwarf minke whales to human observers is influenced by two major factors: (1) the whale group size, and (2) the familiarity of individuals with the stimulus (vessels and swimmers). Individual whales approach swimmers significantly closer when they are part of a larger group (i.e. more than six animals). In addition, familiar (re-sighted) whales pass substantially closer in subsequent interactions and approach humans on average 2.5m closer than non-resighted whales (Mangott, 2010). The whales are approaching a stationary stimulus of their own volition and the apparent increased level of confidence could also be the result of a positive experience with the stimulus and/or the perceived safety among a larger group.

PROTECTING FRIENDLY WHALES

Cetaceans which lose their natural wariness to human activity are potentially more prone to harassment (Samuels and Bejder, 2004, Simmonds this volume), boat strikes (Stone and Yoshinaga, 2000; Spradlin et al, 2001; Bejder and Samuels, 2003), and entanglement in fishing gear or marine debris (Lien, 1994; Knowlton and Kraus, 2001). Habituated dwarf minkes could even become an easy target for the whaling industry in the Southern Ocean (potential feeding grounds). However, their migration paths and destinations are still unknown and there is evidence that their extreme interactivity is confined to the peak of the GBR winter season and is only maintained for a maximum of three years.

Depending on the magnitude of human interactions and the importance of the displaced behaviour to the animals (for example resting, courtship, socializing) there is potential to alter the fitness of the population (see for example, Mann et al, 2000; Frohoff, 2004). Although all sizes of whales are encountered from new-born calves (albeit rarely) to large animals well over 7m in length, the majority of interacting dwarf minke whales are adolescents (Dunstan et al, 2007). Immature animals are often regarded with particular concern as the documented increased familiarity with human activity could alter their normal course of behavioural development, potentially resulting in altered patterns of social behaviour (Samuels et al, 2003) but such permanent habituation has not been documented in minkes.

Close and repeated interactions with marine mammals also pose potential public safety concerns (Spradlin et al, 2001; Samuels and Bejder, 2004). Aggression towards swimmers has been reported in various baleen whale species (Hall, 2000; Rose et al, 2007). In most cases, the aggressive act by the animal was triggered by inappropriate human behaviour (for example, harassing their calf or touching). Although potential agonistic displays by dwarf minke whales are present during interactions with humans (for example, *jaw gapes, bubble releases*), these displays are very rare and none have ever been followed by any apparent increased level of aggression, either between whales or towards swimmers (Mangott, 2010) in many thousands of interactions.

CONCLUDING COMMENTS

After 15 years, it is rather daunting how little we know about these beautiful but enigmatic little whales. It still amazes us that we can be studying a new subspecies (possibly even a new species) of baleen whale in the waters of a developed country such as Australia at this point in the 21st century and it is quite alarming that our knowledge is so limited.

We still know very little about their population size, why they are here and where they are for most months of the year. We know nothing about where and when they feed. What we do know is that they are very curious about us, that they provide us with truly remarkable experiences and a clear responsibility to manage our interactions with them in an ecologically sustainable way that does no harm to these marvellous little whales.

REFERENCES

Arnold, P. (1997) 'Occurrence of dwarf minke whales (*Balaenoptera acutorostrata*) on the northern Great Barrier Reef', *Report of the International Whaling Commission*, vol 47, pp419–424

Arnold, P. W. and Birtles, A. (1999) 'Towards sustainable management of the developing dwarf minke whale tourism industry in northern Queensland', *CRC Reef Research Tech. Rep. 27*, James Cook University of North Queensland (Amended version of SC/50/WW1)

Arnold, P., Marsh, H. and Heinsohn, G. (1987) 'The occurrence of two forms of minke whales in east Australian waters with a description of external characters and skeleton of the diminutive or dwarf form', *Scientific Reports of the Whales Research Institute, Tokyo*, vol 38, pp1–46

Arnold, P., Birtles, A., Dunstan, A., Lukoschek, V. and Matthews, M. (2005) 'Colour patterns of the dwarf minke whale *Balaenoptera acutorostrata* sensu lato: General description, cladistic analysis and taxonomic implications', *Memoirs of the Queensland Museum*, vol 51, no 2, pp 165–195

Aw, M. (1996) 'Meeting with minkes', *Scuba Diver*, March–April, 1996, pp40–43

Bejder, L. and Samuels, A. (2003) 'Evaluating the effects of nature-based tourism on cetaceans', in N. J. Gales, M. A. Hindell and R. Kirkwood (eds) *Marine Mammals: Fisheries, Tourism and Management Issues*, CSIRO Publishing, Collingwood, Australia

Birtles, R. A. and Arnold, P. (2002) *Dwarf Minke Whales in the Great Barrier Reef, Current State of Knowledge*, CRC Reef Research Centre, Townsville, Australia

Birtles, A., Arnold, P., Curnock, M., Valentine, P. and Dunstan, A. (2001) *Developing Ecologically Sustainable Dwarf Minke Whale Tourism (1999–2001)*, Final Report to the Commonwealth Department of Environment and Heritage (Environment Australia), November, James Cook University, Townsville

Birtles, R. A., Arnold, P. W. and Dunstan, A. (2002a) 'Commercial swim programs with dwarf minke whales on the northern Great Barrier Reef, Australia: Some characteristics of the encounters with management implications', *Australian Mammalogy*, vol 24, pp23–38

Birtles, A., Valentine , P., Curnock, M., Arnold, P. and Dunstan, A. (2002b) 'Incorporating visitor experiences into ecologically sustainable dwarf minke whale tourism in the northern Great Barrier Reef', CRC Reef Research Centre Technical Report No. 42, CRC Reef Research Centre, Townsville

Birtles, A., Arnold, P., Curnock, M., Salmon, S., Mangott, A., Sobtzick, S., Valentine, P., Caillaud, A. and Rumney, J. (2008) *Code of Practice for Dwarf Minke Whale Interactions in the Great Barrier Reef World Heritage Area*, Great Barrier Reef Marine Park Authority, Townsville, Australia

Birtles, A., Valentine, P., Curnock, M., Mangott, A., Sobtzick, S. and Marsh, H. (2010) 'Dwarf minke whale tourism monitoring program (2003–08)', Final Report to the Great Barrier Reef Marine Park Authority, 16 April

Caldwell, D. K. and Caldwell, M. C. (1972) *The World of the Bottlenosed Dolphin*, Lippincott, Philadelphia

Compagno, L. J. V., Dando, M. and Fowler, S. (2005) *Sharks of the World*, Collins, London

Dahlheim, M. E., Schempp, J. D., Swartz, S. L. and Jones, M. L. (1981) 'Attraction of gray whales, *Eschrichtius robustus*, to underwater outboard engine noise in Laguna San Ignacio, Baja California Sur, Mexico', *Journal of Acoustical Society of America*, vol 70, ppS83–S84

Dawson, W. W. (1980) 'The cetacean eye', in L. M. Herman (ed) *Cetacean Behaviour: Mechanisms and Functions*, Robert E. Krieger Publishing Company, Malabar, Florida, pp53–100

Dunstan, A., Sobtzick, S., Birtles, R. A. and Arnold, P. W. (2007) 'Use of videogrammetry to estimate length to provide population demographics of dwarf minke whales in the northern Great Barrier Reef', *Journal of Cetacean Research and Management*, vol 9, pp215–223

Fairbanks, L. A. (1993) 'Risk-taking by juvenile vervet monkeys', *Behaviour*, vol 124, pp57–72

Fox, R. (1978) 'Binocularity and stereopsis in the evolution of vertebrate vision', in S. J. Cool and E. L. Smith (eds) *Frontiers in Visual Science*, Springer-Verlag, New York, pp316–327

Frohoff, G. T. (2004) 'Behavioral indicators of stress in odontocetes during interactions with humans: A preliminary review and discussion', International Whale Watching Conference, South Africa

Gedamke, J., Costa, D. P. and Dunstan, A. (1997) 'New vocalisation definitely linked to the minke whale', *J. Acoust Soc. America*, vol 102, pp3121–3122

Hall, H. (2000) 'Wildlife photography', *Natural History*, vol 109, no 10, p107

Herman, L. M. and Tavolga, W. N. (1980) 'The communication systems of cetaceans', in L. M. Herman (ed) *Cetacean Behaviour: Mechanisms and Function*, Robert E. Krieger Publishing Company, Malabar, Florida, pp149–209

Herzing, D. L. (1999) 'Minimizing impact and maximizing research during human/dolphin interactions in the Bahamas', in M. Dudzinski, G. T. Frohoff and T. R. Spradlin (eds) *Wild Dolphin Swim Program Workshop*, Maui, Hawaii

Hinde, R. A. (1966) *Animal Behaviour*, McGraw-Hill, London

Knowlton, A. R. and Kraus, S. D. (2001) 'Mortality and serious injury of northern right whales (*Eubalaena glacialis*) in the western North Atlantic Ocean', *Journal of Cetacean Research and Management*, Special Issue 2, pp193–208

Lien, J. (1994) 'Entrapments of large cetaceans in passive inshore fishing gear in Newfoundland and Labrador (1979–1990)', *Reports of the International Whaling Commission*, Special Issue 15, pp149–157

Mangott, A. (2010) 'Behaviour of dwarf minke whales (*Balaenoptera acutorostrata*) associated with a swim-with industry in the northern Great Barrier Reef', unpublished PhD thesis, James Cook University, Townsville, Queensland, Australia

Mangott, A., Birtles, A. and Marsh, H. (2011) 'Attraction of dwarf minke whales (*Balaenoptera acutorostrata*) in the Great Barrier Reef World Heritage Area: Management challenges of an inquisitive whale', *Journal of Ecotourism*, vol 10, no 1, pp64–76

Mann, J., Connor, R. C., Barre, L. M. and Heithaus, M. R. (2000) 'Female reproductive success in bottlenose dolphins (*Tursiops* sp.): Life history, habitat, provisioning, and group-size effects', *Behavioral Ecology*, vol 11, pp210–219

Mason, W. A. (1973) 'Regulatory functions of arousal in primate psychosocial development', in C. R. Carpenter (ed) *Behavioral Regulators of Behavior in Primates*, Bucknell University Press, Lewisburg, Lewisburg, pp19–33

Morgan, D. and Sanz, C. (2003) 'Naive encounters with chimpanzees in the Goualougo Triangle, Republic of Congo', *International Journal of Primatology*, vol 24, pp369–381

Norris, K. S. and Dohl, T. P. (1980) 'Behavior of the Hawaiian spinner dolphin *Stenella longirostris*', *Fishery Bulletin*, vol 77, pp821–849

Roden, C. L. and Mullin, K. D. (2000) 'Sightings of cetaceans in the northern Caribbean Sea and adjacent waters, winter 1995', *Caribbean Journal of Science*, vol 36, pp280–288

Rose, N. A., Parsons, E. C. M. and Sellares, R. (2007) 'Swim-with-whale tourism: An update on development of a questionnaire', International Whaling Commission, Anchorage

Samuels, A. and Bejder, L. (2004) 'Chronic interaction between humans and free-ranging bottlenose dolphins near Panama City Beach, Florida, USA', *Journal of Cetacean Research and Management*, vol 6, pp69–77

Samuels, A., Bejder, L., Constantine, R. and Heinrich, S. (2003) 'Swimming with wild cetaceans in the southern hemisphere', in N. Gales, M. Hindell and R. Kirkwood (eds) *Marine Mammals: Fisheries, Tourism and Management Issues*, CSIRO Publishing, pp277–303

Sobtzick, S. (in prep) 'Dwarf minke whale biology and implications for tourism management', unpublished PhD thesis, James Cook University, Townsville, Australia

Spradlin, T. R., Barre, L. M., Lewandowski, J. K. and Nitta, E. T. (2001) 'Too close for comfort: Concern about the growing trend in public interactions with wild marine mammals', *Marine Mammal Society Newsletter*, vol 9, pp3–5

Stone, G. S. and Yoshinaga, A. (2000) 'Hector's dolphin *Cephalorhynchus hectori* calf mortalities may indicate new risk from boat traffic and habituation', *Pacific Conservation Biology*, vol 6, pp162–170

Tyack, P. L. (2000) 'Functional aspects of cetacean communication', in J. Mann, R. C. Connor, P. L. Tyack and H. Whitehead (eds) *Cetacean Societies: Field Studies of Dolphins and Whales*, The University of Chicago Press, Chicago, pp270–307

Valentine, P. S., Birtles, R. A., Curnock, M., Arnold, P. and Dunstan, A. (2004) 'Getting closer to whales: Passenger expectations and experiences, and the management of swim with dwarf minke whale interactions in the Great Barrier Reef', *Tourism Management*, vol 25, pp647–655

16

The Cultures of Whales and Dolphins

Hal Whitehead

Despite the difficulties of studying cetacean behaviour, and of identifying cultural processes in studies of wild animals, there is ample evidence for the significance of culture among whales and dolphins, including the baleen whales. The presence of culture makes the animals' lives considerably richer, but makes the tasks of conserving and managing them much more complex. We need to conserve both genetic and cultural diversity, but also to consider the wide range of potential ways that cultural animals may respond to human-induced changes in their environments. The sustainable exploitation of such cultural animals may be impossible to ensure. The significance of culture among the whales and dolphins indicates that they should be included with humans and a few other species in an extended moral community.

Introduction

Culture means a great deal to humans. We largely define ourselves as a species, and as communities within the human species, by our cultures. Culture is cuisine, art, literature, fashion, architecture, music and technology. It is the way we greet each other, the side of the road we drive on, our languages, agricultural methods, medicine, scientific knowledge and religions. It includes passing fads such as the music of a particular 'boy band' as well as peculiar behavioural patterns conserved over millennia by religious strictures. Culture has shaped our evolution; the ability of adults to absorb lactose was driven by the cultural practice of dairy farming, and scientists have argued that the evolution of our large and complex brains owes much to the need to acquire the best cultural information, and to use it effectively. Our cultures have allowed us to use planet Earth in extraordinary ways, as well as to abuse it. And now, the consequences of our culture threaten the planet. For humans, culture is essential (Richerson and Boyd, 2005).

Culture also means a great deal to whales and dolphins. Culture is song, dialect and foraging strategies, greeting ceremonies and ways of moving through the ocean. It includes passing fads, such as everyone pushing dead salmon around for a few weeks, and, in the same group of killer whales, dialects that retain their stability over generations. Cetacean cultures evolve, sometimes very quickly and sometimes slowly, and occasionally there are revolutions. And there is much more that we do not know. I argue in this chapter that culture is an essential part of what cetaceans are, how they take their place in ecosystems, and thus how we conserve, manage and treat them.

I begin with a consideration of what culture is, why it is important and how it can be studied. I then summarize the evidence for culture in whales and dolphins, comparing this with results for other species. The last part of the chapter is about the importance of cetacean culture for the whales and dolphins themselves, for the other species that interact with them, and for humans.

The study of cetacean culture is both difficult and in its infancy. We know very little, and little of the evidence available is based upon designed, replicated experiments. Given the natures of whales and culture it cannot be. This makes for uncertainty, as many have pointed out. In what follows I note areas of uncertainty but follow a path of reasoning that I believe best fits what we do know of these animals.

What is Culture and How May it be Studied?

Culture has been defined in many ways (see for example, Mundinger, 1980; Laland and Galef, 2009). Academics from different disciplines – and several academic disciplines specifically study aspects of human culture – have different concepts of culture. Some of these definitions are of no use to those interested in nonhuman culture because they explicitly or implicitly restrict culture to humans, or define it based on concepts, such as 'values', that we cannot reach in studies of whales or chimpanzees or horses.

However, there is a concept of culture that has emerged independently from several traditions, such as macro-sociobiology (Lenski et al, 1995), theoretical work on human cultural evolution (Richerson and Boyd, 2005), and laboratory studies of culture in animals (Laland and Hoppitt, 2003). It is simple, can be operationalized for studies of nonhumans, and corresponds to most humans' informal view of culture, 'the way we do things' (McGrew, 2003). A more formal version of the definition is 'group-typical behaviour patterns shared by members of a community that rely on socially learned and transmitted information' (Laland and Hoppitt, 2003). Such definitions have three principal components. Culture is behaviour, it is community specific, and it is transmitted through social learning in which behavioural patterns are the result of social relationships. Social learning includes mechanisms such as imitation, emulation and teaching (Whiten et al, 2004).

From this definition it follows that the raw material of studies of culture in nonhumans is behaviour, so these studies are part of the academic discipline known as ethology (Hinde, 1982), and that they focus on transmission mechanisms and group-specific patterns. Transmission mechanisms and group-specific patterns have usually been studied by scientists from different traditions, experimental psychology and field ethology respectively. Especially in studies of chimpanzees (*Pan troglodytes*), those studying social learning in the laboratory have frequently argued against much social learning and thus culture (see for example, Galef, 1992; Tomasello 1994), whereas field researchers have uncovered a great deal of behavioural variation that they attribute to culture (Whiten et al, 1999). These and other disputes became heated, leading to the 'chimpanzee culture wars' (McGrew, 2003). In recent years, the field evidence has mounted and the experimental psychologists, using improved methods, have recanted some, but certainly not all, of their scepticism towards chimpanzee culture (Tomasello, 2009).

The evidence for cetacean culture follows the same two lines, laboratory studies of social learning and field data on behavioural variation, and has attracted the same types of criticism as the research on chimpanzees and other primates (Rendell and Whitehead, 2001, and commentaries). However, there are important differences between what we know of the cultures of whales and primates, as well as how culture is studied in the two groups of species. These contrasts stem from the very different biologies and habitats of apes and whales. Compared with ape culture, whale culture is less material, more vocal, less geographically based, but much larger scale.

Culture in Whales and Dolphins

The evidence for culture in whales and dolphins comes from both studies of social learning, in other words the transmission of culture, as well as from observations of patterns of behavioural variation (Rendell and Whitehead, 2001).

The transmission of culture

Although there has been rather more emphasis on patterns of behavioural variation, it is very clear that dolphins, at least of those species that have been tested, learn socially. The preferred method of training a dolphin to perform particular behaviour for shows or other purposes is to expose it to another dolphin who knows the routine (Pryor, 2001). Not only do dolphins in captivity imitate both actions and sounds of humans or other dolphins in experimental settings, but they do so profligately and understand the concept of imitation (Herman, 2002). Dolphins, like humans and other apes but not monkeys, can follow a command that means 'imitate what I do next', playing the 'Simon says' game. According to Whiten (2001), who studies primates, dolphins imitate humans better than apes. A drawback of the direct laboratory evidence for imitation in cetaceans is that nearly all of it comes from just one species, the bottlenose dolphin (*Tursiops truncatus*), the species that has been the subject of almost all research on captive cetaceans.

Types of social learning other than imitation have received much less attention. However, there are strong indications that wild killer whales (*Orcinus orca*) teach their offspring the difficult and dangerous behaviour of self-stranding on beaches in order to catch seals (Guinet and Bouvier, 1995). Additionally, many of the patterns of behavioural variation among cetaceans discovered by the field ethologists, and which I summarize in the next few paragraphs, have to be transmitted by social learning – there is no other conceivable mechanism – but we do not know which form of social learning (imitation, emulation, teaching, etc.) is operating.

Patterns of behaviour

The first form of cetacean culture to be discovered was the song of the humpback whale (*Megaptera novaeangliae*). During the breeding season, male humpback whales make long, loud, structured and elaborate sequences of vocalizations, spanning a wide frequency range (Payne and McVay, 1971). At any time, all the males in any ocean basin sing nearly the same song, but the song evolves a little over months and more substantially over years, as 'themes' change, drop out or are added (Payne, 1999). The songs in different oceans have different contents and evolutionary trajectories but follow the same rules (Payne, 1999). An exception to the gradual evolution of humpback song occurred off the east Pacific coast of Australia in 1997, when the humpbacks adopted a radically different song, the Indian Ocean song from the west coast of Australia, which seems to have been introduced by Indian Ocean animals migrating the wrong side of Australia into the Pacific (Noad et al, 2000). This was the first-known nonhuman cultural revolution. Although most scientists assume that the humpback's song, sung just by males in the breeding season, is involved in breeding, despite considerable research the precise function of the song is far from clear (see for example, Darling and Bérubé, 2001). However, even the fiercer critics of nonhuman culture admit that the patterns of humpback song must be the result of social learning, humpbacks listening to each others' songs and adjusting their own accordingly, and are a community-wide phenomenon, and thus culture (Laland and Janik, 2006).

It is now becoming clear that males of many of the other baleen whale species, such as bowheads (*Balaena mysticetus*), blue whales (*Balaenoptera musculus*), finbacks (or fin whales) (*Balaenoptera physalus*) and minke whales (*Balaenoptera acutorostrata*) also sing during the breeding season (Würsig and Clark, 1993; Gedamke et al, 2001; Croll et al, 2002; McDonald et al, 2006). Although not as complex as the humpback song, these songs are also structured and cyclical, can evolve, are geographically patterned and may be extremely loud, having enormous power (ca. 186dB re 1μPa) and range (thousands of kilometres) (Payne and Webb, 1971; McDonald et al, 2001).

There has been rather little study of the non-acoustic behaviour of the baleen whales, and most of that has been on the humpback. Humpbacks have traditional feeding grounds, which they probably learn from their mothers while accompanying her during their first pole-wards migration at a few months of age (Rendell and Whitehead, 2001). This social learning results in communities of whales making similar migrations, and so humpback migrations are a form of culture by our definition. Humpback whales have enormously varied, and sometimes complex, foraging strategies including the use of bubbles and other behaviour to herd prey, as well as communal foraging methods (Clapham, 2000). This variation is primarily spatial, with animals in different areas using characteristic foraging behaviour, but also temporal. Particular attention has been paid to the development of 'lobtail feeding' by humpbacks in the Gulf of Maine in the 1980s. This behaviour, which seems to be an amalgamation of two other types of behaviour, was first observed from one animal in 1981, and then was observed being performed by more and more individuals over the next ten years (Weinrich et al, 1992). It involves slamming the tail flukes onto the water before diving under a prey school and also exhaling bubbles beneath the school. This seemed to represent the spread of a new innovation through the population by social learning, and so culture (Rendell and Whitehead, 2001). But, as with so much apparent cultural foraging behaviour, there are doubts. Could changing conditions in the Gulf of Maine have gradually favoured the individual adoption of this type of lobtail feeding, without any learning from other humpbacks (Janik, 2001)? It seems to me much more likely that the humpbacks learned the quite complex foraging technique from each other, and also that the geographical variation in humpback foraging techniques are culture, but we do not know for sure.

There are similar doubts about almost all accounts of foraging cultures, whether in apes, birds or cetaceans (Laland and Janik, 2006). The problem stems from the standard method of inferring culture in field studies of animals. Scientists describe patterns of variation in behaviour, usually spatial variation, but sometimes temporal variation or other sorts of variation. They then rule out genetic determination – that the variation in behaviour is determined by different sets of genes – and ecological determination – that differences in the environmental conditions could have caused the variation, for instance a particular kind of tool use could be dependent on the availability of certain materials, or that the spread of lobtail feeding in humpback whales of the Gulf of Maine was directly the result of changing oceanographic conditions. Sometimes scientists also consider ontogenetic determination – changes in behaviour with age – as a potential cause of behavioural variation. If there are no genetic, ecological or ontogenetic factors that could possibly cause the patterns of variation, then, using the process of elimination, they conclude that social learning must be the cause, and if there are group-specific patterns of behaviour, the animals have culture. This is an unsatisfactory methodology for several reasons. Critics can try to sink culture by coming up with ecological or genetic explanations, and they do this. See, for instance, Janik's (2001) extraordinarily tortuous genetic explanation for sperm whale (*Physeter macrocephalus*) coda dialects, when the cultural alternative is that young sperm whales learn appropriate patterns of sounds from their mothers. The exclusionary methodology makes culture an explanation of last resort for nonhumans, whereas with humans culture is usually assumed to be driving behavioural variation, and it must be ruled out before genetics is considered. Human and nonhuman cultures cannot be compared when the standard of proof for the underlying data is reversed. The exclusionary method of identifying culture is particularly problematic with foraging cultures, as foraging is an interaction with the environment. So environmental variation is almost always present, and is hard to exclude as a potential contributor towards behavioural variation. Yet foraging cultures are undoubtedly very important in many species. New techniques of studying culture are needed, methods that do not exclude but rather apportion behavioural variation between genetic, environmental, cultural and possibly other causes. Such methods are under development (Laland et al, 2009; Whitehead, 2009), but have yet to be applied to any cetacean behaviour.

Because of these methodological problems there is uncertainty about what is potentially one of the most exciting datasets on the cultures of whales and dolphins (Sargeant and Mann, 2009). A very detailed study of bottlenose dolphins in Shark Bay, Western Australia, has been underway for over 20 years. The dolphins use a number of quite complex foraging techniques, including putting sponges on their beaks to ferret prey from crevices, driving fish onto beaches and exploiting human provisioning (Connor, 2001; Sargeant and Mann, 2009). Only subsets of the population use each technique, and at least some of the techniques seem to be mostly passed down from mother to offspring. So these look like cultures. But perhaps the mother–offspring transmission is partially genetic, or perhaps the behavioural variation is at least partially due to differences in microhabitat caused by the different ranges of the dolphins (Connor, 2001; Sargeant and Mann, 2009). Given the abundant evidence for imitation in this species, the coexistence of dolphins with different foraging strategies, and imperfect genetic correlations with behaviour (Krützen et al, 2005), I believe that culture is much the most parsimonious explanation for these diverse foraging strategies, but it does not emerge clearly from the standard 'exclusion' methodology.

Studies of bottlenose dolphins have found many other cases of within-population and between-population behavioural variation (Rendell and Whitehead, 2001; Sargeant and Mann, 2009), much to do with foraging strategies. The exclusion methodology is particularly ineffective in such cases, but, in my opinion, they are probably driven by culture. There are also non-foraging differences between bottlenose dolphin populations that are simpler to assign to culture, as the environment is unlikely to play a role in their generation. A very good candidate is the in-air head butt made by pairs of male bottlenose dolphins in New Zealand, which has been seen nowhere else (Lusseau, 2003). The fact that this relies on two animals coordinating their activities considerably strengthens the case that it is the result of social learning.

Another type of dolphin behaviour that involves coordination among animals is the human–cetacean fishing cooperative. Bottlenose dolphins off Laguna, Brazil, cooperate with each other and local fishers to enhance the catches of both the dolphins and the fishers (Simöes-Lopes et al, 1998). Only some of the dolphins in the area work with the fishers. Each species uses signals to trigger behaviour from the other, following a strict protocol, and both appear to benefit. The behaviour has been transmitted socially within the community of dolphins, within the community of humans, and between the humans and dolphins since at least 1847 (Pryor et al, 1990). There are other multigenerational human–cetacean fishing cooperatives involving other cetacean species in various parts of the world (Pryor et al, 1990).

So there is considerable evidence for cetacean culture in both the baleen whales and small odontocetes (toothed cetaceans), but it is in two species of larger odontocetes that the evidence is particularly strong, and the cultures are especially interesting (Rendell and Whitehead, 2001). Both these features spring from the social systems of the larger odontocetes. Among the baleen whales and smaller odontocetes, as well as the northern bottlenose whale (*Hyperoodon ampullatus*), females have loose and flexible networks of relationships, with the strong bonds being between mother and suckling offspring, or, in a few species, within coalitions of mature males (Connor, 2000; Gowans et al, 2001). But in the killer whales, sperm whales, pilot whales (*Globicephala* spp.), and probably a few other larger cetaceans with teeth, the strongest and most important relationships are among related females. Females form tight permanent or nearly permanent social units with other females, primarily their relatives. There are variants on this. Among 'resident', fish-eating, killer whales off the west coast of North America, mature males, as well as mature females, remain in their mother's unit (Bigg et al, 1990), whereas in sperm whales, mature males lead separate lives and social units may contain unrelated females (Whitehead, 2003). However, the two key attributes of these societies for the study of culture are that most females remain in their social units for their lifetimes, and that social units share their habitat with other social units. This means that the method of identifying culture by the standard method of exclusion works more simply. If social units using the same waters behave differently, then ecological causes for the variation can usually be discarded, and genetic determination can be tested quite simply when genetic markers are available. If the behaviour is performed by all members of the social unit, and some of them, as with sperm whales, are unrelated, then social learning must have a role, and the behaviour is, by our definition, cultural. These same attributes, that behaviour is communal and that social units with different behaviour have overlapping ranges, have important consequences ecologically and evolutionarily, as well as for management and conservation, as I discuss later. Essentially, they produce multicultural societies.

First sperm whales. Sperm whales are found in all oceans and move widely. In the Pacific, a social unit containing about a dozen females and their offspring roams through a home range spanning about 1500km of ocean (Whitehead, 2003). They share this range with other units. Some of these units are from the same clan, while others may be from two to four of the other clans that use the waters. The clans have distinct dialects made up of Morse code-like patterns of clicks: 'click-click-click-click' go members of what we call the 'regular' clan; 'click-click-PAUSE-click' members of the 'plus-one' clan (Rendell and Whitehead, 2003). The social units get together with other units for a few days to form cohesive groups, but only with units from their own clan. The clans, even though using the same area, have distinctive movement patterns, feeding success and even rates of producing calves (Whitehead and Rendell, 2004; Marcoux et al, 2007). They are affected quite differently by the El Niño oceanographic/climatic phenomenon that dramatically changes their habitat every seven years or so (Whitehead and Rendell, 2004). Although some alternative explanations have been proposed (Janik, 2001), there can be little doubt that sperm whale clans are cultural structures – different clans use the same waters, and show no differences in the autosomal genes (i.e. non-sex linked) that might code for behaviour (Whitehead, 2003). They are very large cultural structures. A sperm whale clan may contain 10,000 females and span thousands of kilometres of the Pacific Ocean (Rendell and Whitehead, 2003).

The 'resident' fish-eating killer whales off the west coast of North America are the cetaceans whose culture is best mapped. They share their habitat with 'transient' mammal-eating killer whales, but the two forms, which may be subspecies, rarely interact (Ford et al, 2000). Resident killer whales live, and remain, in matrilineal social units, which are themselves parts of pods, which are members of clans, and one to three clans form a community (Bigg et al, 1990). There are cultural attributes at all levels of this pyramid. Most notable are vocalizations. The pulsed call vocalizations are sufficiently stereotyped that the pod of killer whales can be reliably identified by humans from a short recording (Ford et al, 2000). Dialect similarity decreases as one compares matrilineal groups within pods, pods within clans, clans within communities, and communities (Ford, 1991; Deecke et al, 2000; Yurk et al, 2002). But there are other differences in how they live their lives, such as the elaborate greeting ceremony performed when pods of the Southern Resident community (but not Northern Resident) meet (Osborne, 1986), and 'beach-rubbing' by some Northern Resident but not Southern Resident pods (Ford et al, 2000). Some of this culturally determined behaviour is very stable over generations – an example being the call dialects (Deecke et al, 2000) – but there are also passing 'fads'. For a few weeks in the summer of 1987 the Southern Residents took up pushing dead salmon around, but then dropped the behaviour (Whitehead et al, 2004).

Comparative Cultures

How does what we know, or suspect, of cetacean culture compare with that of other species? There is something of a paradox as we look for culture across species. Culture is based on social learning and social learning, in the broad sense, does not seem to be a particularly demanding activity. If animals can learn from their environments, then why not learn from each other? They do, and there is considerable evidence for social learning in a range of animals from fish to rats to songbirds (Laland and Hoppitt, 2003). Then, if we define culture as group-specific socially learned behaviour, we find culture widely distributed through at least the vertebrates. Culture, as so defined, is not very special.

Despite this, studies of nonhuman culture have focused overwhelmingly on primates (in the recent book, *The Question of Animal Culture* (Laland and Galef, 2009), 6 of 12 chapters were contributed by those studying monkeys or apes). There are two reasons for the scientific focus on primate culture. Primarily, there is an overwhelming fascination with human culture – culture is essential for understanding humans and our place in the world (Richerson and Boyd, 2005). So what are its roots? An obvious place to look is in our closest relatives. And they have culture, at least by some definitions (Whiten et al, 1999), but how much, and how does it compare with human culture? These are the difficult and controversial questions that are so heavily debated (Laland and Galef, 2009). The second reason for the focus on primates is that, in many respects, they seem to have among the most advanced nonhuman cultures. Similarly, a book on bird flight might concentrate on hummingbirds and albatrosses at the expense of chickens, which can fly, but not often or very well.

And this is where cetaceans come into the picture. The evidence, which I have summarized above, is patchy, but it strongly suggests that cetaceans have well-developed cultures that are important to them. How does whale culture compare to ape and monkey culture? There are difficulties with this evaluation, as cetologists and primatologists have quite different methods and foci of their research. However, some contrasts are clear. First, at the root of culture is social learning. Bottlenose dolphins (and probably other cetaceans, but only bottlenose dolphins have been tested) are considerably more adept social learners than nonhuman primates (Whiten, 2001). There are theoretical reasons why we might expect social learning to be more important in the ocean (Whitehead, 2007). Looking across types of culture, the evidence goes both ways. Chimpanzees use material tools for a wide range of purposes and in many ways. Much, if not all, of this variation seems to be cultural (McGrew, 1992). In contrast, there is only one suggestion of a cetacean culture of tool use, the sponge feeding of the Shark Bay bottlenose dolphins (Krützen et al, 2005), and that this is culture is disputed (Laland and Janik, 2006). Conversely, cetacean vocal cultures are prominent and varied, while there is little evidence for vocal learning or vocal cultures in primates (Janik and Slater, 1997). Such contrasts may be partially the result of different foci of research on land (material tools) and in the ocean (sounds), but only partially. The ocean is an excellent medium for sound propagation, but contains few potentially useful tools, or even functional tasks for simple tools, while in chimpanzee habitat these attributes are reversed.

More profound contrasts lie in what might be called the second-order properties of culture. For instance, Tomasello (1994, 2009) discusses 'ratcheting', the accumulation of modifications to build greater functionality. Without cultural ratcheting in human cultures, we could not have shirts or language, let alone computers. Tomasello (2009) considers that there is very little, if any, chimpanzee behaviour that can be considered to ratchet. Although cultural ratcheting in cetaceans cannot match the extraordinary products of the cultures of modern humans, the humpback whale song must be a result of ratcheting, and it is likely that their lobtail feeding is as well (Rendell and Whitehead, 2001).

A second major difference between the cultures of humans and nonhumans is claimed to be that the nonhumans 'do not have moral systems and do not reinforce social rules with symbolic display or signal adherence to specific sets of norms' (Hill, 2009). This is a strong claim and only based upon a perceived lack of evidence to the contrary. It seems entirely probable that cetaceans, as well as many other animals, including nonhuman primates, do have moral systems (Broom, 2003). For instance, consider sperm whales. Sperm whales spend most of their time foraging using the most powerful sound in the animal world, their highly directional echolocation click (Møhl et al, 2000). They also live in tight social groups, frequently very close to, or touching, one another (Whitehead, 2003). A sperm whale who directed its echolocation clicks at a social partner's ears would at the very least incapacitate the recipient's ability to forage for some time, and quite likely cause permanent damage. The situation is akin to a close-knit group of hunters each with a machine gun that he is operating half the time. Without strong norms about gun use the longevity of the hunters would be low. Sperm whales live a long time. Barring the seemingly extremely unlikely possibility that safe use of the echolocation system is completely genetically determined, there must be norms of use, i.e. moral systems or social rules, for any social animal that is using a potentially debilitating weapon close to social partners for most of its life.

The second part of Hill's (2009) claim, the lack of reinforcement of these social rules with cultural displays, is, in a somewhat more general form, the focus of other considerations of the differences between human and nonhuman cultures. For instance, Perry (2009) discusses the lack of evidence for ethnic markers in nonhuman primates. She suggests that these are most likely to exist in species with 'intense cooperative relationships, fission–fusion structures, symbolic capacities, group-specific social norms, and large home ranges', and thinks that cetaceans, as well as birds and bats, are more likely to have ethnic markers than nonhuman primates. I believe that Perry's (2009) conditions are met in the case of the matrilineally based societies of the large odontocetes. For instance killer whales live in tight cooperative social units; they often meet other killer whales with whom they have a variety of social relationships; they may have symbolic capacity in their group-specific dialects; and they do have very large home ranges (Ford et al, 2000). The parallel evolution of dialects in neighbouring pods strongly suggests that the dialects *are* ethnic markers (Deecke et al, 2000). Going back to

Hill's claim, it also seems probable that the greeting ceremony used by Southern Resident killer whale pods (Osborne, 1986) may function, in part, to reinforce the social norms of this community.

So, cetacean cultures are probably closer to human culture than those of nonhuman primates in significant ways. While cetaceans have nothing even comparable with human technologies, the nature and evolution of their social cultures may have important parallels, as well as contrasts, with ours.

Culture and Ecology

Culture controls behaviour, and behaviour is the link between an animal and its environment. Therefore, as is obvious from the human experience, culture can have a major role on the ecology of a species. Social learning can quickly bring a new foraging method to a population, much more quickly than if each individual has to work it out for themselves. Although, as noted earlier, the proof is not fully in, it seems (to me at least!) almost certain that culture had a major role in the development of the cetaceans' diverse and complex foraging methods, allowing complex collaborations with each other (Similä and Ugarte, 1993; Gazda et al, 2005) and with us (Pryor et al, 1990, Simöes-Lopes et al, 1998). This leads to cultural species potentially having wider, more profound and more unexpected influences on their environments than species for which social learning has little role in foraging.

But culture can also have almost the opposite effect. Cultural conformism and conservatism can evolve in species where social learning is important, making animals resistant to changing behavioural norms (Richerson and Boyd, 2005). In the face of a changing environment, such conformism to old and no longer appropriate norms can have dire consequences for the individuals and their populations (Whitehead and Richerson, 2009). Killer whales seem to have strong streaks of cultural conformism (Osborne, 1999). In addition to its negative effects when the environment changes, conformism can lead to maladaptive behaviour, such as human cultures of celibacy and suicide. It seems entirely probable that some of the maladaptive behaviour of cetaceans, such as mass strandings, may have cultural roots (Rendell and Whitehead, 2001). For instance, cultural group conformity in movement patterns may override an individual's survival instincts when the group gets into trouble.

Culture and Conservation

The ecologies of the whales and dolphins have been altered by humans, once we worked out how to use their environments. This ecological change has interacted with their cultures, confounding attempts to manage and conserve populations (Kasuya, 2008).

Most directly, when whales and dolphins make part of their living from human activities, then the interaction between human and whale culture is central, and may be complex. For instance, in Moreton Bay, Australia, a community of bottlenose dolphins regularly feeds from trawler discards, whereas another, largely socially distinct, community using the same waters does not (Chilvers and Corkeron, 2001). In such circumstances, changes in fishery regulations may have complex implications for the dolphins.

From the perspective of human fishers, the interactions between cetaceans and human fisheries can be positive, as in the fishing cooperatives in Brazil and elsewhere, neutral, as with the Moreton Bay dolphins who feed from discards, or perceived as negative. Long-line fishers in many parts of the world believe that killer whales, sperm whales and other species take fish from their lines. These practices can spread through cetacean populations fast, presumably by social learning, i.e. culture, but the consequences on the fishers' attitudes, and indirectly on the cetaceans (through actions such as shooting, ramming and calling for culls), are negative (Whitehead et al, 2004).

As shown by the Moreton Bay dolphins, when populations are culturally structured, then the different segments may be differentially affected by human activities (Whitehead et al, 2004). As another example, the different clans of sperm whales in the eastern Pacific are differentially affected by El Niño oceanographic/climatic phenomenon (Whitehead and Rendell, 2004). With global warming, conditions are likely to become more like El Niño, and El Niños themselves may become more frequent, in which case all sperm whales in the region are likely to be negatively affected, but groups of the 'plus-one' clan will be better adapted (Whitehead and Rendell, 2004). This highlights the potential significance of maintaining cultural diversity for adaptation to changing environments. The goal of conservation is generally the maintenance of biodiversity, but the diversity of life can be a result of both genetic and cultural processes, and we should conserve both (Whitehead et al, 2004; Ryan, 2006). The intensity of whaling on the most heavily targeted whale populations may well have led to a loss of the knowledge aggregate in the population, just as the genetic diversity was diminished. As an example, Whitehead et al (2004) suggest that the North Atlantic right whale, *Eubalaena glacialis*, which was reduced perhaps to a few tens of animals in the early 20th century, may have lost the knowledge of how to use some of the species' pre-whaling feeding grounds. The species is now virtually absent from the waters off Newfoundland and Labrador and the eastern Atlantic. This reduction in the available diversity of foraging opportunities makes the species more vulnerable to natural, and human-caused, variation in the quality of those feeding grounds that they now depend upon.

Other instances in which the cultures of whales and dolphins interact with human activities are discussed by Whitehead et al (2004). The bottom line is that conservation and management become considerably more complex when culture is an important determinant of behaviour and thus fitness. The goal of our conservation/management actions is now to maintain cultural, as well as genetic, diversity. The effects of human-made changes to the environment of the whales and dolphins become extremely unpredictable. Sometimes new behaviour spreads rapidly through social learning. At others, the extreme conservatism that can sometimes result from conformist cultures leaves animals using inappropriate behaviour when conditions change. Kasuya (2008), after a lifetime's experiences in cetacean science, considers that 'it will be extremely difficult to develop an exploitation scheme for such highly social species of cetaceans without disrupting their cultural traits', and that current patterns of whaling and hunting of smaller cetaceans off Japan are only possible because cetacean individuality, sociality and culture are ignored.

A possible rejoinder to Kasuya's perspective is that while some cetaceans, such as killer and sperm whales, may have complex social structures, and knowledge bases – cultures – built upon them, there are others, for instance the baleen whales, and in particular the minke, where social life is much simpler, and culture is therefore not an issue. However, in all areas where there has been substantial study of baleen whale behaviour (principally vocal, migratory and foraging behaviour), we have found culture. And, although culture may be more prominent in species with complex social structures, such as chimpanzees and killer whales, there are species, such as orang-utans (*Pongo pygmaeus*) where individuals are much more solitary but important cultures have evolved (van Schaik et al, 2003). It seems likely that the same is true of the baleen whales.

Are Cultural Animals Special?

Culture is a vital element of the modern human, and is often presented as the defining difference between 'us' and 'them'. It can then be argued (for example, Fox, 2001; McGrew, 2003) that species that also have advanced cultures should be included, with humans, in 'an extended moral community'. A key element here is 'advanced'. By my definition of culture, many species have it. But only in a very few has culture become a major determinant of many forms of behaviour. When this happens, processes that are important, but rare or absent in the standard genetically evolved species, begin to operate (Richerson and Boyd, 2005): cultural group selection, conformism, cultural ethnicity with symbolic markers, and so on. These processes change the nature of society as well as ecology. There is strong selection within such species to use this culture effectively. Perhaps this 'cultural-drive' is a principal or contributing cause as to why cetaceans, humans and a few other species have evolved self-awareness, large brains and astute intelligences (van Schaik, 2006). Thus there are good reasons to give highly cultural species special considerations. They are 'more like us' not only because of the culture itself, but also because the advanced culture is at least a marker, and perhaps a cause, of other attributes that we think of as particularly human.

REFERENCES

Bigg, M. A., Olesiuk, P. F., Ellis, G. M., Ford, J. K. B. and Balcomb, K. C. (1990) 'Social organization and genealogy of resident killer whales (*Orcinus orca*) in the coastal waters of British Columbia and Washington State', *Reports of the International Whaling Commission*. Special issue, vol 12, pp383–405

Broom, D. (2003) *The Evolution of Morality and Religion*, Cambridge University Press, Cambridge

Chilvers, B. L. and Corkeron, P. J. (2001) 'Trawling and bottlenose dolphins' social structure', *Proceedings of the Royal Society of London, B*, vol 268, pp1901–1905

Clapham, P. J. (2000) 'The humpback whale: Seasonal breeding and feeding in a baleen whale', in J. Mann, R. C. Connor, P. L. Tyack and H. Whitehead (eds) *Cetacean Societies*, University of Chicago Press, Chicago, IL, pp173–196

Connor, R. C. (2000) 'Group living in whales and dolphins', in J. Mann, R. C. Connor, P. L. Tyack and H. Whitehead (eds) *Cetacean Societies*, University of Chicago Press, Chicago, IL pp199–218

Connor, R. C. (2001) 'Individual foraging specializations in marine mammals: Culture and ecology', *Behavioral and Brain Sciences*, vol 24, pp329–330

Croll, D. A., Clark, C. W., Acevedo, A., Tershy, B., Flores, S., Gedamke, J. and Urban, J. (2002) 'Only male fin whales sing loud songs' *Nature*, vol 417, p809

Darling, J. D. and Bérubé, M. (2001) 'Interactions of singing humpback whales with other males', *Marine Mammal Science*, vol 17, pp570–584

Deecke, V. B., Ford, J. K. B. and Spong, P. (2000) 'Dialect change in resident killer whales: Implications for vocal learning and cultural transmission', *Animal Behaviour*, vol 40, pp629–638

Ford, J. K. B. (1991) 'Vocal traditions among resident killer whales (*Orcinus orca*) in coastal waters of British Columbia', *Canadian Journal of Zoology*, vol 69, pp1454–1483

Ford, J. K. B., Ellis, G. M. and Balcomb, K. C. (2000) *Killer Whales*, UBC Press, Vancouver, British Columbia

Fox, M. A. (2001) 'Cetacean culture: Philosophical implications', *Behavioral and Brain Sciences*, vol 24, pp333–334

Galef, B. G. (1992) 'The question of animal culture', *Human Nature*, vol 3, pp157–178

Gazda, S. K., Connor, R. C., Edgar, R. K. and Cox, F. (2005) 'A division of labour with role specialization in group-hunting bottlenose dolphins (*Tursiops truncatus*) off Cedar Key, Florida', *Proceedings of the Royal Society of London, B*, vol 272, pp135–140

Gedamke, J., Costa, D. P. and Dunstan, A. (2001) 'Localization and visual verification of a complex minke whale vocalization', *Journal of the Acoustical Society of America*, vol 109, pp3038–3047

Gowans, S., Whitehead, H. and Hooker, S. K. (2001) 'Social organization in northern bottlenose whales (*Hyperoodon ampullatus*): Not driven by deep water foraging?' *Animal Behaviour*, vol 62, pp369–377

Guinet, C. and Bouvier, J. (1995) 'Development of intentional stranding hunting techniques in killer whale (*Orcinus orca*) calves at Crozet Archipelago', *Canadian Journal of Zoology*, vol 73, pp27–33

Herman, L. H. (2002) 'Vocal, social, and self-imitation by bottlenosed dolphins', in K. Dautenhahn and C. L. Nehaniv (eds) *Imitation in Animals and Artifacts*, MIT Press, Cambridge, MA, pp63–108

Hill, K. (2009) 'Animal "culture"', in K. N. Laland and B. G. Galef Jr (eds) *The Question of Animal Culture*, Harvard University Press, Cambridge, MA, pp269–287

Hinde, R. A. (1982) *Ethology: Its Nature and Relation with Other Sciences*, Oxford University Press, Oxford

Janik, V. M. (2001) 'Is social learning unique?', *Behavioral and Brain Sciences*, vol 24, pp337–338

Janik, V. M. and Slater, P. J. B. (1997) 'Vocal learning in mammals', *Advances in the Study of Behavior*, vol 26, pp59–99

Kasuya, T. (2008) 'The Kenneth S. Norris lifetime achievement award lecture: Presented on 29 November 2007 Cape Town, South Africa', *Marine Mammal Science*, vol 24, pp749–773

Krützen, M., Mann, J., Heithaus, M. R., Connor, R. C., Bejder, L. and Sherwin, W. B. (2005) 'Cultural transmission of tool use in bottlenose dolphins', *Proceedings of the National Academy of Sciences of the United States of America*, vol 102, pp8939–8943

Laland, K. N. and Galef B. G. Jr, (2009) *The Question of Animal Culture*, Harvard University Press, Cambridge, MA

Laland, K. N. and Hoppitt, W. (2003) 'Do animals have culture?', *Evolutionary Anthropology*, vol 12, pp150–159

Laland, K. N. and Janik, V. M. (2006) 'The animal cultures debate', *Trends in Ecology and Evolution*, vol 21, pp542–547

Laland, K. N., Kendal, J. R. and Kendal, J. R. (2009) 'Animal culture: Problems and solutions', in K. N. Laland and B. G. Galef Jr (eds) *The Question of Animal Culture*, Harvard University Press, Cambridge, MA, pp174–197

Lenski, G., Nolan, P. and Lenski, J. (1995) *Human Societies*, McGraw-Hill, New York

Lusseau, D. (2003) 'The emergence of cetaceans: Phylogenetic analysis of male social behaviour supports the Cetartiodactyla clade', *Journal of Evolutionary Biology*, vol 16, pp531–535

Marcoux, M., Rendell, L. and Whitehead, H. (2007) 'Indications of fitness differences among vocal clans of sperm whales', *Behavioural Ecology and Sociobiology*, vol 61, pp1093–1098

McDonald, M. A., Calambokidis, J., Teranishi, A. M. and Hildebrand, J. A. (2001) 'The acoustic calls of blue whales off California with gender data', *Journal of the Acoustical Society of America*, vol 109, pp1728–1735

McDonald, M. A., Mesnick, S. L. and Hildebrand, J. A. (2006) 'Biogeographic characterization of blue whale song worldwide: Using song to identify population', *Journal of Cetacean Research and Management*, vol 8, pp55–65

McGrew, W. C. (1992) *Chimpanzee Material Culture: Implications for Human Evolution*, Cambridge University Press, Cambridge

McGrew, W. C. (2003) 'Ten dispatches from the chimpanzee culture wars', in F. B. M. de Waal and P. L. Tyack (eds) *Animal Social Complexity: Intelligence, Culture, and Individualized Societies*, Harvard University Press, Cambridge, MA, pp419–439

Møhl, B., Wahlberg, M., Madsen, P. T., Miller, L. A. and Surlykke, A. (2000) 'Sperm whale clicks: Directionality and source level revisited', *Journal of the Acoustical Society of America*, vol 107, pp638–648

Mundinger, P. C. (1980) 'Animal cultures and a general theory of cultural evolution', *Ethology and Sociobiology*, vol 1, pp183–223

Noad, M. J., Cato, D. H., Bryden, M. M., Jenner, M. N. and Jenner, K. C. S. (2000) 'Cultural revolution in whale songs', *Nature*, vol 408, p537

Osborne, R. W. (1986) 'A behavioral budget of Puget Sound killer whales', in B. Kirkewold and J. S. Lockard (eds) *Behavioral Biology of Killer Whales*, A. R. Liss, New York, pp211–249

Osborne, R. W. (1999) 'A historical ecology of Salish Sea "resident" killer whales (*Orcinus orca*): With implications for management', PhD dissertation, University of Victoria, Victoria, British Columbia

Payne, K. (1999) 'The progressively changing songs of humpback whales: A window on the creative process in a wild animal', in N. L. Wallin, B. Merker and S. Brown (eds) *The Origins of Music*, MIT Press, Cambridge, MA, pp135–150

Payne, R. and McVay, S. (1971) 'Songs of humpback whales', *Science*, vol 173, pp587–597

Payne, R. S. and Webb, D. (1971) 'Orientation by means of long-range acoustic signalling in baleen whales', *Annals of the New York Academy of Sciences*, vol 188, pp110–142

Perry, S. (2009) 'Are non-human primates likely to exhibit cultural capacities like those of humans?', in K. N. Laland and B. G. Galef Jr (eds) *The Question of Animal Culture*, Harvard University Press, Cambridge, MA, pp247–268

Pryor, K. W. (2001) 'Cultural transmission of behavior in animals: How a modern training technology uses spontaneous social imitation in cetaceans and facilitates social imitation in horses and dogs', *Behavioral and Brain Sciences*, vol 24, p352

Pryor, K., Lindbergh, J., Lindbergh, S. and Milano, R. (1990) 'A dolphin-human fishing cooperative in Brazil', *Marine Mammal Science*, vol 6, pp77–82

Rendell, L. and Whitehead, H. (2001) 'Culture in whales and dolphins', *Behavioral and Brain Sciences*, vol 24, pp309–324

Rendell, L. and Whitehead, H. (2003) 'Vocal clans in sperm whales (*Physeter macrocephalus*)', *Proceedings of the Royal Society of London, B*, vol 270, pp225–231

Richerson, P. J. and Boyd, R. (2005) *Not by Genes Alone: How Culture Transformed Human Evolution*, Chicago University Press, Chicago

Ryan, S. J. (2006) 'The role of culture in conservation planning for small or endangered populations', *Conservation Biology*, vol 20, pp1321–1324

Sargeant, B. L. and Mann, J. (2009) 'From social learning to culture: Intrapopulation variation in bottlenose dolphins', in K. N. Laland and B. G. Galef Jr (eds) *The Question of Animal Culture*, Harvard University Press, Cambridge, MA, pp152–173

Similä, T. and Ugarte, F. (1993) 'Surface and underwater observations of cooperatively feeding killer whales in northern Norway', *Canadian Journal of Zoology*, vol 71, pp1494–1499

Simöes-Lopes, P. C., Fabián, M. E. and Menegheti, J. O. (1998) 'Dolphin interactions with the mullet artisanal fishing on southern Brazil: A qualitative and quantitative approach', *Revista Brasileira de Zoologia*, vol 15, pp709–726

Tomasello, M. (1994) 'The question of chimpanzee culture', in R. W. Wrangham, W. C. McGrew, F. B. M. de Waal and P. G. Heltne (eds) *Chimpanzee Cultures*, Harvard University Press, Cambridge, MA, pp301–317

Tomasello, M. (2009) 'The question of chimpanzee culture, plus postscript (chimpanzee culture, 2009)', in K. N. Laland and B. G. Galef Jr (eds) *The Question of Animal Culture*, Harvard University Press, Cambridge, MA, pp198–221

van Schaik, C. (2006) 'Why are some animals so smart?', *Scientific American*, vol 294, no 4, pp64–71 van Schaik, C. P., Ancrenaz, M., Borgen, G., Galdikas, B., Knott, C. D., Singleton, I., Suzuki, A., Utami, S. S. and Merrill, M. (2003) 'Orangutan cultures and the evolution of material culture', *Science*, vol 299, pp102–105

Weinrich, M. T., Schilling, M. R. and Belt, C. R. (1992) 'Evidence for acquisition of a novel feeding behaviour: Lobtail feeding in humpback whales, *Megaptera novaeangliae*', *Animal Behaviour*, vol 44, pp1059–1072

Whitehead, H. (2003) *Sperm Whales: Social Evolution in the Ocean*, Chicago University Press, Chicago, IL

Whitehead, H. (2007) 'Learning, climate and the evolution of cultural capacity', *Journal of Theoretical Biology*, vol 245, pp341–350

Whitehead, H. (2009) 'How might we study culture? A perspective from the ocean', in R. N. Laland and B. G. Galef Jr (eds) *The Question of Animal Culture*, Harvard University Press, Cambridge, MA, pp125–151

Whitehead, H. and Rendell, L. (2004) 'Movements, habitat use and feeding success of cultural clans of South Pacific sperm whales', *Journal of Animal Ecology*, vol 73, pp190–196

Whitehead, H. and Richerson, P. (2009) 'The evolution of conformist social learning can cause population collapse in realistically variable environments', *Evolution and Human Behavior*, vol 30, pp261–273

Whitehead, H., Rendell, L., Osborne, R. W. and Würsig, B. (2004) 'Culture and conservation of non-humans with reference to whales and dolphins: Review and new directions', *Biological Conservation*, vol 120, pp431–441

Whiten, A. (2001) 'Imitation and cultural transmission in apes and cetaceans', *Behavioral and Brain Sciences*, vol 24, pp359–360

Whiten, A., Goodall, J., McGrew, W. C., Nishida, T., Reynolds, V., Sugiyama, Y., Tutin, C. E. G., Wrangham, R. W. and Boesch, C. (1999) 'Cultures in chimpanzees', *Nature*, vol 399, pp682–685

Whiten, A., Horner, V., Litchfield, C. and Marshall-Pescini, S. (2004) 'How do apes ape?', *Learning and Behaviour*, vol 32, pp36–52

Würsig, B. and Clark, C. W. (1993) 'Behavior', in J. J. Burns, J. J. Montague and C. J. Cowles (eds) *The Bowhead Whale*, The Society for Marine Mammalogy, Lawrence, Kansas, pp157–199

Yurk, H., Barrett-Lennard, L., Ford, J. K. B. and Matkin, C. O. (2002) 'Cultural transmission within maternal lineages: Vocal clans in resident killer whales in southern Alaska', *Animal Behaviour*, vol 63, pp1103–1119

Part III

New Insights – New Challenges

Figure III *Minke whale surfaces for air*

Source: Rob Lott

17

Whales and Dolphins on a Rapidly Changing Planet

Mark Peter Simmonds and Philippa Brakes

Perceptions and Timelines

Innovative technologies are increasingly allowing us to affect the seas and the animals that inhabit them, in ways that would not have been possible in the past. We are now entering even the deepest and remotest seas to exploit their resources. As a result, our activities are having an impact on cetaceans in new ways across more of their ranges. Human activities are known to affect cetaceans and to degrade, or destroy, their habitats (Kemp, 1996).

Human impacts on the oceans are often difficult for us to see. One exception being massive oil spills, such as that in the Gulf of Mexico in 2010, and in such instances images can be rapidly transmitted around the world and viewed by millions of people. Such events are exceptions and even in the case of such vast and dramatic incidents, the effects on cetaceans are often far from clear. They are also hotly disputed. Not only are our activities in the sea typically far removed from the sight of most of us but the fundamental differences between cetaceans and humans can create another barrier to us fully appreciating our impacts on their world.

The mammalian order of primates, including humans, is primarily distinguished by certain features (although not all have all of these features), include relatively large brains, frontally placed eyes with binocular vision (protected by bony sockets), grasping hands and feet (with opposable thumbs and large toes), nails rather than claws, small numbers of offspring and slowly maturing young (Sorenson, 2009). Primates are also mainly vegetarian and tend to climb trees. Whales and dolphins do not climb trees, are not vegetarian, and have no hands or opposable thumbs. For them – and this is a key issue – vision is a secondary sense to hearing. This is reflected very obviously in their anatomy, with eyes placed on the sides of their heads and far less well protected than ours. They also have unique structures in their heads for the production, focusing and reception of sound. These anatomical differences reflect a very different evolution and biology to our own. They reflect adaptations to an aquatic world very unlike ours, where different laws of physics apply.

Like the apes, cetaceans have diversified across a wide range of habitats. Some species are only found in restricted areas, for example the dolphins that live exclusively in big tropical rivers. Some others exploit the resources only found in the deepest oceans. Many more live out in the open sea. Some appear to have solitary behaviour, but the majority are known to specialize in collaborating to find scattered prey and are believed to use shared knowledge to help them survive. Cetaceans also produce few offspring and look after them for extended periods compared to most other animals. These long periods of maternal care allow the young cetacean, just like the young primate, to learn from its parent and social group the skills that will allow it not only to survive but also to integrate into its society. The differences and similarities between cetaceans and primates in functional anatomy and lifestyles help point to their likely vulnerabilities. In particular, we can expect them to be especially sensitive to noise pollution and, like other social mammals, the removal of one individual can affect others within the social unit. Furthermore, the shared knowledge that allows particular populations to exploit some areas may be lost if that particular group is totally removed. This may explain why some areas previously colonized by whales have not been recolonized even after the main whaling activity that removed them has ceased.

There have been a number of expert reviews which have started to unravel the implications for cetaceans of particular human activities. In this short chapter we will highlight some of these in-depth analyses, focusing on those human activities that incidentally affect cetaceans (hunting being dealt with in other chapters).

Whales in an Increasingly Noisy Ocean

Light rapidly disappears with increasing depth. At best, vision may be useful over tens of metres in the sea, whereas sound travels so well underwater it may be useful for communication over hundreds, if not thousands, of kilometres (Weilgart, 2007). Many marine animals, particularly marine mammals, are known to be very sensitive to sound, using it for almost all important aspects of their life, including feeding, reproduction and navigation. If whales cannot hear each other, it impairs their ability to find each other, for example, for breeding purposes and to collaborate to find prey. Not surprisingly, there is some evidence that cetaceans move away from noisy and disturbed environments. This displacement can have serious implications if they are driven away from a habitat that includes an important food source or is a suitable breeding ground. However, it is a challenge for us to quantify this impact adequately and provide evidence that its significance is sufficient for mitigating actions to be warranted.

In fact, addressing noise pollution in the oceans has been controversial since it first emerged as an important issue in the late 1990s (Weilgart, 2007). Loud noise sources include military sonar, the seismic survey activities of the oil industry and shipping. In some areas, underwater background noise levels have doubled every decade for the last few decades, most likely due to commercial shipping (Hildebrand, 2005; McDonald et al, 2006). For the whales, an acoustic 'fog' has now descended in the seas. At one time some of the great whales probably communicated across the better part of entire ocean basins. However, our noise, mainly from shipping, has reduced their ability to do this and contracted the range of their senses from hundreds (maybe sometimes thousands) of kilometres down to just a few tens. Some of our more powerful noises also cause distress, confusion and sometimes even stranding and death. Reflecting such concerns, the Standing Working Group on Environmental Concerns of the IWC's Scientific Committee has included an item on underwater sound on its agenda each year since 2004. In 2010, it focused on 'masking' noise and considered a mechanistic model that demonstrated the significant reduction in the 'communication space' of baleen whales that now occurs, especially near shipping lanes and busy ports (IWC, 2010). It also reviewed other possible effects of noise pollution on whales, including the altered calling patterns and frequency shown by fin whales in the western Mediterranean Sea and humpback whales off the coast of northern Angola in the presence of low-frequency sound from shipping and seismic airguns. The Working Group noted particular concerns for some populations, including the chronic exposure of the small population of humpback whales in the Arabian Sea to low-frequency sound from construction, shipping and seismic surveys. The IWC Scientific Committee duly reviewed this information and called for better regulation of seismic surveys, better collection and analyses of baseline data and other research.

WHALES IN A TAINTED OCEAN

Because they are at, or close to, the apex of marine food chains, the tissues of cetaceans tend to accumulate a concentrated and potentially unhealthy cocktail of certain xenobiotics (chemicals not previously known in nature). These can affect the health of the individual cetaceans and their populations (and also potentially those humans who eat them (see for example, Simmonds et al, 1994)). This is not news. In fact there is a vast and growing literature concerning contaminant levels in cetaceans (for example, recent accounts by Isobe et al, 2008, for dolphins sampled in Japan; Law et al, 2010, concerning European harbour porpoises, *Phocoena phocoena*; and, for a review, see Reijnders, 1996).

The risks from this acquired cocktail have been recognized for some decades and, in several case studies, immunological and reproductive disorders in marine mammals have been linked to them (Reijnders, 1996). The smaller toothed cetaceans that sit higher up the food chain, especially those living and feeding in near-shore polluted waters, are generally viewed as being most at risk. Jepson et al (2009) recently highlighted the likelihood that the disappearance of bottlenose dolphins from much of Europe's shores could have been caused by pollution – a hypothesis arising from the association that Jepson and his co-workers had found between poor health status (expressed as mortality due to infectious disease) and chemical contamination for a large sample of harbour porpoises stranded in the UK since 1990. This association exists for blubber concentrations of polychlorinated biphenyls (PCBs) above 17 parts per million total. (The PCBs are a persistent group of industrial organic chemicals with similar properties to some pesticides that were used in a number of industries prior to bans being implemented.) Bottlenose dolphins (*Tursiops truncatus*) in the same region and time period show up to one order of magnitude higher PCB levels in blubber; and the decline in their numbers coincides with the peak of PCB presence in the environment, strongly indicating a causal link.

It is not all bad news. There has been a gradual temporal decline in PCB levels in UK stranded harbour porpoises since 1990 (Jepson et al, 2009) and some other contaminants of concern are also decreasing (see, for example, Law et al, 2010), demonstrating that pollution controls can work. Uncertainty remains about how low residue levels need to fall until they are of no health consequence. Obviously we should continue to be vigilant, especially with new generations of xenobiotics entering the environment; and we must also be mindful of the related issue of emerging diseases in marine mammals. There appears to have been an increase in major disease events in marine mammals in recent years. Pollution-compromised immune systems could facilitate major mortalities where pathogens are the proximate cause but contaminants are playing a significant role (Simmonds and Mayer, 1997).

Whales in a Warming Ocean

Human activities have released what have become known as 'greenhouse gases' into the planet's atmosphere. Since the Industrial Revolution, this has primarily been carbon dioxide (CO_2). As a result we have changed the planet's heat-retaining abilities and we find ourselves in a rapidly changing world. Essentially we have broken the planet's homoeostatic mechanism. A mechanism that was allowing the coexistence of all the life forms currently present on our planet; a planet that is now sailing into uncharted waters because '*when you change the climate you change everything*' (WWF, 2010) and there are immediate and longer-term consequences for all species, including our own.

For the cetaceans, climate change is primarily expected to affect them via loss of habitat because each species has a distinct thermal range that it has evolved to exploit and because of potential changes in their prey distribution and numbers

(Learmonth et al, 2006; Simmonds and Elliot, 2009). However, changes in human behaviour, for example, resulting from increased flooding, fisheries collapse or other food security issues arising as a result of climate change, may also have an influence (Alter et al, 2010). For example, reduced ice cover in the Arctic is projected to lead to increased shipping, oil and gas exploration, and fishing, which is likely to result in additional noise and chemical pollution and increased cetacean bycatch. In more tropical areas, as fish become more difficult to find, so directed takes of cetaceans may increase.

Whales in an Increasingly Acidic Ocean

There is another problem relating to excess CO_2 in the atmosphere – marine acidification. The oceans absorb approximately one quarter of the CO_2 added to the atmosphere from human activities each year, and this has greatly reduced the speed of human-induced climate change. However, when CO_2 dissolves in seawater, carbonic acid is formed, a process causing ocean acidification. This is now having an influence on many organisms (Royal Society, 2005; Doney et al, 2009). Particularly vulnerable are those organisms that form shells, including corals, many marine plankton and shellfish. Some of the smaller shell-bearing organisms are important food sources for higher marine organisms, and reef-building corals form rich habitats for many other organisms. Acidification also appears to adversely affect the survival of larval marine species, including commercial fish and shellfish.

Acidification may also have a direct impact on the deep-sea schooling squid that form the key prey for many deep-diving whale species (Royal Society, 2005). Squid are reported to be very sensitive to increases in external CO_2 due to their energetically demanding 'jet propulsion', which requires a good supply of oxygen from the blood. An increasing CO_2 concentration lowers the pH of the blood reducing its ability to carry oxygen.

Whales in a Busy Ocean

Our physical presence in cetacean habitats is increasing as we build out into them, produce bigger and faster boats, and intensify our fisheries. For example, the proliferation of vessels, ranging from jet-skis up to the super tankers, is increasing the risk of physical collisions with cetaceans. Despite the misconception that 'they would get out of the way', there is ample grisly evidence, including the huge carcasses of struck whales that can appear splayed across the bows of ocean-going vessels as they come into port, that this is not always the case.

Injuries from ship strikes are often classified as either 'sharp trauma', which is usually caused by contact with a propeller, or 'blunt trauma', which usually results from being struck by the hull. Propellers can cause significant and very rapid damage when they come into contact with cetaceans. Such injuries may or may not be immediately life threatening. Even when victims survive in the short term, their longer-term survival may be significantly compromised. Propeller wounds can reopen and become infected, sometimes years later (Campbell-Malone et al, 2008). Injuries associated with blunt trauma include broken bones (often including the jaw or skull) and internal haemorrhage.

Sadly it is often the most threatened species that fall victim to ship strikes. Speed is not always a factor, as demonstrated by a report on 3 August 2010 of a southern right whale being struck by a container ship whilst the vessel was moving very slowly into dock (Martinez, 2010). Being struck by a vessel is a major cause of mortality for right whales, which like to 'log' at the surface. Between 1970 and 2006, over 50 per cent of North Atlantic right whale carcasses for which a post mortem examination was conducted were killed as a result of vessel collision (Campbell-Malone et al, 2008).

Besides marine transport, we are also increasing our activities in the marine environment in other dramatic ways. Fish farming now occupies vast swathes of coastline in some areas. The quest for renewable energy sources is driving our engineering activities into the seas in the form of wave, tidal and wind power generation. There are many potential issues associated with scaling up our commercial use of the marine environment, including disruption of the seabed and cetacean thoroughfares. Perhaps the issue that causes the greatest concern at present is the noise associated with pile driving – the process typically used to fix structures to the seabed (Simmonds and Brown, 2010).

Whales in a Netted Ocean

A significant part of human business at sea remains, of course, our fishing activities, and incidental capture in fishing nets and gear is adversely affecting many cetacean populations. These incidental takes, also referred to as bycatch, are widely regarded as the most immediate threat to many populations and even some entire species. Large numbers of animals are killed each year in fishing operations. In the US, for example, the annual bycatch of marine mammals was more than 6000 individual animals between 1990 and 1999, mainly taken in gillnets (Read et al, 2006). This suggests an annual global removal rate of tens of thousands. Andrew Read, an authority on such matters, comments that 'immediate action is needed to assess the magnitude of bycatch, particularly in many areas of Africa and Asia where little work has been conducted' and he concludes that 'new and innovative solutions to this problem are required that take account of the socioeconomic conditions experienced by fishermen and allow for efficient transfer of mitigation technology to fisheries of the developing world' (Read, 2008). Another problem generated by fisheries may be prey depletion arising from fisheries activities. This is something that is very difficult to investigate and that may be exacerbated by climate change.

Whales in the 21st Century and Beyond

The various threats to whales were reviewed in a 1996 anthology (Simmonds and Hutchinson, 1996) and there is an opportunity here to highlight changes in the intervening decade and a half since that text was published. While this is to some extent subjective, we believe the following to be true:

- There is a greater acceptance of noise pollution as a substantive threat and some movement to address this. However, there are also new types of loud noise coming into the oceans, notably the pile driving associated with marine wind farms and other renewable energy sources.
- Bycatch remains a substantive issue and, while placing pingers on nets seems to act as a deterrent for some species and gear types, their deployment has not generally gained support from the fishing communities around the world.
- Levels of some of the more infamous pollutants have fallen (with regional variations) but probably not sufficiently yet to eliminate them as a threat. New substances are now being found in marine animals, indicating we need to maintain vigilance.
- There has been considerable new research into marine mammal diseases, including epizootics, and there is a growing awareness of the vulnerability of marine mammal populations to disease events and the potential for human activities to contribute to them.
- Climate change is now an accepted phenomenon and cetaceans are beginning to change their distribution in response to changes in the oceans and their ecosystems – but the long-term implications are unclear.
- As we look to the future, it is important to appreciate that particular threats do not act in isolation.

A hypothetical whale travelling between her breeding and feeding grounds could well be exposed to various man-made stresses. She will certainly be carrying pollutants in her tissues with perhaps a small chronic effect on her health. She might encounter an oil spill on her travels. Breathing in the fumes, she could become intoxicated and move into the path of a ship. A ship that she cannot hear approaching, because the level of background noise masks its engines until it is almost on top of her. But then she dives, and survives.

However, when she arrives at the feeding grounds, the prey she needs in suitably high abundance to recharge her energy stores may not be there. Oceanic conditions may have changed as a result of global warming and the planktonic bloom is now out of step with the arrival of the whales. So she feeds very poorly and will be in no condition to breed this season, especially as her return journey is longer than usual because its path is displaced by loud noise emanating from a pile-driving operation close to her return route.

Figure 17.1 *Marine conservation efforts are sometimes too little, as well as too late*

Source: Cartoon provided by H., S. and M. P. Simmonds

The story could go on, but hopefully the point is made. Many factors may impact the individual, with only minimal attempts being made to mitigate them (see Figure 17.1) probably because for many of us they remain 'out of sight and out of mind'. These factors do not necessarily need to result in death to have some influence at the population level. For example, a strong link has been made between climatic conditions and the calving success of southern right whales, *Eubalaena australis* (Leaper et al, 2006).

The great conservationist Sir Peter Scott famously said 'If we can't save the whales ... we have little chance of saving anything else.' Simmonds and Hutchinson (1996) added 'and to save the whales you need to save the seas'. Others have spoken on the need to tread lightly on our planet. We now suggest that it should be obvious that our planet is overburdened by our presence, we are exhausting its resilience and great efforts need to be made by all to address this. But how can we weigh something like cetacean protection against the urgent need to address climate change? One response is that maintaining biodiversity is the only way to maintain robust ecosystems, and robust ecosystems are the foundations for maintaining life on our crowded planet. Each population or species is a 'brick' in the ecosystem structure; each one removed threatens to destabilize and destroy its architecture.

We could also posit that just as whales were the iconic call to arms for the environmental movement of the 1970s, the potential loss of these flagship species could once more provide the motivation for action on climate and other matters. Then, of course, there are also important ethical arguments for protecting cetaceans. These we return to in the closing chapter of this volume.

References

Alter, S. E., Simmonds, M. P. and Brandon, J. R. (2010) 'Forecasting the consequences of climate-driven shifts in human behavior on cetaceans', *Marine Policy*, vol 34, pp943–954

Campbell-Malone, R., Barco, S. G., Daoust, P.-Y., Knowlton, A. R., McLellan, W. A., Rotstein, D. S. and Moore, M. J. (2008) 'Gross and histological evidence of sharp and blunt trauma in North Atlantic right whales (*Eubalaena glacialis*) killed by vessels', *Journal of Zoo and Wildlife Medicine*, vol 39, no 1, pp37–55

Doney, S. C., Fabry, V. J., Feely, R. A. and Kleypas, J. A. (2009) 'Ocean acidification: The other CO_2 problem', *Annu. Rev. Mar. Sci.*, vol 1, pp169–192

Hildebrand, J. A. (2005) 'Impacts of anthropogenic sound', in J. E. Reynolds, W. F. Perrin, R. R. Reeves, S. Montgomery and T. J. Ragen (eds) *Marine Mammal Research: Conservation beyond Crisis*, The Johns Hopkins University Press, Baltimore, MD, pp101–124

Isobe. T., Ochi, Y., Ramu, K., Yamamoto, T., Tajima, Y., Yamada, T. K., Amano. M., Miyazaki, N., Takahashi, S. and Tanabe, S. (2008) 'Organohalogen contaminants in striped dolphins (*Stenella coeruleoalba*) from Japan: Present contamination status, body distribution and temporal trends (1978–2003)', *Marine Pollution Bulletin*, vol 58, pp396–401

IWC (International Whaling Commission) (2010) *Report of the Scientific Committee of the International Whaling Commission*, IWC/62/Rep 1, International Whaling Commission, Cambridge, UK

Jepson, P. D., Tregenza, N. and Simmonds, M. P. (2009) 'Disappearing bottlenose dolphins (*Tursiops truncatus*) – is there a link to chemical pollution?', submitted to the IWC Scientific Committee SC/60/E7

Kemp, N. J. (1996) 'Habitat loss and degradation', in M. P. Simmonds and J. D. Hutchinson (eds) *The Conservation of Whales and Dolphins: Science and Practice*, John Wiley and Sons, Chichester, UK, pp263–280

Law, R. J., Barry, J., Bersuder, P., Barber, J. L., Deaville, R., Reid, R. J. and Jepson, P. D. (2010) 'Levels and trends of brominated diphenyl ethers in blubber of harbor porpoises (*Phocoena phocoena*) from the UK, 1992–2008', *Environ. Sci. Technol.*, vol 44, pp4447–4451

Leaper, R., Cooke, J., Trathan, P., Reid, K., Rowntree, V. and Payne, R. (2006) 'Global climate drives southern right whale (*Eubalaena australis*) population dynamics', *Biology Letters*, vol 2, pp289–292

Learmonth, J. A., Macleod, C. D., Santos, M. B., Pierce, G. J., Crick, H. Q. P. and Robinson, R. A. (2006) 'Potential effects of climate change on marine mammals', *Oceanogr. Mar. Biol.*, vol 44, pp429–456

Martinez, D. P. (2010) 'Cargo vessel strikes a right whale', Ballenas de las Bahias, http://bioreporte.com/bahia/08/2010/cargo-vessel-strikes-a-southern-right-whale-in-argentina

McDonald, M., Hildebrand, J. A. and Wiggins, S. M. (2006) 'Increases in deep ocean ambient noise in the Northeast Pacific west of San Nicolas Island, California', *J. Acoust. Soc. Am.*, vol 120, no 2, pp711–718

Read, A. J. (2008) 'The looming crisis: Interactions between marine mammals and fisheries', *Journal of Mammalogy*, vol 89, no 3, pp541–548

Read, A. J., Drinker, P. and Northridge, S. (2006) 'Bycatch of marine mammals in US and global fisheries', *Conservation Biology*, vol 20, pp163–169

Reijnders, P. J. H. (1996) 'Organohalogen and heavy metal contamination in cetaceans: Observed effects, potential impact and future prospects', in M. P. Simmonds and J. D. Hutchinson (eds) *The Conservation of Whales and Dolphins: Science and Practice*, John Wiley and Sons, Chichester, pp205–217

Royal Society (2005) *Ocean Acidification due to Increasing Atmospheric Carbon Dioxide*, The Royal Society, London

Simmonds, M. P. and Brown, V. C. (2010) 'Is there a conflict between cetacean conservation and marine renewable energy developments?', *Wildlife Research*, vol 37, pp688–694

Simmonds, M. P. and Elliot W. J. (2009) 'Climate change and cetaceans: Concerns and recent developments', *Journal of the Marine Biological Association of the United Kingdom*, vol 89, no 1, pp203–210

Simmonds, M. P. and Hutchinson, J. D. (1996) 'Preface', in M. P. Simmonds and J. D. Hutchinson (eds) *The Conservation of Whales and Dolphins: Science and Practice*, John Wiley and Sons, Chichester, UK, ppxiii–xv

Simmonds, M. P. and Mayer, S. J. (1997) 'An evaluation of environmental and other factors in some recent marine mammal mortalities in Europe: Implications for conservation and management', *Environ. Rev.*, vol 5, no 2, pp89–98

Simmonds, M. P., Johnston, P. A., French, M. C., Reeve, R. and Hutchinson, J. D. (1994) 'Organochlorines and mercury in pilot whale blubber consumed by Faroe Islanders', *Sci. Total Environ.*, vol 149, pp97–111

Sorenson, J. (2009) *Ape*, Reaktion Books Ltd, London, UK

Weilgart, L. S. (2007) 'The impacts of anthropogenic ocean noise on cetaceans and implications for management', *Can. J. Zool.*, vol 85, pp1091–1116

WWF (World Wide Fund for Nature) (2010) 'The changes that climate change brings. A series of briefing on the impacts of climate change from the World Wide Fund for Nature', http://wwf.panda.org/about_our_earth/aboutcc/problems

18

From Conservation to Protection: Charting a New Conservation Ethic for Cetaceans

Philippa Brakes and Claire Bass

Traditionally, governments tend to manage our interactions with cetaceans from a conservation perspective, i.e. focusing management activities on the protection of species or populations. In contrast, much less effort is focused on protecting cetaceans at the individual level. Some exceptions exist, such as the attempts to protect solitary sociable dolphins from harassment and specific legislation prohibiting captivity.

In this chapter we challenge the traditional management approach and discuss the links between recent behavioural research, the capacity for cetaceans to suffer and the management of our direct and indirect interactions with these species.

Conservation: A Numbers Game

It is estimated that over 300,000 cetaceans are incidentally caught or killed in fishing nets worldwide each year (Read et al, 2006). Traditionally, governments have looked to scientists to tell them whether such removals from populations are sustainable. Similarly, some national governments endorse and encourage the hunting of whales, dolphins and porpoises on the basis that certain populations have 'surplus' animals – animals that can be removed without endangering the long-term viability of the species or population. The lexicon of cetacean management, for example in intergovernmental forums such as the IWC, demonstrates the simplistic, numeric nature of current conservation approaches – cetaceans are 'stocks' with 'yield rates' to be 'harvested'.

This 'numbers game' is not unfamiliar; we also tend to manage many other species in this relatively crude manner. The IUCN species listings are based on the current and projected health of a population, using data on ecology, population size, threats, conservation actions and utilization (IUCN, 2010). Thus, at the international, institutional level, common practice is to ask relatively simple questions such as 'How many are there?' and 'What threats do they face?'. The application of this approach to cetacean management is also perhaps unsurprising in light of the expertise, effort and money invested in counting cetaceans and determining how to extrapolate survey data into population estimates. Much less consideration has been given to the implications of fishing activities and other anthropogenic impacts on cetaceans at the individual level (Soulsbury et al, 2008; WDCS, 2008).

Thus, for most governments, modern marine mammal management tends to focus only on managing the 'numbers', i.e. evaluating the sustainability of certain human activities on cetacean populations. This 'numbers game' has also played out in recent years at the IWC, where nations barter over acceptable whaling quotas on the purported basis of sustainable removals (an issue on which governments and scientists do not always agree), with the political overlay of whaling proponents attempting to ensure lucrative quotas, now and in the future (WDCS, 2010).

The worldwide conservation movement – itself partially launched by the 'Save the Whale' movement of the 1960s and 1970s – has been very successful in instilling in many governments a sense of moral responsibility towards maintaining viable populations of wild animals. Internationally, despite the significant divide between the pro- and anti-whaling nations, there is often common ground on the principle that cetacean management should be broadly risk averse and that anthropogenic removals should not deplete populations to levels that may risk extinction. Whether a government approaches cetacean management from the perspective of species conservation, or whether the objective is to cull cetaceans that are deemed to be interfering with human fisheries,[1] the commonly shared objective is to 'manage the numbers' through an increase, in the former case, or decrease, in the latter.

However, is this almost exclusive interest in managing numbers sufficiently sophisticated to adequately cater for socially complex and highly migratory animals such as cetaceans? There is a growing body of evidence that suggests we should instead be seeking to manage cetaceans at lower orders of organization than species or geographically discrete populations, for example, as cultural units (Kasuya, 2008; Whitehead, this volume).

Conversely, despite the inconsistency in approach between conservation and animal welfare considerations, 'numbers', and the related genetic diversity, are undoubtedly important in conservation terms, particularly to ensure that gene pools are adaptive. Another important, but often overlooked issue, is that the genetic health of a population is also of importance to the welfare of generations of individuals within that population, for example, to ensure sufficient diversity for adaptation to environmental pressures and population fitness, so that individuals can thrive.

Social Complexity and Modern Conservation Mandates

Recent research suggests that in some bottlenose dolphin populations certain individuals may have specific roles (Lusseau, 2007). Such roles may be of importance both to the welfare of others within the group and to conservation. Further evidence that all individuals within a population are not necessarily equal from a conservation perspective is provided by investigations into orca societies, where it has been proposed that some matrilines may play a more central role than others in the wider social network (Williams and Lusseau, 2006). Where cultural transmission between individuals is established (as described in detail by Whitehead in this volume), ensuring such transmission of information may also be important to the fitness of the group in the longer term. For example, in 2005 scientists found that the use of sponges as foraging tools by wild bottlenose dolphins in Australia's Shark Bay displayed strong matrilineal heritability (Krützen et al, 2005). Such foraging strategies may have selective advantage.

It follows, therefore, that the removal of just a single individual displaying an advantageous adaptive trait (such as foraging tool use or knowledge of a productive feeding ground) could have longer-term implications for a population's health and survival. Thus, for those species where there is evidence of cultural transmission of knowledge, or where individuals have specific roles, the importance of the individual within the cetacean society may also be of significance to the wider conservation status of the group. The question is, are we doing enough to protect cetaceans at lower orders of organization?

Often when scientists discuss the conservation of cetacean populations they look to 'recovery rates'. Of course, many baleen whale populations are recovering from severely depleted states as a result of commercial whaling, therefore, a recovery rate alone may not be indicative of whether a population is 'healthy' or nearing carrying capacity. Since the vitality of certain individuals may be important to the wider population health, overlaying traditional conservation approaches with consideration of the 'conservation value' of specific individuals to the cetacean society, and thus potentially the population, may help us to develop an increasingly sophisticated means of evaluating the true projected health of populations and smaller subunits.

Perhaps then it is time to develop a new, more refined approach for evaluating the status of populations and cultural groups or, giving consideration in management and conservation plans to the potential existence of such subdivisions. Such a marriage between traditional conservation measures, behavioural ecology and arguably also animal welfare, could provide a new approach that would tell us not only 'how many' individuals there are, but additionally provide a means of understanding cetaceans at the social and cultural levels. In other words, policy-makers could use such a method to consider not only whether a population is genetically viable (the traditional conservation measure), but also whether there are discrete social/cultural subpopulations (which could, for example, influence their likelihood to breed, based on cultural status). In addition, a more comprehensive approach to monitoring the wellbeing of cetacean groupings would entail gaining an understanding of the 'quality' of the lives of the individuals within the population in terms of their group and individual welfare, and their ability to express natural behaviours and transmit culture. One possibility would be to deconstruct the welfare of the population or subpopulation in question into physical and behavioural components, which could perhaps include aspects such as their ability to share in cultural exchange of knowledge (an ability that can be adversely affected by human activities, such as noise pollution).

Measuring many of these parameters to provide a qualitative and quantitative analysis of population health, beyond size and genetic diversity, which would also take into consideration aspects such as individual welfare and cultural transmission, would not be without considerable challenges. However, like us, cetaceans are complex, social mammals and, in comparison, it is difficult to imagine that an anthropologist or social geographer would approach the projection of any human population on the simplistic basis of the size of the population, how many breeding females there are and what the environmental threats to the population are likely to be. There would be important additional questions, such as, do the females have access to everything they need, both physically and socially, in order to breed successfully?

Understanding Cetacean Welfare

If such an approach, which extended to individual welfare, were developed there is much more we would need to learn. Cetaceans live in an environment entirely alien to our own. Over the decades we have endeavoured to understand what life must be like in their aquatic world, but there are many aspects of their lives that are very difficult for us to imagine, such as 'seeing' with sound. Indeed, there are even some fairly basic facts about cetaceans' lives that have eluded our scrutiny. For example, until very recently it was believed that olfaction in baleen whales is only rudimentary. But new research has revealed that bowhead whales have a functioning olfactory system that suggests they have a sense of smell. Scientists speculate that they may use smell to help them locate schools of krill (Thewissen et al, 2010).

In this volume Thomas White describes the possibility that echolocation data that are intercepted by another individual 'could generate objects that are experienced in more nearly the same way by different individuals than ever occurs in communal human experiences when we are passive observers of the same external environment', and Hal Whitehead describes the concept that some cetacean species may have a collective consciousness; a sense of 'us' rather than just 'I'. White cites Jerison, who suggests that in dolphins 'communal experience might actually change the boundaries of the self to include several individuals' (Jerison, 1986).

If one were to compare cetacean research with research on large terrestrial mammals, such as the great apes and elephants, it is clear that the physical challenges associated with studying these animals in their natural environment have hindered equivalent progress in the field of understanding cetacean behaviour and their social systems (see Slooten, this volume). This is simply because many cetacean behaviours and interactions occur far from the view of humans. Nevertheless, some excellent research has been conducted, but practical limitations often mean that progress is slow and a great deal of patience and financial resources are often required.

Thomas White, in this volume, poses the question 'What's it like to be a dolphin?' and highlights the many similarities between ourselves and dolphins, but also discusses the many differences. Such comparisons are also important for our consideration of the lives of whales. Cetaceans are far-ranging, long-lived mammals and modelling the expected welfare status of individual cetaceans under specific circumstances is complex. The welfare of cetaceans cannot easily be compared with the welfare of terrestrial animals, for which many welfare scales and metrics have been developed. Recently a method for estimating the relative exposure of wild cetaceans to certain anthropogenic threats, such as purse seine fisheries, has been developed (Archer et al, 2010), but the measurement of exposure is only a first step in attempting to evaluate welfare status.

The different medium in which cetaceans live and the different focus of their senses, in comparison to our own, pose many questions related to how we can accurately assess their welfare. As marine mammals there are many environmental factors that can influence their welfare such as temperature, salinity, turbidity and prey availability, as well as social considerations. In addition, disease, which can also influence behaviour (Broom, 2006), may play an important role in both individual and group welfare, specifically where disease interferes with behaviours related to the wellbeing of the group.

To date, largely due to practical limitations, most attempts to measure cetacean welfare have been focused on the welfare of cetaceans in captivity or *in extremis* during stranding. Such simulations do not provide an accurate picture of wild cetacean welfare; we cannot, for example, easily obtain baseline data from wild, healthy cetaceans for the blood parameters used to evaluate stress levels. Furthermore, cetaceans often have rich and complex societies (Mann et al, 2000) and in captivity there is usually little or no interaction with their wild counterparts – both conspecifics and other species. Therefore, using behavioural and welfare data gathered in captivity to extrapolate to wild situations is problematic: the 'system' being measured has been significantly altered by the measurement process (i.e. captivity).

As for most species, cetacean welfare issues can be both acute and chronic. Bycatch and hunting are acute welfare issues and are often the focus of public attention and outrage. Nevertheless, chemical and noise pollution, ingestion of debris, harassment etc. may present more chronic and insidious threats to cetacean welfare, whose effects are equally significant but harder to quantify. It is logical that maintaining cetaceans' natural environment, in all its aspects, is most likely to be best for their long-term welfare; thus welfare and habitat protection are inextricably linked.

Many of the current approaches to cetacean welfare consider the physical aspects of harm and how these influence the wellbeing of the individual and the expectation for short- or long-term survival. Harder to predict are the psychological effects of anthropogenic activities – since measuring such influences on behaviour are complex. As a result there is a tendency to shy away from these trickier questions. Nevertheless, with our increasing knowledge of the complexity of cetacean societies and their individual intelligence, it is perhaps time to consider that some of the psychological impacts of human activities may be *at least* as profound as the physical impacts. In essence, preventing cetaceans from expressing their natural behaviours may have important implications for both their conservation and their individual wellbeing.

The 'Special Case' of Cetacean Bycatch

The cetacean bycatch issue provides a stark example of both the numeric context under which cetaceans are currently managed and the cost to individual welfare. While many governments take a strong position on issues such as whaling and harassment of individual cetaceans (on the basis of animal welfare concerns), a 'blind eye' is often turned to the animal welfare issues associated with the protracted death of cetaceans accidentally caught in fishing gear. Instead, policy-makers tend to focus on whether or not such removals from populations are sustainable.

The inconsistency of this approach is undoubtedly the result of the political pressures associated with tackling domestic fishing industries and associated vested interests among the voting public. But this contrasts profoundly with the fact that many countries have legislation that is intended to protect individual cetaceans from harassment and, increasingly, to prohibit cetacean captivity.

Nevertheless, fisheries bycatch is often treated as if it is exempt from animal welfare considerations (and even welfare legislation), despite the fact that cetacean deaths in fishing gear are often brutal and protracted. Post mortem evidence shows that dolphins typically suffocate over many minutes in the nets, rather than drown, a fact that surely qualifies as 'unnecessary' suffering by any definition. The harsh reality for cetaceans is that it may be easier for government officials to moderate the conduct of an overzealous jet-skier, keen to interact with a solitary dolphin, or to make vitriolic speeches at the IWC about whaling, than it is to temper the activities of their own fishing industry, where harm to cetaceans takes place beyond public purview.

Towards 'Protection'

Our increased understanding of the transmission of culture between cetaceans and the knowledge that certain individuals may have specific roles within populations and subgroupings, bestows upon us a responsibility to consider how we can protect cetacean populations beyond the normal geographic and genetic boundaries that have traditionally been used to delineate, manage and protect populations. We must now consider how we can protect cetaceans as individuals and communities, as well as species and populations. Protecting cetaceans at the individual and society level would clearly have significant welfare benefits for the individuals concerned as well as their social or familial conspecifics, and potentially have conservation significance at higher levels of organization.

Future 'protection' of cetaceans should include consideration of all their interests, such as ensuring they have healthy environments, enough food to eat, space and safety to exhibit their range of natural behaviours, as well as freedom from hunting and capture. Equipped with the knowledge that some cetacean species are self-aware and that there is a growing body of research that suggests that some cetacean brains contain structural facets suggestive of advanced cognitive capacities (Eriksen and Pakkenberg, 2007; Marino et al, 2007), it is arguably incumbent upon us to include in our management criteria aspirations to protect these individuals from suffering.

If the global community acknowledges – as science is telling us that we should – that these are intelligent, sentient beings, often with complex social networks, some of which can transmit cultural knowledge between individuals and generations, then we must do more than simply protect 'numbers'. We should also seek to understand and protect the physical and psychological wellbeing of cetaceans as individuals and societies when we consider their welfare, conservation and protection.

Finally, there is a third tier of protection that we should begin to embrace: the importance of protecting cetaceans and their cultures for their own sake. A time may come when future generations, whom we hope will have a better understanding of the nonhuman animal cultures with which we share the planet, will wonder at the folly of our grand-scale interspecies cultural imperialism and will marvel that despite all the indicators that we needed to do better, we resorted to just 'counting the numbers'.

Note

1 For a review of the 'whales eat fish' debate see Holt (2006) and for an analysis of the trophic role of whales off the coast of Northwest Africa, a region where this debate has momentum, see Morissette et al (2010).

References

Archer, F. I., Redfern, J. V., Gerrodette, T., Chivers, S. J. and Perrin, W. F. (2010) 'Estimation of relative exposure of dolphins to fishery activity', *Marine Ecology Process Series*, vol 410, pp245–255

Broom, D. M. (2006) 'Behaviour and welfare in relation to pathology', *Applied Animal Behaviour Science*, vol 97, pp73–83

Eriksen, N. and Pakkenberg, B. (2007) 'Total neocortical cell number in the mysticete brain', *Anat. Rec.*, vol 290, no 1, pp83–95

Holt, S. J. (2006) 'Whales competing? An analysis of the claim that some whales eat so much that they threaten fisheries and the survival of other whales', International League for the Protection of Cetaceans, 27 May

IUCN (International Union for Conservation of Nature) (2010) 'Red List Overview' www.iucnredlist.org/about/red-list-overview#assessment_process

Jerison, H. J. (1986) 'The perceptual world of dolphins', in R. J. Schusterman, J. A. Thomas and F. G. Wood (eds) *Dolphin Cognition and Behavior: A Comparative Approach*, Lawrence Erlbaum Associates, Hillsdale, New Jersey, pp141–166

Kasuya, T. (2008) 'The Kenneth S. Norris lifetime achievement award lecture: Presented on 29 November 2007 Cape Town, South Africa', *Marine Mammal Science*, vol 24, pp749–773

Krützen, M., Mann, J., Heithaus, M. R., Connor, R. C., Bejder, L. and Sherwin, W. B. (2005) 'Cultural transmission of tool use in bottlenose dolphins', *Proceedings of the National Academy of Sciences of the United States of America*, vol 102, no 25, pp8939–8943

Lusseau, D. (2007) 'Evidence for social role in a dolphin social network', *Evol. Ecol.*, vol 21, pp357–336

Mann, J., Connor, R. C., Tyack, P. L. and Whitehead, H. (eds) (2000) *Cetacean Societies: Field Studies of Dolphins and Whales*, The University of Chicago Press, Chicago

Marino, L., Connor, R. C., Fordyce, R. E., Herman, L. M., Hof, P. R., Lefebvre, L., Lusseau, D., McCowan, B., Nimchinsky, E. A., Pack, A. A., Rendell, L., Reidenberg, J. S., Reiss, D., Uhen, M. D., Van der Gucht, E. and Whitehead, H. (2007) 'Cetaceans have complex brains for complex cognition', *Public Library of Science (PLoS) Biology*, vol 5, no 139, pp966–972

Morissette, L., Kaschner, K. and Gerber, L. R. (2010) 'Ecosystem models clarify the trophic role of whales off Northwest Africa', *Marine Ecology Progress Series*, vol 404, pp289–302

Soulsbury, C. D., Iossa, G. and Harris, S. (2008) 'The animal welfare implications of cetacean deaths in fisheries', School of Biological Sciences, University of Bristol, UK

Read, A. J., Drinker, P. and Northridge, S. (2006) 'Bycatch of marine mammals in US and global fisheries', *Conservation Biology*, vol 20, no 1, pp163–169

Thewissen, J. G. M., George, J., Rosa, C. and Kishida, T. (2010) 'Olfaction and brain size in the bowhead whale (*Balaena mysticetus*)', *Marine Mammal Science*, doi:10.1111/j.1748-7692.2010.00406.x

WDCS (Whale and Dolphin Conservation Society) (2008) 'Shrouded by the sea: The animal welfare implications of cetacean bycatch in fisheries – a summary document', Whale and Dolphin Conservation Society, Chippenham, UK

WDCS (2010) 'Reinventing the whale: The whaling industry's development of new applications for whale oil and other products in pharmaceuticals, health supplements and animal feed', Whale and Dolphin Conservation Society, Chippenham, UK

Williams, R. and Lusseau, D. (2006) 'A killer whale social network is vulnerable to targeted removals', *Biol. Lett.*, vol 2, pp497–500

19

What is it Like to Be a Dolphin?

Thomas I. White

In the last 20 years, marine scientists have discovered that advanced abilities traditionally thought to be unique to humans are present in a variety of the marine mammals. Research on dolphins, however, has provided the richest picture of sophisticated intellectual and emotional abilities in cetaceans. The scientific evidence in a variety of areas is striking: self-awareness, dolphins' abilities to understand artificial human language, their capacity to think abstractly and solve problems, cetacean emotional abilities and social intelligence. Humans are no longer the only animals on the planet with such sophisticated intellectual and emotional abilities (Cavalieri and Singer, 1993; Reiss and Marino, 2001; White, 2007).

At the same time, fundamental differences between humans and dolphins have also surfaced. The dolphin brain has an older architecture than the human brain, and dolphin and human brains have features not found in the other. Dolphins possess a sense that humans lack (echolocation). And humans and dolphins have profoundly different evolutionary histories (Glezer et al, 1988; Ridgway, 1990).

This juxtaposition of important similarities and differences has significant ethical implications. The similarities suggest that dolphins qualify for moral standing as individuals – and therefore are entitled to treatment of a particular sort (White, 2007). The differences, however, suggest that dissimilar standards may apply when it comes to determining something as basic as 'harm'. Thousands of dolphins are killed by humans each year. Hundreds are kept captive for entertainment, therapy and military use. Is such treatment ethically defensible? Any attempt to develop an interspecies ethic about human–dolphin interaction, however, brings with it a variety of important issues that demand careful attention.

First, any such ethic will be logically dependent on our understanding of the nature of dolphins. The answers to such questions as 'What constitutes ethical and unethical treatment of dolphins by humans?' depend upon what dolphins are like, what they need to flourish, and so forth.

Second, research on many dolphin species suggests that dolphins are intellectually and emotionally sophisticated, self-aware beings. Some of the most important answers to the ethical issues, then, will depend on assumptions about the subjective dimensions of a dolphin's experience. For example, the answer to the question of whether certain treatment of dolphins inflicts traumatic, emotional pain depends a great deal on the character of the subjective experience of the dolphins involved.

Third, in investigating these issues, there is the constant danger of species bias. Anthropocentrism tempts us to think that in order for us to conclude that nonhumans have abilities similar to our own, they must demonstrate these abilities in the same way that we do. (For example, to be truly 'intelligent', the dolphin brain should have the same 'advanced' traits as the human brain, for example, a prefrontal cortex.) However, the ways that dolphins differ from humans are at least as important – if not *more* important – than the ways they are like us. So, in trying to determine what counts as appropriate treatment of dolphins, it is critical that we take into account and recognize the significance of these differences.

This chapter makes a preliminary attempt to address some of these issues by asking what we can currently say about 'what it's like to be a dolphin'. It will suggest a three part answer:

1. To be a dolphin is to be similar to humans.
2. To be a dolphin is to be different from humans.
3. To be a dolphin is to be the victim of unintentional anthropocentrism.

The chapter then concludes with some brief reflections on the ethical implications of these three claims.

To Be a Dolphin is to Be Similar to Humans

Historically, humans have seen our species as the 'gold standard' for sentience, intelligence, emotional sophistication and moral standing on Earth. We 'count' in a way that other beings do not. We're 'subjects'; they're objects. We're 'people'; they're 'animals'. We think it is kind and 'humane' of us to treat other beings in a considerate fashion, but, strictly speaking, this is nothing they're owed. We assume that the only reason we'd have to treat 'animals' any better is if and only if they demonstrate the same intellectual and emotional abilities that we have – and demonstrate them in the same way. In other words, only beings 'just like us' deserve moral consideration.

The discoveries of modern cetacean science, however, fundamentally challenge humanity's claim to uniqueness. Most importantly, a survey of the highlights of dolphin research in recent decades uncovers key data that suggest that dolphins, like humans, are 'persons'.

To be a 'person' is to be a 'who', not a 'what'. The sophisticated intellectual and emotional abilities characteristic of persons produce self-conscious, unique individuals (with distinctive personalities, life-long memories and personal histories) who are vulnerable to a wide range of physical and emotional pain and harm, and who have the power to reflect upon and choose their actions (Herzing and White, 1998; White, 2007). Dolphins join the great apes as nonhumans who demonstrate a nexus of advanced cognitive and affective abilities traditionally thought to surface only in humans (Cavalieri and Singer, 1993). Research shows that:

- The dolphin brain has a variety of advanced traits. It has a large cerebral cortex and a substantial amount of associational neocortex. Most anatomical ratios that assess cognitive capacity (brain weight/spinal cord, Encephalization Quotient) place it second only to the human brain. From this perspective, the dolphin brain appears to be, at the very least, one of the most complicated and powerful brains on the planet (Marino, 1995, 2004).
- On the subject of 'mirror self-recognition', scientists have made a strong case for the idea that dolphins possess not simply consciousness but self-consciousness. The dolphins studied were able to pass a standard test of self-awareness that we use with humans. It involves how we behave when we see our reflection in a mirror. The pioneer in this area was Ken Marten who worked with five dolphins at Sea Life Park in Hawaii (Marten and Psarakos, 1995). The most rigorous study was done by Diana Reiss and Lori Marino with two dolphins at the New York Aquarium. When a removable mark was put on the dolphins' bodies, the dolphins behaved in front on the mirror in a way that suggests that they were using it to examine themselves and to look at the mark. They appeared to understand that they were looking at a reflection of *their own* bodies (Reiss and Marino, 2001).
- Like humans, dolphins can solve problems by reasoning, as argued by John Gory, Stan Kuczaj and Rachel Walker (Gory and Kuczaj, 1999; Kuczaj and Walker, 2006). The dolphins in these studies were able to 'create a novel and appropriate solution in advance of executing the solution' – something, they argue, that 'can only be achieved if an animal has an ability to represent the causal structure of its environment' (Gory and Kuczaj, 1999). Other researchers have also observed novel and imaginative behaviours that appear to proceed from this same cognitive skill. The behaviour at issue may seem prosaic – blowing bubbles, for example – but it appears to require at least an intuitive grasp of the basic physical laws that govern the situation and a working knowledge of hydrodynamics (Marten et al, 1996).
- There is also evidence of problem solving among wild dolphins in a variety of hunting strategies that have been devised by members of the community, and then continued via cultural transmission to the next generation. This includes not only the use of sponges and bubble nets as tools but also cooperative fishing strategies with humans (Pryor et al, 1990; Smolker, 2001; Whitehead et al, 2004).

- Dolphins share with us an impressive level of cognitive sophistication – the ability to handle words, syntax, grammatical rules and the like, as suggested by Lou Herman's famous work with artificial human languages (Herman et al, 1984, 1993, 1999; Richards et al, 1984; Herman, 1986; Kako, 1999). Even more important, however, is that Herman's dolphins were able to perform so well while operating in what Denise Herzing has called 'a foreign conceptual environment' (private communication). It is clear that dolphins communicate with each other in the wild, but there is, as yet, no evidence that dolphins have a language equivalent to ours. The amount of cognitive flexibility demonstrated by Herman's dolphins, then, is remarkable.
- It is in social intelligence that dolphins likely excel, however. Rachel Smolker claims that the dolphins she studied 'spend most of their time and mental energy sorting out their relationships' (Smolker, 2001). Smolker highlights this fact even further in her description of the mind of one of the dolphins in the group. She writes, 'Her mind is a social mind, her intellectual skills lie in the realm of relationships, politics, social interaction' (Smolker, 2001). Richard Connor's discovery of political alliances among dolphins is particularly important in this regard (Connor and Peterson, 1994).
- Emotions ranging from anger to grief are a standard part of a dolphin's inner world, as suggested by observations by Denise Herzing and Ronald Schusterman (Herzing, 2000; Schusterman, 2000).
- Centuries of stories document dolphin altruism towards one another and to humans; cetaceans may be the only nonhumans on the planet to demonstrate enough curiosity about another intelligent being to seek out interaction with that species; and, like humans, dolphins appear to engage in sex simply for pleasure (Norris, 1991).

The catalogue of similarities between dolphins and humans is striking, and, in my view, it is enough to grant what philosophers call 'moral standing' to dolphins. That is, the advanced level of dolphin cognitive skills and affective abilities (that is, the capacity to experience, reflect upon and manage a wide range of emotional states) implies that, in terms of subjective experience, dolphins have similar 'individual consciousness' as humans. Therefore, they should 'count' in a moral calculation on similar grounds that individual humans count.

This conclusion derives from the fact that, traditionally, it is the distinguishing features of human consciousness that underpin the human claim that we are entitled to special treatment at the hands of one another – and treatment significantly different from how we're obligated to treat other animals. Advanced cognitive and affective traits give us the capacity for a rich subjective experience that, we argue, demands respect. These include:

- our ability to be aware of ourselves, our past, and to imagine ourselves in the future;

- our vulnerability to emotional as well as physical pain;
- our ability to empathize with and assist those around us – even if we have no close emotional bond with them;
- our rich intellectual and emotional lives;
- our ability to control our actions, and hence to be held responsible for what we do.

The richness of our abilities is so considerable that we consider the subjective experience of each human to be unique – and so valuable as to be beyond measure. As the philosopher Immanuel Kant expressed it, '[E]verything has either a price or a dignity. Whatever has a price can be replaced by something else as its equivalent; on the other hand, whatever is above all price, and therefore admits of no equivalent, has a dignity' (Kant, 1993). Kant is plain in asserting the fundamental and critical difference between people, on the one hand, and objects, on the other. Chairs and tables have a price, but we have a dignity. And our actions towards each other should reflect that.

In at least one fundamental and critical way, then, the subjective experience of 'being a dolphin' is remarkably similar to the subjective experience of 'being a human'. To be a dolphin is to be similar to humans in that we both experience life as 'persons'.

To Be a Dolphin is to be Different from Humans

As intriguing as the similarities between dolphins are humans are, however, the differences between our two species are probably more important. One of humanity's traditional weaknesses in dealing with nonhumans is to assume that other animals deserve consideration only to the extent that they are 'just like us'. The risk of anthropocentrism is ever present, so it's important to remind ourselves that evolution and adaptation may produce different types of consciousness and intelligence in different species. As Diana Reiss observes:

> *It is important to realize that in some cases the borders between species are not real but are, rather, assumptions based on a lack of evidence or data. Historically it has been a tacit assumption that we are the only symbol- and tool-using species and are at the pinnacle of evolution. The view that physical evolution is pyramidal has been replaced by a view of evolution as a spreading structure with diverse life forms from different phyla. It is clear that there are both convergent and divergent processes and a variety of strategies operating throughout the biological world that enable different species to survive and flourish in their own environments. Perhaps in the near future we will view the evolution of intelligence in a similar way.* (Reiss, 1990)

Reiss considers the differences in the outcome of evolutionary processes experienced by humans and dolphins to be so profound that she has suggested that we regard dolphins as an example of an 'alien intelligence' (Reiss, 1990).

In view of the many ways that the outer world (physical environment, anatomy) probably shapes the inner world (consciousness, self-consciousness, intelligence, etc.), we have, then, the dual challenge of: (1) identifying appropriately 'alien' factors in a dolphin's natural environment and the features they lead to, and (2) determining their significance for 'what it's like to be a dolphin'.

A good place to start is simply common-sense observations that underscore just how fundamentally different it is for members of our two species to experience some of the most basic dimensions of life:

- First, because dolphins live in the water, they are virtually always moving. Even when they rest, they move. Coastal dolphins may sometimes station themselves on the bottom, but many pelagic (deep water) dolphins probably go through their entire lives without ever seeing a stable ocean floor. Humans, by contrast, require stillness and steadiness in virtually everything we do. Even when we travel, we need cars, trains and planes to be as stable as possible. We seem to be able to think better and focus more sharply on problems when we sit down. The dynamic and fluid quality of a dolphin's life would be disorienting to us.
- Because vision is such a central sense to humans, we require light for an extraordinary amount of what we do – to the extent that we've even invented artificial light. However, dolphins who live in the oceans have adapted to operating in darkness. Even on a sunny day, light does not penetrate very far into the ocean. And dolphins can, in fact, be far more active by night than during the day because more prey is available to hunt at this time.
- Because dolphins don't sleep the way that humans do, their awareness of life seems not to be punctuated by the stretches of unconsciousness that we experience (Wells et al, 1999). Ideally, we sleep and dream for about a third of each day. Because every breath is a conscious action for a dolphin, any significant stretch of unconsciousness would mean death. Uninterrupted awareness, then, is as normal for dolphins as periodic unconsciousness is for humans.
- In humans and other primates, the face is a critical tool for communication. Indeed, the face has no small number of muscles dedicated exclusively to producing the expressions associated with one emotion or another. And this is supported by as much as one third of the motor cortex of the human brain. In assessing the mood of another human, we likely 'read' his or her face first. Dolphins, by contrast, have no 'face', in the human sense. The famous dolphin 'smile' does not express an inner emotional state. It is simply the result of millions of years of adaptation for feeding and optimal hydrodynamics. External signals of dolphin emotional states are something other than 'facial' movements. Open mouth behaviours, tail slaps and bubbles are signs of aggression that are often apparent even to humans. But the more subtle pectoral touches, postures

and the like may be the equivalent of the small changes in facial expression that are understandable only to people who know one another very well.

- Another significant difference between the everyday experience of humans and dolphins is in how we perceive the objects on which we focus our attention. Because sound passes through tissue and other materials, when dolphins scan objects with echolocation clicks, their brains construct representations that are not only three dimensional, but transparent (or at least translucent). Without an artificial technology, however, human perception is opaque, and we see only the surface of objects. When we tap objects, the sound produced and the sensation we feel may tell us something about the density of objects. But we get only a tiny fraction of the information dolphins receive about the inner composition or structure of objects.

As intriguing as such a common-sense, unscientific list of differences is, it is difficult to identify the significance of each entry. However, the sum of the entire list clearly suggests that dolphins have adapted to an environment that is 'foreign' to humans and that this has probably led to some important differences in the inner worlds of the two species.

Fortunately, as cetacean research has progressed, details of key differences between humans and dolphins are beginning to surface. And two specific findings point to at least one fundamental difference in how our respective species experience life.

First, scientists have determined that dolphins can 'eavesdrop' on one another's echolocation clicks (Harley et al, 1995). By listening to another dolphin's echoes, then, one dolphin might actually share the other dolphin's experience. As a result, psychologist Harry Jerison makes the fascinating suggestion that this produces a kind of 'social cognition' – something for which there is no human equivalent. He writes:

> *Intercepted echolocation data could generate objects that are experienced in more nearly the same way by different individuals than ever occurs in communal human experiences when we are passive observers of the same external environment. Since the data are in the auditory domain the 'objects' that they generate would be as real as human seen-objects rather than heard-'objects,' that are so difficult for us to imagine. They could be vivid natural objects in a dolphin's world.* (Jerison, 1986)

Second, because of differences in how the human and dolphin brains evolved, it is possible that the limbic system in the dolphin brain has more of an impact on the processing of information than is the case in the human brain (Morgane et al, 1986; Jerison, 1986). This has led Jerison to yet another interesting claim – that dolphins may have deeper emotional attachments than humans experience.

The combination of shared perception of echolocation clicks and deep social attachments, then, makes it reasonable to suggest that the dolphin 'sense of self' differs from what humans experience. Indeed, Jerison suggests that dolphin 'social cognition' produces a 'sense of self' that is qualitatively different – and more social – from what humans experience. He writes that 'the processes underlying decisions might be shared by several dolphins as a group when facing the same task' and that this 'communal experience might actually change the boundaries of the self to include several individuals' (Jerison, 1986). He suggests, for example, that in dolphin 'reciprocal altruism', 'the "individual" (at least during the altruistic episode) [is] not one animal but a group of dolphins sharing communally in the experience as well as the behavior' (Jerison, 1986). Such a 'social self' might help explain certain dolphin behaviours during mass strandings – that is, why healthy members of a community will not abandon their sick companions. It could explain why dolphin aggression against each other is far less mortal than is the case with humans. (In contrast to the many daily reports of humans murdering humans, there are far fewer accounts of dolphins killing other dolphins.) Perhaps dolphins experience conspecifics as less of a threat than humans do. And it might also explain why individual dolphins will not escape from being encircled by nets they could easily leap over. Cetacean scientist Kenneth Norris observed that when the Hawaiian spinner dolphins that he studied were confronted by something unfamiliar, or when the school was under attack, 'individuality is reduced close to zero' (Norris, 1991). Norris noted that the impulse to act primarily for what's good for the group is so strong that in threatening situations, dolphins won't try to escape separately as individuals, even though the option is readily available. They will refuse to move in ways that the entire group can't.

In a fundamental way, then, the subjective experience of 'being a dolphin' is remarkably different from the subjective experience of 'being a human'. To be a dolphin is to be different from humans in that the dolphin experience of 'selfhood' is probably social and shared with others.

To Be a Dolphin is to Be the Victim of Unintentional Anthropocentrism

While this chapter has been speculating about the subjective experience of dolphins by exploring the implications of scientific research, there is one final, very different dimension of 'what it's like to be a dolphin' that we need to examine. That is, 'to be a dolphin' is to be vulnerable to humans' beliefs about who dolphins are and what kind of treatment dolphins deserve. Thousands of dolphins die or are injured each year as a result of human fishing practices, and hundreds of dolphins are kept captive in entertainment facilities for shows or dolphin swim programmes. However, virtually all of these practices are sanctioned by the laws of one country or another and, in theory, supported by the best available science. That is, we are not talking about the blind, unthinking predation of one species by another. The way that humans treat dolphins is supposedly the result of intelligent dialogue and formal processes that aspire to fairness, compassion and objectivity.

Despite humanity's best intentions, however, prejudices of all sorts (racial, sexual, religious, cultural, etc.) regularly play a role in shaping laws and policies. One of the public benefits of science, then, is to uncover and examine facts related to such matters so that prejudice is at least minimized. Yet for this to be successful, scientific investigation itself must be unbiased. And in the case of factual claims about nonhumans on which human practices rest, this means that scientific investigation must be free of unintentional anthropocentrism.

It's no secret that humans have a vested interest in being able to claim that we have the best brains on the planet. For thousands of years, this belief has allowed us to justify the idea that we are the only 'intelligent' species on Earth, and that the rest of creation lies at our feet, waiting for us to do with it as we please. As long as no other beings 'think' or 'feel', we don't need to have any moral qualms about how we treat them. Accordingly, in light of the obvious temptation to protect the primacy of humans, it's important for us to ask ourselves if we're being as objective as we should be. That is, is it possible that when we discuss topics of the sort we're involved in, could we – even unintentionally or unconsciously – interpret data about the rest of nature so that it supports a preconceived picture of reality and doesn't challenge our privileged status? In particular, is it possible for *scientists* to do this?

Ideally, of course, we would want to answer this question with an unequivocal 'no'. Yet humans have a long history of citing 'objective facts' in defence of practices that are either firmly rooted in irrational prejudice or at least conveniently self-serving (Tuana, 1993; Gould, 1996). And science has not been as unbiased as it should have been – particularly when it comes to assessing the claims of groups who have traditionally been labelled 'inferior'. As Steven Jay Gould has detailed, the work of two 19th-century craniologists, Samuel George Morton and Paul Broca, is worth special note because both men paid careful attention to detail and attempted to follow disciplined, scientific methodology. And yet, both came to the conclusion that the 'facts' irrevocably proved the intellectual superiority of white males (Gould, 1996).

Moreover, there are two critical points to realize about Morton and Broca. First, they were not rabid racists and sexists. They truly believed that they were objective scientists who were simply reporting data and following the facts wherever they led. Second, despite their stated, conscious desire to be objective, both men were apparently unaware of the fact that they were arguing for conclusions *that their data did not support*. In reviewing Morton's data, Gould discovered that although Morton 'finagled' and 'juggled' his data to back up his claims, he apparently was not consciously aware of this. (And when recalculated in an objective light, Morton's data reveal no significant differences between the races or sexes.) Gould acquits Morton of fraud, but points to a more insidious process. He writes:

> *Yet through all this juggling, I detect no sign of fraud or conscious manipulation. Morton made no attempt to cover his tracks and I must presume that he was unaware he had left them. He explained all his procedures and published all his raw data. All I can discern is an a priori conviction about racial ranking so powerful that it directed his tabulations along pre-established lines. Yet Morton was widely hailed as the objectivist of his age, the man who would rescue American science from the mire of unsupported speculation.* (Gould, 1996)

Broca, in particular, had a sophisticated understanding of statistics and knew how to correct for various factors that could colour the outcome: differences in body size, age, health and the like. However, like Morton, Broca was a man of his time – and in his time, it was self-evident that non-white people and women were perceived to be intellectually inferior. As Gould explains:

> *I spent a month reading all of Broca's major work, concentrating on the statistical procedures. I found a definite pattern in his methods. He traversed the gap between fact and conclusion by what may be the usual route – predominantly in reverse. Conclusions came first and Broca's conclusions were the shared assumptions of most successful white males during his time – themselves on top by the good fortune of nature, and women, blacks, and poor people below. His facts were reliable (unlike Morton's), but they were gathered selectively and then manipulated unconsciously in the service of prior conclusions. By this route, the conclusions achieved not only the blessing of science, but the prestige of numbers. Broca and his school used facts as illustrations, not as constraining documents. They began with conclusions, peered through their facts, and came back in a circle to the same conclusions.* (Gould, 1996)

Of course, when confronted with the examples of 19th-century scientists such as Morton and Broca, most of us are probably tempted to find one reason or another to think that this kind of thing couldn't happen in our century. Perhaps we think that contemporary science is more sophisticated and objective than in the past, so questionable data are more readily and effectively challenged. Or maybe we'd say that we live in more egalitarian times with greater sensitivity to human rights. Perhaps we'd even claim if there's any preconceived idea in our society that exerts unconscious pressure on scientists, it's the belief in the equality of races and the sexes. However, there are two reasons why we should take the example of Morton and Broca very seriously.

First, the second half of the 20th century saw more than one attempt to use quantifiable data and rigorous methodology in a way that suggests irrevocable racial differences in intelligence:

- In 1969, Arthur Jensen argued that differences in IQ scores between white and non-white people in America were largely the result of genetic, not environmental, factors (Jensen, 1969).
- In 1971, H. Eysenck argued that African and black American babies develop sensorimotor skills more quickly than Caucasians do, and he then claimed that such speedy development as an infant correlates with lower IQ later in life. 'These findings', he observes, 'are important because of a very general view in biology according to which the more prolonged the infancy the greater in general are the cognitive or intellectual abilities of the species' (Eysenck, 1971).
- In 1994, Richard Herrnstein and Charles Murray echoed Jensen's earlier claim that differences in IQ scores between white and non-white people were mainly genetically based (Herrnstein and Murray, 1994).

Moreover, Jensen, Herrnstein and Murray clearly link their scientific findings to recommendations regarding social policy. Jensen (1969) begins his article by writing, 'Compensatory education has been tried, and it apparently has failed.' And his subsequent study allegedly shows why additional compensatory education programmes would also fail. Herrnstein and Murray (1994) similarly use their data to argue that a variety of traditional programmes designed to eradicate inequalities can be nothing but fruitless.

But the most important reason to take the examples of Morton and Broca to heart is that it is as 'obvious' today that nonhuman beings are 'just animals' as it was in the 19th century that non-white people and women were inferior. In our culture, it is self-evident that 'animals' are completely different from humans. They are not thought to have self-awareness, and they are seen to have very limited cognitive and affective abilities. And because they're so different from us, we really don't have to worry too much about harming them. Can we say with certainty that the 'objective' research of contemporary science can't be affected by these beliefs?

Consider one more comment by Gould (1996) about the troubling possibility that scientists may not be as free from the attitudes that predominate in the societies in which they live as we would like to think:

> *Clearly, [in the 19th and early 20th centuries], science did not influence racial attitudes ... Quite the reverse: an* a priori *belief in black inferiority determined the biased selection of 'evidence.' From a rich body of data that could support almost any racial assertion, scientists selected facts that would yield their favoured conclusion according to theories currently in vogue. There is, I believe, a general message in this sad tale. There is not now and there never has been any unambiguous evidence for genetic determination of traits that tempt us to make racist distinctions (differences between races in average values for brain size, intelligence, moral discernment, and so on). Yet this lack of evidence has not forestalled the expression of scientific opinion. We must therefore conclude that this*

> *expression is a political rather than a scientific act – and that scientists tend to behave in a conservative way by providing 'objectivity' for what society at large wants to hear.* (Gould, 1996)

The point to recognize is that 19th-century scientists could just as easily have taken the position that differences between races and the sexes did not prove anything about the superiority or inferiority of one group over the other. And yet, instead of concluding nothing, they took a position that not only was unsupported by the facts, but that clearly reflected the dominant attitudes and prejudices operating in their society. Is it possible that contemporary scientists in some way do the same thing when studying dolphins?

Anthropocentrism and cetacean science

But is this any more than an idle fear? Do we have any reason to think that contemporary scientists might draw conclusions about dolphins that are unintentionally affected by species bias? Most scientists are appropriately cautious in the conclusions they draw. They recognize that what we might know about brain structure does not settle the issue of general cognitive ability or behavioural flexibility, and they regularly call for more research. And contemporary science has progressed in a way that it would be considered bad science to make the sort of heavy-handed pronouncements about the relationship between a single feature (for example, skull size) and a property such as 'intelligence' that we find in 19th-century science.

However, this does not mean that species bias is impossible – only that it would be expressed in more subtle ways. The language of contemporary science is both more technical and the conclusions more carefully crafted, so any unintentional species bias would come through tone, for example, or in what facts are *not* mentioned. And occasionally we do find scientists discussing matters in a way that lets us ask whether species bias is inadvertently creeping in to the interpretation of the data:

1 In a short article that argues that dolphins fail to show evidence of advanced intelligence, Margaret Klinowska (1989) gives the following account of the dolphin brain:

> *The newest studies of dolphin brains show that they have not developed the latest stage in the evolution of the brain. Their cortex seems to be lacking some features that are characteristic of primates and many other mammals. It seems that these structures started to evolve among land mammals about 50 million years ago, while the ancestors of modern cetaceans returned to the water a few million years earlier. Even the most advanced cetacean brains seem to be stuck at a stage*

> *called the paralimbic-parinsular, which is the most primitive stage in land mammals.*
>
> *In many respects, then the cetacean brain is actually quite primitive. It retains all the structures found in primitive mammals, such as hedgehogs and bats. It shows none of the structural differences from area to area typical of advanced brains like those of primates. The regions of the cortex are not separated by so-called associative areas, as they are in most other mammals, but they do seem to be arranged in much the same order as we imagine they were in the ancestor of all mammals. (Klinowska, 1989)*

The basic facts that this scientist cites about the cortex and the brain structure are correct. However, in the context of her entire article, the unmistakable implication (that these facts suggest that dolphins could not have advanced cognitive abilities) is questionable. Missing, of course, are provisos about the relevance of the different evolutionary histories between humans and dolphins (Marino, 2002).

2 In the course of their commentary on the main study that argues for the 'initial brain' hypothesis, Lester Aronson and Ethel Tobach appeal to a variety of grounds to challenge John Lilly's claim that the size of the dolphin brain suggests remarkable similarities with the human brain. They rely on brain measurements in the following way:

> *Through the efforts of [Glezer, Jacobs and Morgane], we are now able to correlate the behavioural level with the anatomical level of the neocortex and probably with the physiological level as well. We see at once that the anatomical level is considerably below that of the higher primates, and far below the human level. Those who favour the hypothesis of a high level of cetacean intelligence almost always emphasize the large, highly convoluted cortical surface area which is larger in* Homo *and which forms a vast array of sulci and gyri. But [the authors of this study] show paradoxically that the corticalization index in* Tursiops *(volume of cortex over volume of brain x 100) is even below that of the basal insectivore which is their extant model of the hypothetical 'initial' mammalian ancestor.* (Aronson and Tobach, 1988)

Again, there is a factual basis for these claims. However, the facts show only that the anatomical level of the dolphin brain is 'different from', not 'considerably below', the human brain. In addition, Aronson and Tobach ignore significant facts. The 'corticalization index' is not the only measure that Glezer et al note in their research. Three other ratios are cited (for 'encephalization' and 'neocorticalization') that show rough equivalence between humans and dolphins.

At the very least, this argues for a more cautious conclusion than the one that Aronson and Tobach indulge in:

> *We think that Gaskin put it well: 'If I may borrow and embellish a phrase from a paper by the Caldwells, there is abundant evidence that dolphins communicate information about "what", "where" and "who". There is no substantive evidence that they transmit information about "when", "how" or "why". So, no matter what some might wish to believe, with respect to Kipling's "six honest serving men" of learning and intellect ["What", "Why", "When", "How", "Where" and "Who"], the dolphin appears to be three servants short'.* (Aronson and Tobach, 1988)

3 And even though Glezer et al offer a carefully worded caution in a reply to Aronson and Tobach's comments about intelligence, note how they conclude their comments:

> *Relative to certain points brought up by Aronson and Tobach, it is likely that the behavioural status of the dolphin is exaggerated in the literature. Obviously, caution is needed in comparing intelligence among different species living in various ecological niches. Our investigations do not suggest any direct correlations between neocortical morphology and behaviour, but they point out the obvious morphological fact that dolphin neocortical organization bears a close resemblance to that of the hedgehog.* (Glezer et al, 1988)

What they give with one hand ('caution is needed in comparing intelligence'), they take away with the other ('close resemblance to [the brain] of the hedgehog').

In each of these three cases, no scientist explicitly says, 'specific features of the dolphin brain prove that dolphins cannot have advanced cognitive abilities'. However, the authors strongly imply what they think is *probably* the case – despite the existence of a large body of data that essentially precludes any judgement based on structure alone.

Why does the conflicting data not simply lead them to an absolutely noncommittal stance? Perhaps we have the same phenomenon that Steven Gould described above. In other words, it's possible that these scientists – even though operating in good faith – might nonetheless be influenced in how they view and interpret data by their society's overwhelming belief that only humans have advanced intellectual and emotional abilities.

To be a dolphin, then, is to be the victim of unintentional anthropocentrism in that there is reason to fear that contemporary cetacean science is not as unbiased as it should be.

Conclusion: The Ethical Implications

In exploring the question, 'What is it like to be a dolphin?' this essay makes three claims:

1. To be a dolphin is to be similar to humans in that we both experience life as 'persons'.
2. To be a dolphin is to be different from humans in that the dolphin experience of 'selfhood' is likely social and shared with others.
3. To be a dolphin is to be the victim of unintentional anthropocentrism in that there is reason to fear that contemporary cetacean science is not as unbiased as it should be.

If the three claims of this essay are correct, what are the ethical implications? A full account of all of the ethically problematic aspects of contemporary human–dolphin interaction is beyond this chapter, but the following are at least the most obvious conclusions.

First, the deaths of hundreds of thousands of dolphins each year at the hands of humans are without question ethically indefensible. The most dramatic slaughter comes from the annual Japanese dolphin 'drive hunts'. However, over 300,000 cetaceans are thought to die annually around the world as a result of fisheries bycatch (WDCS, 2008). Dolphin self-awareness implies that the death of each individual dolphin is the ethical equivalent of the death of an individual human. Moreover, the suffering that accompanies these deaths – from suffocation, extreme panic, drowning, bleeding to death, witnessing the deaths of members of one's community and the like – is surely extreme.

Second, Jerison's (1986) claims about dolphin 'social cognition', a 'social self' and the powerful social bonds among dolphins imply that the number and quality of their relationships are important factors in their welfare. (The average human, by contrast, can flourish with a much thinner social network.) This means that all versions of dolphin captivity – for entertainment, therapy and military use – are probably depriving dolphins of conditions that are critical to their wellbeing. Captivity is characterized by a small number of relationships, all of which are managed by humans, in sterile conditions that can hardly be characterized as constituting an authentic dolphin community. Given the intense social nature of dolphins, it is difficult to see how captive situations can nourish this dimension of their being – thus reducing dolphins to commodities being used primarily to benefit humans.

Third, the fundamental differences between humans and dolphins suggest that – as is the case with concepts such as 'intelligence' and 'self' – what constitutes 'harm' is to a large degree specific to the nature of each species. This is particularly true with beings with such advanced cognitive and affective capacities. The development of an interspecies ethic that can apply to human–dolphin interaction, then, requires more research into the lives of wild dolphins for the sake of identifying the basic conditions that foster the welfare of both individual dolphins and their community.

It is important to realize, however, that this essay has focused on the smaller dolphins primarily because of the large amount of research available on them – not because they are the only cetaceans with advanced intellectual and emotional abilities. It is obviously easier to do research on small cetaceans than on large ones, but this shouldn't suggest that dolphins are the only cetacean candidates for nonhuman personhood or moral standing – or that dolphins are the only cetacean under human assault.

Large spindle cells, which are thought to play a significant role in human self-awareness, social cognition and possibly even communication, have been found in the brains of humpback whales, fin whales, sperm whales and orcas (Hof and van der Gucht, 2007). Cultural transmission has been discovered in dolphin, orca, sperm whale and humpback societies (Whitehead et al, 2004). A variety of larger cetaceans (orcas, sperm and humpback whales) demonstrate abilities to communicate, form coalitions, cooperate and use tools (Connor et al, 1992; Clapham and Mead, 1999; Clapham, 2000; Rendell and Whitehead, 2001, 2003; Valsecchi et al, 2002). Such emotions as grief and parental love appear to have been observed in orcas (Rose, 2000a, 2000b). And, surveying the most relevant literature on cetaceans, Simmonds argues that for at least some cetacean species, there is unequivocally evidence for significant social structure and highly developed social behaviour (Simmonds, 2006).

Similarly, the larger cetaceans face their own ethically problematic treatment at the hands of humans. The current large-scale commercial whaling practised mainly by Japan, Norway and Iceland is disturbing. So too are the deaths of many whales taken in subsistence hunts (Brakes et al, 2004). But if commercial whaling is fully revived (by the lifting of the IWC's moratorium), the slaughter that would ensue would be awful and, in this one gesture, humankind would reaffirm its ancient worldview on the nature of cetaceans and our exploitative relationship with them. Even the current, incomplete state of research on the intellectual and emotional abilities of the large cetaceans implies that the systematic hunting of whales should be seen more as 'genocide' than 'hunting for food' or 'preserving a cultural tradition'.

Whether such a slaughter will happen depends on whether policy-makers and the public are willing to listen to scientists and cetacean advocates with an open mind. But, at this point, the jury is still out on which will carry the day: vested interest and speciesism or the ethical implications of the objective facts of science. We can hope for the latter. But the history of our species is a grim reminder of how difficult it will be to prevent the former.

References

Aronson, L. R. and Tobach, E. (1988) 'Conservative aspects of the dolphin cortex match its behavioral level', *Behavioral and Brain Sciences*, vol 11, no 1, pp89–90

Brakes, P., Butterworth, A., Simmonds M. P. and Lymbery, P. (2004) *Troubled Waters: A Review of the Welfare Implications of Modern Whaling Activities*, World Society for the Protection of Animals, London

Cavalieri, P. and Singer, P. (eds) (1993) *The Great Ape Project: Equality beyond Humanity*, St. Martin's Press, New York

Clapham, P. J. (2000) 'The humpback whale: Seasonal breeding and feeding in a baleen whale', in J. Mann, R. C. Connor, P. L. Tyack and H. Whitehead (eds) *Cetacean Societies: Field Studies of Dolphins and Whales*, University of Chicago Press, Chicago, pp173–196

Clapham, P. J. and Mead, J. G. (1999) 'Megaptera novaeangliae', *Mammal Species*, vol 604, pp1–9

Connor, R. C. and Peterson, D. M. (1994) *The Lives of Whales and Dolphins*, Henry Holt, New York

Connor, R. C., Smolker, R. A. and Richards, A. F. (1992) 'Dolphin alliances and coalitions', in A. H. Harcourt and F. B. M. de Waal (eds) *Coalitions and Alliances in Humans and Other Animals*, Oxford University, Oxford, pp415–443

Eysenck, H. J. (1971) *The IQ Argument: Race, Intelligence and Education*, Library Press, New York

Glezer, I. I., Jacobs, M. S. and Morgane, P. J. (1988) 'Implications of the "initial brain" concept for brain evolution in Cetacea', *Behavioral and Brain Sciences*, vol 11, no 1, pp75–116

Gory, J. D. and Kuczaj, S. A. (1999) 'Can bottlenose dolphins plan their behavior?', paper presented at the Biennial Conference on the Biology of Marine Mammals, Wailea, Maui, Hawaii, November–December

Gould, S. J. (1996) *The Mismeasure of Man*, revised and expanded edition, Norton, New York, NY

Harley, H. E., Xitco, M. J. and Roitblat, H. L. (1995) 'Echolocation, cognition, and the dolphin's world', in R. A. Kastelein, J. A. Thomas and P. E. Nachtigall (eds) *Sensory Systems of Aquatic Mammals*, De Spil Publishers, Woerden, The Netherlands, pp529–542

Herman, L. M. (1986) 'Cognition and language competencies of bottlenose dolphins', in R. J. Schusterman, J. A. Thomas and F. G. Wood (eds) *Dolphin Cognition and Behavior: A Behavioral Approach*, Lawrence Erlbaum Associates, Hillsdale, New Jersey, pp221–252

Herman, L. M., Richards, D. G. and Wolz, J. P. (1984) 'Comprehension of sentences by bottlenosed dolphins', *Cognition*, vol 16, pp129–219

Herman, L. M., Pack, A. A. and Morrel-Samuels, P. (1993) 'Representational and conceptual skills of dolphins', in H. L. Roitblat, L. M. Herman and P. E. Nachtigall (eds) *Language and Communication: Comparative Perspectives*, Erlbaum Associates, Hillsdale, NJ, pp403–442

Herman, L. M. and Uyeyama, R. K. (1999) 'The dolphin's grammatical competency: Comments on Kako (1999)', *Animal Learning & Behavior*, vol 27, no 1, pp18–23

Herrnstein, R. J. and Murray, C. (1994) *The Bell Curve: The Reshaping of American Life by Difference in Intelligence*, Free Press, New York

Herzing, D. L. (2000) 'A trail of grief', in M. Bekoff (ed) *The Smile of a Dolphin: Remarkable Accounts of Animal Emotions*, Discovery Books, New York, pp138–139

Herzing, D. and White, T. I. (1998) 'Dolphins and the question of personhood', *Etica & Animali*, Special Issue on Nonhuman Personhood, vol 9/98, pp64–84

Hof, P. R. and van der Gucht, E. (2007) 'Structure of the cerebral cortex of the humpback whale, *Megaptera novaeangliae* (Cetacea, Mysticeti, Balaenopteridae)', *The Anatomical Record*, vol 290, pp1–31

Jensen, A. (1969) 'How much can we boost IQ and scholastic achievement?', *Harvard Educational Review*, vol 39, pp1–123

Jerison, H. J. (1986) 'The perceptual world of dolphins', in R. J. Schusterman, J. A. Thomas and F. G. Wood (eds) *Dolphin Cognition and Behavior: A Comparative Approach*, Lawrence Erlbaum Associates, Hillsdale, New Jersey, pp141–166

Kako, E. (1999) 'Elements of syntax in the systems of three language-trained animals', *Animal Learning & Behavior*, vol 27, no 1, pp1–14

Kant, I. (1993) *Grounding for the Metaphysics of Morals*, translation by J. E. Ellington, Hackett, Indianapolis, IN

Klinowska, M. (1989) 'How brainy are cetaceans?', *Oceanus*, vol 32, no 1, pp19–20

Kuczaj, S. A. and Walker, R. (2006) 'Problem solving in dolphins', in T. Zentall and E. A. Wasserman (eds) *Comparative Cognition: Experimental Explorations of Animal Intelligence*, MIT Press, Cambridge, MA, pp580–601

Marino, L. A. (1995) 'Brain-behavior relationships in cetaceans and primates: Implications for the evolution of complex intelligence', PhD thesis, State University of New York at Albany

Marino, L. (2002) 'Convergence of complex cognitive abilities in cetaceans and primates', *Brain, Behavior and Evolution*, vol 59, pp21–32

Marino, L. (2004) 'Cetacean brain evolution: Multiplication generates complexity', *International Journal of Comparative Psychology*, vol 17, pp1–16

Marten, K. and Psarakos, S. (1995) 'Evidence of self-awareness in the bottlenose dolphin (*Tursiops truncatus*)', in S. T. Parker, R W. Mitchell and M. L. Boccia (eds) *Self-Awareness in Animals and Humans: Developmental Perspectives*, Cambridge University Press, New York, pp361–379

Marten, K. Shariff, K., Psarakos, S. and White, D. J. (1996) 'Ring bubbles of dolphins', *Scientific American*, August, pp83–87

Morgane, P. J., Jacobs, M. S. and Galaburda, A. (1986) 'Evolutionary morphology of the dolphin brain', in R. J. Schusterman, J. A. Thomas and F. G. Wood (eds) *Dolphin Cognition and Behavior: A Comparative Approach*, Lawrence Erlbaum Associates, Hillsdale, New Jersey, pp5–30

Norris, K. S. (1991) *Dolphin Days: The Life and Times of the Spinner Dolphin*, W. W. Norton, New York

Pryor, K., Lindbergh, J., Lindbergh S. and Milano, R. (1990) 'A dolphin-human fishing cooperative in Brazil', *Marine Mammal Science*, vol 6, pp77–82

Reiss, R. (1990) 'The dolphin: An alien intelligence', in B. Bova and B. Preiss (eds) *First Contact: The Search for Extraterrestrial Intelligence*, Penguin Books, New York, pp31–40

Reiss, D. and Marino, L. (2001) 'Mirror self-recognition in the bottlenose dolphin: A case of cognitive convergence', *Proceedings of the National Academy of Science*, vol 98, no 10, pp5937–5942

Rendell, L. and Whitehead, H. (2001) 'Culture in whales and dolphins', *Behavioral and Brain Sciences*, vol 24, pp309–382

Rendell, L., and Whitehead, H. (2003) 'Vocal clans in sperm whales (*Physeter macrocephalus*)', *Proceedings of the Royal Society: Biological Sciences*, vol 270, pp225–231

Richards, D. G., Wolz, J. P. and Herman, L. M. (1984) 'Vocal mimicry of computer-generated sounds and vocal labelling of objects by a bottlenose dolphin, *Tursiops truncatus*', *Journal of Comparative Psychology*, vol 98, no 1, pp10–28

Ridgway, S. H. (1990) 'The central nervous system of the bottlenose dolphin', in S. Leatherwood and R. Reeves (eds) *The Bottlenose Dolphin*, The Academic Press, San Diego, pp69–97

Rose, N. A. (2000a) 'A death in the family' in M. Bekoff (ed) *The Smile of a Dolphin: Remarkable Accounts of Animal Emotions*, Discovery Books, New York, pp32–33

Rose, N. A. (2000b) 'Giving a little latitude', in M. Bekoff (ed) *The Smile of a Dolphin: Remarkable Accounts of Animal Emotions*, Discovery Books, New York, pp144–145

Schusterman, R. (2000) 'Pitching a fit' in M. Bekoff (eds) *The Smile of a Dolphin: Remarkable Accounts of Animal Emotions*, Discovery Books, New York, pp106–107

Simmonds, M. P. (2006) 'Into the brains of whales', *Applied Animal Behaviour Science*, vol 100 pp103–116

Smolker, R. (2001) *To Touch a Wild Dolphin*, Doubleday, New York

Tuana, N. (1993) *The Less Noble Sex: Scientific, Religious and Philosophical Conceptions of Woman's Nature*, Bloomington, Indiana University Press

Valsecchi, E., Hale, P., Corkeron, P. and Amos, W. (2002) 'Social structure in migrating humpback whales (*Megaptera novaeangliae*)', *Molecular Ecology*, vol 11, pp507–518

WDCS (Whale and Dolphin Conservation Society) (2008) 'Shrouded by the sea… The animal welfare implications of cetacean bycatch in fisheries – a summary document', WDCS, Chippenham, UK

Wells, R. S., Boness, D. J. and Rathburn, G. B. (1999) 'Behavior' in E. Reynolds III and S. A. Rommel (eds) *Behavior of Marine Mammals*, Smithsonian Institution Press, Washington, pp324–422

White, T. I. (2007) *In Defense of Dolphins: The New Moral Frontier*, Blackwell, Oxford

Whitehead, H., Rendell, L., Osborne, R. W. and Wursig, B. (2004) 'Culture and conservation of non-humans with reference to whales and dolphins: Review and new directions', *Biological Conservation*, vol 120, no 3, pp427–437

20

Thinking Whales and Dolphins

Philippa Brakes and Mark Peter Simmonds

O what an astonishing mass of animated substance!
(The Reverend Alexander Fletcher describing 'the whale' in *Natural History Scripture*, published circa 1870)

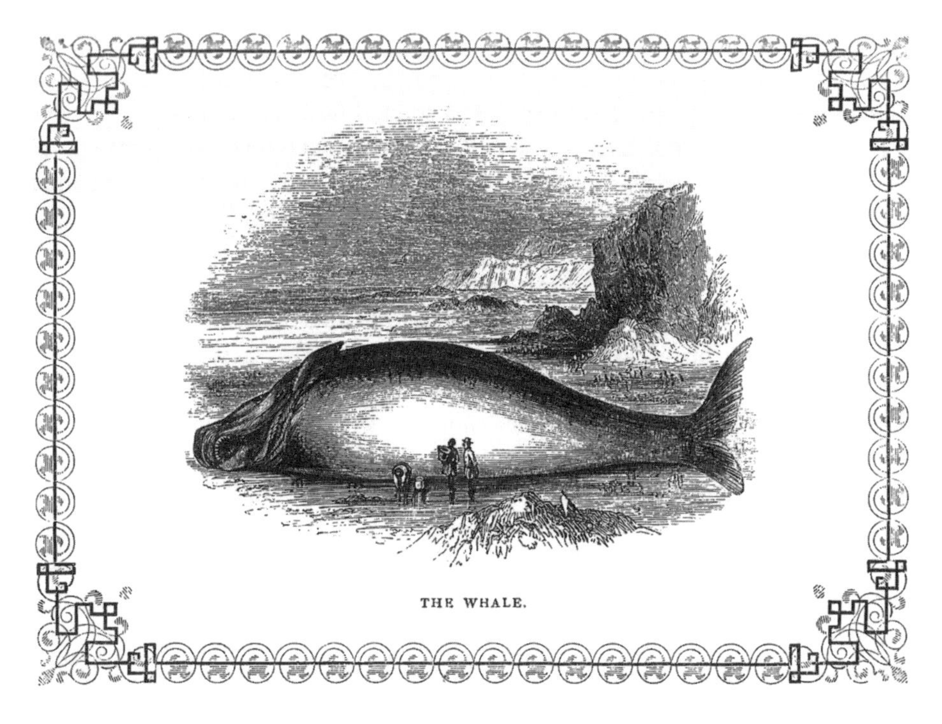

Figure 20.1 *'The whale' as portrayed in* Natural History Scripture

Source: Fletcher (1870)

Perspectives Over Time

The relationships between humans and cetaceans are typically both complex and long-standing, with the lives of many coastal peoples and cetaceans in many different and widely dispersed lands bound together in various ways over millennia. The nature of these relationships may vary with the locality and species concerned and, as several chapters in this volume attest, may also depend upon the point in history you choose to view them. However, whether you have an ancestral spiritual connection with whales, see cetaceans as food, use dolphins to guide you in your fishing activities, view such mammals as integral components of their ecosystems, or simply regard them with awe and wonder, cetaceans clearly have an importance to many of us. In many parts of the world cetaceans invoke strong feelings as well as disparate views, as illustrated in this volume by the relevant chapters. There is of course much more that could be said on these issues but here, in this closing chapter, we highlight some of the key points made by the authors that have contributed to this volume and present our own view.

We (the editors of this book) live in Western society and this will have helped to shape our views. Many of our ancestors would have subscribed to the Judaeo-Christian tradition of 'dominion' over other living things, including whales. The tradition of dominion places a superior humankind in control of all nature. Yet, paradoxically, just as we seem in some ways to have achieved 'dominion' over much of nature, taming and converting it to our needs with our machines and technologies, our activities have also driven some things beyond our control. For example, some fish populations devastated and decimated by overfishing may never be able to recover. In addition, our climate is now changing in an unprecedented manner, with no indication that the cause of these changes will be remedied in a timely fashion.

We are now at a point in our history as a species where we have changed much of the world around us (as illustrated by Simmonds and Brakes, this volume) and humans are, arguably, uniquely the single most important factor in shaping the future of all Earth's inhabitants. There have, of course, been other times in the planet's history when conditions have suddenly shifted. For example, there was a mass extinction of most of the dinosaurs at the end of the Cretaceous period 65 million years ago. The compelling difference between other such catastrophic historic events and our current situation is that now the force for change is humans, and we are entirely aware of the likely outcome of our actions and have the ability to act to change the consequences.

Bearing in mind our accumulated knowledge about whales and dolphins, including the many threats affecting them, we are compelled to challenge the notion that cetaceans are simply there for us to 'use'. Our improving knowledge and insights into these remarkable creatures, should now – we urge – hail a new era in our relationship with them. At the end of his chapter, Stuart Harrop notes that in relation to the many pressures on these animals and their habitats, 'the impeccable choice may well be to leave the whales free to live out their last days'. We also hope that we can do better than this for at least some cetaceans.

With knowledge comes responsibility. We now know that cetaceans are intelligent, social animals that communicate with each other in ways that we are still only just beginning to understand (as, for example, described in the chapter by Paul Spong). Hal Whitehead reminds us in his chapter that some have even been shown to exhibit cultural transmission of knowledge. (For others, less is known of their biology, including their social interactions, but this absence of evidence cannot be taken as evidence of absence of such attributes.) The new knowledge of societies and culture comes at a time when there is also growing acceptance within the scientific community that some species, including the cetaceans, have rich and complex emotional lives (Simmonds, 2006; Bekoff, 2007).

Prior to our expansion into their watery habitats, cetaceans could be regarded as a remarkably well-established and successful group of mammals. Toothed and baleen whales have been in existence for some 35 million years (Roman, 2006). The apes, which include our own species, by contrast are relative newcomers. They separated from other primates a mere 26–25 million years ago, with the divergence that led to chimpanzees, bonobos and ourselves becoming distinct only around 5 million years ago (or perhaps even more recently) (Sorenson, 2009). In fact, 90 million years of separate evolution stand between the primates and the cetaceans, demonstrating that high intelligence and complex societies evolved more than once on Earth and that several species of highly encephalized animals can coexist (Marino et al, 2004). The dramatic changes that we are causing through accelerated industrialization of the oceans, and overexploitation of their resources, have not been encountered over the millions of years of existence of the cetaceans prior to our arrival. As time passes there are increasingly fewer safe refuges for these species; places where they can go about their business unharmed, unharassed and unencumbered by the pollution and detritus of human existence. This situation is exacerbated by the fact that many species travel extensively, some undertaking regular migrations from feeding to breeding grounds. During such journeys they may encounter various threats.

Nevertheless, the 'direct' threats to cetaceans (specifically the various forms of hunting) and the so-called 'indirect' threats (such as habitat degradation and fisheries bycatch) continue unabated in many parts of the world. As the philosopher Thomas White observes in his chapter, 'virtually all of these practices are sanctioned by the laws of one country or another and, in theory, supported by the best available science. That is, we are not talking about the blind, unthinking predation of one species by another. The way that humans treat dolphins is supposedly the result of intelligent dialogue and formal processes that aspire to fairness, compassion and objectivity.'

A View on the 'Whaling Wars'

What constitutes 'best available science' can often be debated and this can be considered in the case of commercial whaling that, inevitably, is one of the reoccurring themes in this book. In his chapter, Richard Cowan gives us a precise and important analysis of the current situation at the relevant international body, the IWC. During the preparation of this book, in 2009 and 2010, the IWC was involved in complex discussions on one of its own reoccurring themes: an attempt to formulate a deal between those member nations that favour commercial whaling's resumption and those that do not. At the 2010 meeting the attempt finally failed. Simmonds and Fisher (2010) and Holt (2010) provide helpful critiques of the deal, although its details were heavily obscured by secrecy. Nonetheless, in a few short weeks leading up to the key decision-making meeting of the Commission in June 2010, as the word finally broke that the global moratorium on commercial whaling agreed in 1982 might actually be lifted, over 200 marine scientists lent their names to a letter calling for the moratorium to be maintained. Many influential and distinguished marine scientists signed this document, including Sylvia Earle, David Suzuki, Sidney Holt, Hal Whitehead, Roger Payne and many others from many different nations, thereby sending an important message to the member nations of the IWC (see Box 20.1). Most significantly, this demonstrates (counter to many claims to the contrary), that science is not simply on the side of those that insist that commercial whaling is sustainable and should be reinvigorated.

Looking to the Future

Whatever your view of cetaceans, it is difficult not to concede that, in the face of the mounting evidence presented here, a fundamental shift is required in our relationship with these animals. What we seek is not a gentle nudge in the direction of improved protection, or small incremental steps towards better policies to protect these animals; time is not on the cetaceans' side. Instead, what we need is a monumental and collective upheaval of the paradigm under which we regard these animals; from commodity to cohabiter.

Similarly, lofty goals have been espoused *and achieved* in the past; the moratorium on commercial whaling is one such example. The 'Save the Whales' movement of the 1960s and 1970s was motivated by emerging scientific facts, such as the revelation that humpback whales could sing, and further by a simple, growing sense that killing whales and dolphins was wrong, based largely on its cruelty and lack of sustainability. Nevertheless, today, some 40 years later, we have the substantiation of decades of cetacean research and some powerful philosophical arguments (see for example, D'Amato and Chopra, 1991; White, 2007) to further support our entreaty. What further evidence do human societies need before we are sufficiently compelled to take the necessary, radical action required to protect cetaceans effectively?

Box 20.1 Marine scientists petition to the IWC delivered June 2010

We the undersigned marine scientists respectfully call on the member nations of the International Whaling Commission (IWC) not to undermine the conservation achievements of the last few decades by again endorsing commercial whaling at their next meeting.

We are aware that at its 62nd meeting in Agadir, Morocco, June 21st–25th, the IWC will consider a proposal to grant catch limits to the three member nations of the IWC – Japan, Norway and Iceland – that continue to take whales for commercial gain, using well known loopholes in the International Convention for the Regulation of Whaling. The proposal will even permit whaling in a Marine Protected Area ('sanctuary' in the terminology of the IWC) created specifically to protect whales in large parts of their ranges. We believe that to do so would be highly inappropriate and untimely and would again risk the future of the whales. Whilst aware that some whale populations are showing signs of increase in the absence of whaling pressure, partly as a successful result of the global 'moratorium' on commercial whaling adopted in 1982, and partly from application of the management procedures agreed in 1975, such increases are not a sufficient rationale to justify the IWC endorsing commercial catches. There is no evidence that any of the few populations and species known to be increasing have reached, or are anywhere near, the levels that might justify non-zero catch limits under the IWC's existing management and conservation policies and procedures. Furthermore, whales inhabit marine ecosystems that are now increasingly impacted by human activities ranging from oil spills to the effects of persistent pollutants, climate change and increased ship traffic and other hazards; these provide further rationale for providing these remarkable animals of the global commons with the highest possible levels of protection, including protecting them from commercial takes.

The lessons of the past show that commercial whaling has always been intractable to sustainable management, and we see no changes in the attitudes of the industry which continues to favour extracting monetary value from the whales as fast as possible and, in the process, evading and obstructing efforts to ensure full compliance with international regulations and transparent supervision. The long-lived and slow-breeding whales are also difficult and expensive to monitor adequately. We are also growing increasingly aware of the complexity of their population structures, behaviour and societies. Given the risks involved and that commercial whaling meets no essential human need, we call on all the IWC governments to abandon experiments in the lethal use of whales and instead refocus their efforts on the conservation of whale populations, on understanding their roles in the marine ecosystems of which they are important parts, and promoting, where appropriate, responsible non-lethal uses of them such as whale-watching.

We can take heart that a movement towards better protection of cetaceans exists and is exemplified by the 'Declaration of Rights for Cetaceans: Whales and Dolphins' agreed at a special meeting of experts in Helsinki during 2010 (see Box 20.2).

Box 20.2 Declaration of Rights for Cetaceans: Whales and Dolphins

Based on the principle of the equal treatment of all persons;
Recognizing that scientific research gives us deeper insights into the complexities of cetacean minds, societies and cultures;
Noting that the progressive development of international law manifests an entitlement to life by cetaceans;
We affirm that all cetaceans as persons have the right to life, liberty and wellbeing.
We conclude that:

1. Every individual cetacean has the right to life.
2. No cetacean should be held in captivity or servitude; be subject to cruel treatment; or be removed from their natural environment.
3. All cetaceans have the right to freedom of movement and residence within their natural environment.
4. No cetacean is the property of any State, corporation, human group or individual.
5. Cetaceans have the right to the protection of their natural environment.
6. Cetaceans have the right not to be subject to the disruption of their cultures.
7. The rights, freedoms and norms set forth in this Declaration should be protected under international and domestic law.
8. Cetaceans are entitled to an international order in which these rights, freedoms and norms can be fully realized.
9. No State, corporation, human group or individual should engage in any activity that undermines these rights, freedoms and norms.
10. Nothing in this Declaration shall prevent a State from enacting stricter provisions for the protection of cetacean rights.

Agreed, 22nd May 2010, Helsinki, Finland

We believe that the groundswell of more positive attitudes towards cetaceans extends around the world. For example, Jun Morikawa reminds us in his conversation with Erich Hoyt in this volume that in 2005 whale meat consumption in Japan was so negligible that it was less than 0.2 per cent of the overall meat consumption of the nation and the price of dolphin meat in Japan dropped from ¥600/kg in 1994, to ¥150/kg in the spring of 2010. Siri Martinsen also notes in her chapter that a 2010 opinion poll showed that 53 per cent of Norwegians would like to go whale watching.

But so much more must be done to make aspirations for the better treatment of cetaceans a reality; even the so-called whale conservation-minded governments turn a conspicuously blind eye to issues such as the incidental capture of cetaceans in fishing gear. Erich Hoyt notes in this volume 'we know too much' not to do better. If today's politicians were projected several hundred years into the future to be put on trial for the maltreatment and loss of some of these unique individuals and species, how would they answer our descendants' accusations of 'How did you let it happen?' and 'Why didn't you do more?'?

But, realistically, what can be done? It is worth bearing in mind that all customs are initiated at some point in history and many evolve over time. In our view it is now time to initiate a new world vision for cetaceans, to call for a grand ascent of the standards for our interactions with, and protection of, cetaceans. Scrutiny of the current 'management' paradigm for cetaceans reveals a somewhat anachronistic approach that does little to take into consideration modern scientific understanding, such as the important role some individuals have within their societies (see Brakes and Bass in this volume) or the existence of discrete cultural units within some populations (as explained by Whitehead) that could be vulnerable and therefore worthy of specific protection. As the Japanese cetacean biologist Toshio Kasuya notes, the current exploitation of cetacean species within Japan 'is only possible by ignoring the individuality of group members and functions of living in a group' (Kasuya, 2008). The same can be said for many other issues that we currently manage with a 'blind eye'.

But why should we bother to make a special effort for the cetaceans? One answer might be that imagining a future without cetaceans in our oceans, seas and rivers is a little like imagining the world in monochrome; another may be that motivating the public about whales and dolphins also appeals at a deep level to wider interest in nature, thus collectively moving us towards greater understand of the importance of protecting ecosystems as well as individual species.

In his chapter, Thomas White asks us to consider the similarities between dolphins and ourselves and concludes that 'The catalogue of similarities between dolphins and humans is striking, and, in my view, it is enough to grant what philosophers call "moral standing" to dolphins … Therefore, they should "count" in a moral calculation on similar grounds that individual humans count.' In our view, it is likely that the same may indeed be true for all whales.

Whales and dolphins remain now, as ever, iconic species – intriguing, enchanting, motivating and increasingly vulnerable. The same things that make them 'special' – their intelligence, their communication, their need for long periods of maternal care, specific habitat requirements and complex societies – can also make them especially vulnerable. As this book helps to detail, humans have many complex and differing relationships with these fellow beings, but one thing is certain: we can do better, and cetaceans *deserve* better.

References

Bekoff, M. (2007) *The Emotional Lives of Animals*, New World Library, Navato, California

D'Amato, A. and Chopra, S. K. (1991) 'Whales: Their emerging right to life', *American Journal of International Law*, vol 85, p21

Fletcher, A. (1870) *Natural History Scripture*, George Virtue, London

Holt, S. (2010) 'Lies and nonsense about whaling', blog posted 24 September, www.mywhaleweb.com/?p=6163

Kasuya, T. (2008) 'The Kenneth S. Norris lifetime achievement award lecture: Presented on 29 November 2007 Cape Town, South Africa', *Marine Mammal Science*, vol 24, pp749–773

Marino, L., McShea, D. W. and Uhen, M. D. (2004) 'Origin and evolution of large brains in toothed whales', *Anatom. Rec.*, Part A, vol 81A, pp1–9

Roman, J. (2006) *Whale*, Reaktion Books, London

Simmonds, M. P. (2006) 'Into the brains of whales', *Applied Animal Behaviour Science*, vol 100, pp103–116

Simmonds, M. P. and Fisher, S. (2010) 'Oh no, not again', *New Scientist*, 10 April, pp22–23

Sorenson, J (2009) *Ape*, Reaktion Books, London

White, T. I. (2007) *In Defense of Dolphins: The New Moral Frontier*, Blackwell, Oxford

Index

aboriginal subsistence (whaling) 9–21, 37, 40–44, 48, 93
acidification, oceanic 2, 4, 173
acoustics 42, 110, 129–134, 142, 171
aggression 110, 145, 193, 195
Agreement on International Humane Trapping Standards 13
Ainu people 93
altruism xii, 136, 191, 195
ancestral connections 10, 27, 208
animal rights 11, 19, 49, 69, 73, 83, 85, 211–212
Antarctica 25, 32–33, 60–61, 90
anthropocentric view xii, 84, 107, 136, 189
apes 5, 117, 124–126, 151–157, 170, 183, 190, 209
aqua-culture 174
archaeocetes 116–118
Arctic 60, 173
Argentina 32–34, 61, 99
Aristotle 131
Arnold, Peter 140
Aronson, Lester 200–201
ASCOBANS 70
Atlantic 32, 59, 63, 110, 160
Australia
  conservation and whaling policy 37, 68, 92
  research 44, 140–145, 152, 154, 160, 181
  whale watching in 90, 96
Azores 50

baiji (Yangtze River dolphin) 10, 11
Baird's beaked whale 92
Baja, California 29, 32, 33, 137
Balaena mysticetus 40, 152
Balaenoptera musculus 33, 152
Balaenoptera physalus 33, 152
baleen 58, 118, 119
Barnum, P. T. 39
Barret-Hamilton, Major G. E. H. 61
Basque whaling 30, 58–59
BBC (British Broadcasting Corporation) 56, 65, 68
BDMLR 56
beach-rubbing 156
beaked whale 29, 92
beluga 39, 137
binocularity 143–144, 169
blue whale xi, 18, 33, 60, 79, 152
Borne Free Foundation 64
boto 10, 34
bottlenose dolphin
  brain and cognition 117, 122–125, 152–157, 181
  captivity 64, 152
  protection 65, 70, 172, 181
  research 65, 109, 152–157, 160
  solitary sociable 65–67
bowhead whale 15, 40, 60, 182
brain size 107, 115–120, 125, 198
brain structure 115–128
Broca, Paul 196
Bryde's whale 98
bubbles 153, 190 193
Buckland, Francis 62–63
bycatch 43, 71, 173–175, 184, 202, 209

Caine, Michael 63
Canada 65, 92, 99
captive research 109, 132, 151, 183
captivity 3, 135, 179, 183, 184, 202, 212
  Free Willy 39
  public opinion 64–65, 69, 184
Cardigan Bay 65
CARIBwhale 50
Chile 30, 33, 61
Chilean dolphin 34
chimpanzees xi–xii, 73, 150, 151, 157, 158, 161, 209
Chukotka 41
clan or clans (human and whale) 23, 112, 156, 160
climate change 1, 4, 20, 44, 100, 172–176, 211.
codas 112, 131, 154
cognition / cognitive abilities xii, 3, 68, 107–126, 136–137, 185, 188–203
cohabiter, cetaceans as 210
Commerson's dolphin 34
commodity, cetaceans as 210
communication, in cetaceans 83, 121, 129–134, 170, 213
Connor, Richard 191
Conservation Committee 43
Convention on the Conservation of Migratory Species of Wild Animals 13, 70
Cook Islands 24
cortex 107, 117, 121–126, 189, 190, 193, 200
The Cove 39, 94
cull (marine mammal) 160, 180,
cultural group 5, 159, 161, 181
cultural transmission (in cetaceans) 4, 181–182, 190, 203, 209

Dall's porpoise 94
Dave (the dolphin) 66–68, 136
de Waal, Frans 135
Declaration of Rights for Cetaceans: Whales and Dolphins 211–212
Delphinoidea 116, 120
dialect 112, 129, 131, 150, 154, 156, 158
dolphinaria / dolphin shows 39, 56, 62–65
Dominica 50, 54
Dominican Republic 50, 54
dominion 208
drive hunts 24, 58, 93, 113, 125, 202
Dunstan, Andy 140

Earl, Sylvia 210
Earth Day 37
eavesdropping in cetaceans 194
echolocation 3, 107–108, 118–124, 129, 158, 182, 188, 194–195
ecosystems 1–2, 12, 21, 44, 150, 175, 176, 208, 211, 213
El Niño 156, 160
elephants 119, 124, 131, 135–137, 183
emotions xi–xii, 121, 125–126, 137, 188–193, 201, 203, 209
empathy 135, 137
emulation 150, 152
encephalization quotient (EQ) 116–117, 190
Endangered Species Act 38
entanglement 14, 43–44, 67, 144
Eschrichtius robustus 29, 41
ethics
  conservation ethic/new ethic 51, 113, 125–126, 179–185
  values 2, 9, 11, 20, 39, 49, 68, 188–189, 202–203
ethology 151
Eubalaena australis 30, 176
Eubalaena glacialis 160
evolution xi, 10, 13, 37, 107, 115–126.
Faroe Islands 20, 58
Fiji 24–26
filter-feeding 119
fin whale 33, 60, 73, 152, 171, 203
fisheries 4, 40, 43, 49, 80, 82, 92, 160, 173–174, 180, 183
fission–fusion 109, 158
Flipper (television series) 39
foraging 109–110, 118, 133, 150, 153–154, 158–161, 181
Foyn, Svend 78, 84

Franciscana 34
Free Willy 39
French Polynesia 24
freshwater dolphins 120
fur farming 77

Galapagos Islands 30–31
Goodall, Jane xi, 73, 135
Gordon, Jonathon 131
Gould, Steven Jay 196
gray whale 29–31, 33, 41, 59, 137
Gray, John Edward 61
great apes 5, 117, 124, 126, 183, 190
Great Barrier Reef 140
Greenland 15, 19–20
greeting ceremony 156, 159
Grenada 49–50
grenade harpoon 60
grief 136, 191, 203
Guadeloupe 50
Guam 25
Gulf of Mexico 169
gyrification index 122

habituation 65, 145
harassment 38, 144, 179, 184
harbour porpoise 63, 70–71, 171–172
head butting 154
Health Studies 94
Hector's dolphin 110–112
hedgehogs 18, 200–201
Herman, Lou 191
Herrnstein, Richard 198
heterodonty 118
Heyerdahl, Thor 79–80
Holt, Sidney 210
homodonty 118
horses 78, 150
human–cetacean fishing cooperative 155, 160
humpback whale
  research 38, 110, 131–132, 152–158, 171, 203, 210
  whale watching 24–26, 47–54, 72, 96–97
  whaling and conservation 25, 33

hydrophone 132–133, 142

Iceland 19–20, 39, 203, 211
iconic status of whales 1, 3, 38, 47, 51, 131, 177, 213
imitation 109, 150–152, 154
Indigenous Peoples (rights of) 13, 44
individual, role of/individuality 109–110, 141–143, 170–176, 188–195, 212–213
individual, value of xv, 33, 181
Inia geoffrensis 10, 34
innovation, in behaviour 107, 109–110, 153
intelligence 4, 107–108, 115–125, 136–137, 189–193
International Convention for the Regulation of Whaling (ICRW) 16, 33, 101–102, 211
International Labour Organization (ILO) Convention 13–14
International Whaling Commission 9, 13–18, 30–34, 39–45, 50–51, 61, 69, 76, 82, 91–94, 100–103, 141, 171, 179–180, 184, 203, 210–211
Inuit 11–12, 93
Isle of Harris 59
IUCN 180
IWC Scientific Committee 100, 102, 141, 171

Japan, whaling 16–18, 30, 39–43, 49–51, 89–99, 102, 161, 171, 202–203, 211, 213
Japanese Fishery Agency 90–93
Japanese Institute of Cetacean Research (ICR) 92
Jensen, Arthur 198

Kant, Immanuel 192
Kasuya, Toshio 94, 112, 159, 161, 180, 213
Keiko 36
killer whale *see* orca
Kinsman, Catherine 137
Klinowska, Margaret 199–200

Kshamenk 29
Kyodo Senpaku Co. 92

Lilly, John 131, 200
limbic system 137, 194, 118, 121–122, 137, 194
Lipotes vexillifer 10
lobtail 109–110, 153, 158
London Natural History Museum 61
Luna 65, 68

Madeira 69
Makah 16, 41–42
Maori 23
Marine Animal Rescue Coalition (MARC) 57, 66
Marine Mammal Protection Act (MMPA) 38, 41–44, 63
Martinique 50
maternal care 170, 213
matriarchal 109
matrilineal 109, 156, 158, 181
McGonagall, William 71–72
McGrew, William 135
McKenna, Virginia 64
McVay, Scott 131
Megaptera novaeangliae 33, 38, 72, 152
Mercury 94
Mexico 29, 31–33, 96, 99, 137, 169
migration 4, 12, 52, 61, 118, 122, 144, 153, 209
military sonar 42–43, 171
minke whale 18, 25, 43, 76–77, 81–83, 86, 140–145, 152, 161, 167
  common or Antarctic species 18, 43, 76–77, 81, 83, 86, 152, 167
  dwarf subspecies 140–148
mirror self-recognition 125, 137, 190
Moby Dick xi, 31, 37
moral community 149, 161
moral standing 188–189, 191, 203, 213
moral system 158
moratorium on commercial whaling 17, 30, 33, 39, 42–43, 45, 61, 69, 76, 93, 100, 103, 203, 210–211
Moray Firth, Scotland 65
Moreton Bay, Australia 160
mortality 43, 71, 135, 172, 174
Morton, Samuel George 196
MPA / Marine Protected Area 31–32, 89, 211
Murray, Charles 198
mysticete (baleen whale) 116–124
myth 3, 9–12, 16, 19–20, 27, 83, 86

nasofacial muscle complex 107, 118
neoceti 116, 118, 120
neocortex 122–125, 190, 200
neuroanatomical trajectory 107, 115
New Caledonia 24
New Zealand 24, 37, 68, 90, 92, 96, 154
Nihon Kyodo Hogei 90, 92
noise xiii, 43–44, 100, 170–175, 182, 184
nonhuman cultural revolution 152
North Atlantic right whale 160, 174
northern bottlenose whale 56–57, 121, 155
Norway 3, 19–20, 43, 45, 61,73, 76–86, 203, 211

ocean acidification 2, 4, 173
Odontocete 107, 116–124, 132, 155, 158
oil spill 169, 175, 211
olfaction 121, 182
omega oils 4
orang-utans (*Pongo pygmaeus*) 161
orca
  brain 203
  captivity 64–65
  human cultures and 29, 39, 65
  research 110, 112, 122, 136, 152, 181
OrcaLab 130, 133
Orcinus orca 29, 65, 122, 152

Pakicetus 116
paralimbic cortex 107, 122
paralimbic lobe 122, 126, 137
Payne, Katy 131
Payne, Roger 131, 210
Peale's dolphin 34

Pelly Amendment 40, 43
persons, personhood 189–192, 202–203, 212
Peru 10, 29, 30, 33
phonation 132
photo-ID 141
Physeter macrocephalus 30, 115, 121, 154
pilot whale 58, 92, 110, 155
pingers 175
pollution, chemical xiii, 3, 44, 94, 100, 172–173, 184, 209
  see also noise
polychlorinated biphenyls (PCB) 172
precautionary principle 38
prey depletion 4, 174
primate 2, 4, 73, 107, 119, 122, 124–125, 135, 143, 151, 157–159, 169–170, 193, 199–200, 209
protectionists/protectionism 11, 16–19, 49

Queen Victoria 17

Reiss, Diana 190, 192–193
renewable energy71, 174–175
reproduction 112, 170
Revised Management Procedure (RMP) 101–102
rights
  animal 11, 19, 49, 69, 73, 83, 85, 211–212
  Indigenous Peoples 13, 44
Risso's dolphin 24
rough-toothed dolphin 24
Rumney, John 140
Russia 41, 61

Salvesen, Christian 60
Samoa 23
Santa Catarina 29–30
sapience xiii, 5
scientific whaling 25, 43–45, 103
Scotland 58, 70–71
sealing 84, 86
Sedna 11–12, 20
sei whale 32
seismic 42, 70, 171
self-awareness 4, 107–108, 125, 137, 161, 185, 188, 189–190, 198, 202–203
sentience xiii, 2, 5, 73, 99, 131, 185,
sequence analysis 110
sexual behaviour 110, 142
Shark Bay, Australia 154, 157, 181
Shetland, Scotland 58–61
signature whistle 132
Smolker, Rachel 191
social cognition 124, 194–195, 202–203
social learning 107–109, 150–161
social unit 2, 112, 155–156, 158, 170, 181
solitary sociable dolphins 65–66, 179
Solomon Islands 24
song xi, 38, 99, 110, 131–132, 150–152, 158,
South Georgia 31–33, 60–61
southern right whale 30, 34–35, 99, 174, 176,
Special Areas of Conservation (SAC) 70
species bias 189, 199
sperm whale
  brain 115, 119, 121, 203
  human cultures and 13, 24,
  hunting 30–31
  research 73, 108, 110, 112, 131, 136, 154–156, 158, 160–161
spindle cells or spindle neurons 4, 124, 137, 203
spinner dolphin 25–26, 195
sponge feeding (or 'sponging') 109, 154, 157, 181, 190
spyhop 105, 144,
St Kilda 59
St Kitts and Nevis 49, 50
St Lucia 49, 50
St Vincent and the Grenadines 15, 49, 50
Stellwagen Bank National Marine Sanctuary 54
stock (whale) xiii, 33, 34, 39, 42–43, 61, 70, 102, 179
stranding 13, 24, 29, 57–58, 95, 152, 159, 171–172, 183, 195

subsistence 9, 12–21, 40–42, 44, 48–49, 54, 203,
super whale (concept) 19–20, 81, 83
Suzuki, David 210
swimming with dolphins 65
swimming with whales 140–145

Tay whale 71–73
teaching xii, 52, 150, 152
Thames whale 56–57, 71–72
Tobach, Ethel 200–201
Tonga 24, 110
tool use / tools 45, 58, 69, 99, 129, 133, 142, 153, 157, 181, 190, 192–193, 203
tourism 24–25, 49–50, 65, 97, 112, 144
tradition xii, 3, 9–21, 24–27, 41, 49, 52, 78,80,91, 93–95, 203, 208
Tucuxi 34
Turks and Caicos 64
Tursiops truncatus 64, 117, 152, 172
Tuvalu 23–27
Tyack, Peter 132, 133

United Kingdom (UK) 3, 7, 17, 30, 37, 56–74, 172
United Nations 2, 39
United Nations Charter for the Rights of Indigenous Peoples 13
United States of America (USA) 16, 33, 37–45
Universal Declaration on Animal Welfare (UDAW) 2
Uruguay 30

vaquita 34
vessel strikes (ship or boat strikes) 67, 100, 144, 173–174
Von Economo neurons 124

Wales 58, 70
Weilgart, Lindy 131, 134
whale watching 24, 32–38, 49–50, 52, 69, 83, 86, 89–99, 140–141, 211, 212
whalebone 29, 60
whaling
  aboriginal 9, 13–16, 18, 37, 40–42, 44, 48, 93
  commercial 14–19, 30, 32, 39, 41–45, 50, 56,61, 69,76, 81, 86, 90–94, 100–103, 181, 203
  scientific 25, 43–45, 103
Whitehead, Hal 4, 73, 110, 112, 131
wind farms 175
World War I 59, 93
World War II 16, 58, 60, 89, 91

xenobiotics 171–172

Yagan 12, 13
Yangtze River dolphin (baiji) 10, 11
yield 38, 179

CW00810492

Collided

Editing by One Love Editing

Proofreading by Rumi

Cover photography by Michelle Lancaster

Becca Steele

www.authorbeccasteele.com

AUTHOR'S NOTE

The author is British, and this story contains British English spellings and phrases.

Enjoy!

For those who march to the beat of their own drum. Never try to fit in. You're perfectly imperfect, just the way you are.

True perfection seems imperfect, yet it is perfectly itself.

LAO TZU

ONE

COLE

"I'm getting married."

I stared at my mum, my jaw dropping. "What the fuck?" I whispered. Since when had she been in a relationship?

Her mouth thinned, her way of letting me know she didn't approve of my language, but I was beyond caring. She was getting *married*? To whom?

"Yes. I'm marrying David."

"David Granger? Your *boss*?" My voice was rising as I clenched and unclenched my fists. "Isn't he married?"

"Not anymore." She smiled then, and I'd never seen her look so happy. "I know this must come as a shock, but he's a wonderful man, and I know you'll grow to love him. We'll be moving into his house next week."

My gaze whipped around our flat. It was small, yeah...we were in London, after all, even if it was a shit part of the city. But it was home, and it had been since my parents had split when I was fifteen. My dad, Matthew Clarke, had since remarried, and he lived up in Aberdeen with his wife. Suffice it to say we didn't speak often. A monthly allowance was deposited into my account, and we exchanged the occasional text about the football scores—the one thing we had in common was that we both supported Arsenal—but that was more or less the extent of our interaction.

But that was irrelevant right now. What was relevant was the fact that out of completely fucking nowhere, my mum had gone and dropped this bombshell, leaving me reeling. I didn't want to leave my home. Technically, I could move out now since I was already eighteen, but realistically, that wasn't going to happen. I was a school student for another two months, and then there were the London prices to consider. No one was going to want to hire me when I was barely scraping through my A levels, and I already had a part-time volunteer job at a charity lined up for the summer, which would be taking up a chunk of my time.

"I'd like you to pack up all your things this weekend. David was kind enough to volunteer his son's help to get everything unpacked at the other end. Did you know he has a son just a little older than you? Huxley, his name is. Bit of a wild card, by David's accounts."

"No. I didn't know he had a son," I said sourly. I barely knew anything about the man other than the fact that he was the manager of a small architectural firm in London, supposedly married, and my mum was his personal assistant. She'd mentioned him on occasion, but I knew nothing whatsoever about him other than that, let alone that they were dating.

"You'll like Huxley," she said and then picked up her

phone, placing it to her ear. I guessed the conversation was over.

Spoiler alert: I didn't like him. But as much as I had an instant dislike for him, he *loathed* me on sight.

I'd just finished dumping the final box of my shit into my new bedroom—the smallest of the four in the Granger household, but still bigger than my bedroom in our flat—when the door flew open with a crash, rebounding against the wall. I didn't even have a chance to take a breath before I was being shoved back against the wall, an arm across my throat as a body roughly the same height as mine but a little more wiry held me immobile.

"Don't make the mistake of thinking you belong here," a voice snarled in my ear. I blinked, recalibrating, and then my attacker's face came into focus. Bleached blond hair, smudged eyeliner and dark brows, a stubbled jaw, pierced ears and septum...

Huxley.

"You really are a walking, talking stereotype," I laughed softly, which made him bare his teeth.

"Explain." His sapphire-blue eyes connected with mine, rage sparking in them.

"The emo bad boy. Let me guess, you sit in your room and play your guitar while you cry about how unfair your privileged life is. Oh, yeah, and you've got a stash of weed hidden from your dad in your bedside drawer."

He gaped at me but quickly recovered. "Fuck you, loser. You don't belong here, and as soon as school finishes, you'd better get the fuck out of my house, otherwise I'll make you regret every single one of your life choices."

I swallowed against the tattooed arm that was pressing into my throat. Time to teach this wanker a lesson. Shoving off the wall, I sent us both crashing across the bedroom. Unfortunately, in the short time I'd been pinned, I'd managed to forget all about the piles of boxes currently scattered across the floor. Catching his foot on one, he fell backwards, taking me down with him, both of us smacking our heads on yet another box on the way down.

Our noses cracked together, and then we were scrambling away from each other and onto our feet, both of us breathing heavily. I bared my teeth at him. "Let's get one thing straight, emo boy. My mum is going to marry your dad. I like it just about as much as you do, but they're adults, and it's not our decision to make. Therefore, that means I have just as much right to be here as you do. You can start by getting the fuck out of my room."

With those words, I pushed him backwards with all my strength, sending him staggering back into the doorway. One more push, and I slammed the door in his face, giving him the middle finger as I did so. When I was finally alone, I examined my nose in the mirror. It didn't appear to be broken, but I decided to head down to the kitchen for some ice, just in case.

In the large, spotless kitchen, I found my mum pouring a glass of wine, humming to herself as something simmered on one of the shiny stainless-steel hobs on the huge range cooker. She glanced up as I entered the room, the smile falling from her face as she took me in.

"Cole! What's the matter?"

"Uh..." Stalling, I went to the freezer, rummaging around inside until I found a bag of frozen peas. That should help with any possible swelling. When I was seated at the kitchen island with the peas pressed to my nose, gradually freezing my skin, I replied as carefully as I could. "Would there be any reason why David's son would hate me on sight? Other than the fact that he's a dic—uh, other than his personality?"

She sighed. "Cole...there's something you should know. David and I...he was still married when we began our relationship. That is to say, as far as Huxley was concerned, the marriage was still real and valid."

What? "Mum! Why would you sleep with a married man?"

"I know. Believe me, that's the part of this I regret the most. It's not that easy. We tried so hard to stay away from one another, but eventually, our...attraction to one another became too great to ignore."

"He still should've left her before he did anything with you," I muttered. "And it still doesn't mean that Huxley should hate me. I didn't even know you were seeing the guy until three days ago."

"Cole, listen. David and Catherine had a very unhappy marriage, from what I could gather. They'd already been sleeping in separate bedrooms for over a year before he and I first...got together, although they'd taken care to hide the breakdown of their marriage from Huxley. He had already been in quite a bit of trouble at school, and they didn't want to contribute to his...issues. From what I could gather, Catherine didn't take the news of our relationship too well, even though she'd fallen out of love with David a long time ago. She...well, there was a lot of animosity on both ends." Picking up her wine glass again, she took a sip, and I noticed the tremble in her hand. "When she left, she wasn't interested in taking Huxley with her. She'd never been the maternal type, I suppose, and David said that Huxley was quite upset when he found out he had to stay with his dad. You have to understand, his world has been upended. He can't direct his anger at me, not in person at least, and unfortunately, that leaves you as the next best target."

"Great. Well, that's just idiotic." I placed the bag of peas down on the kitchen island. "Mum, are you sure it was a good idea for me to be here? I don't want to live somewhere where I'm going to have to be walking on eggshells."

Meeting my eyes, she finally let a small smile cross her face. "Cole. You've never let anyone push you around before. Show him you won't accept his behaviour, and then maybe you can move past everything."

Somehow, I didn't think it would be that easy. But she had one thing right. I'd never let anyone push me around before, and I wasn't going to start now.

I wasn't interested in being friends with an asshole like Huxley Granger. And if he continued to threaten me, he'd soon learn that I wasn't the easy target he thought I was.

TWO

COLE

Huxley gave me a wide berth over the next few days, other than the odd times our paths crossed and we'd both glare at each other until one of us cracked and left the room. But through his dad and my mum, I managed to find out a few things about him. Not by choice but by the fact that David and my mum seemed to want to saturate me with information about the Grangers every time I was in the same room as them. It was like my mum was trying to make up for not telling me anything by oversharing every single detail she could think of.

I couldn't blame them. Regardless of the fact that they'd gone about things the wrong way, they were both trying to make an effort now. And now I was over the shock of being uprooted from my life—at least my school wasn't too far away—I could see there were some benefits to being here. Much more space. Quieter, too, without paper-thin walls and neighbours who liked to blast music at all hours of the day and night. My mum seemed much happier as well, happier than I'd seen her in years. David seemed alright, from what I could tell. He'd taken a particular interest in my upcoming summer voluntary role at Mind You, a mental health charity, offering to research the charity's background and financials. He'd also offered me the opportunity to gain some work experience at his architecture company—something I turned down because working with my mum and my soon-to-be stepdad was not something I had an interest in. And I'd rather get by without handouts if at all possible.

Anyway, back to Huxley. If my brain had a list made up of all the things they'd told me, it would look something like this.

Things I know about Huxley Granger:

1) His middle name is John, the same as his dad's and his grandad's

2) He plays the guitar (I knew it!!!)

3) He's studying A levels in business (same as me), computing, and music (not the same as me—my other subjects are maths and design & technology)

4) David wants him to work at his architecture firm but isn't pushing him into it because he doesn't think he wants to work there

5) He broke his left arm when he was six, falling out of a tree

6) He has a sweet tooth, and his favourite fruit is strawberries (same as me)

7) He was close to his mum when he was younger, but when he became a teenager, they grew apart

8) He's hot, but his wankerish personality cancels it out

Okay, scrap the first part of point number eight, and keep the part about his wankerish personality.

It became especially clear on Tuesday night. My mum and David were out entertaining some clients. I'd set myself up at the kitchen island, switching between revising for my A-level maths exam and scouring the internet for possible part-time jobs that I could fit around my volunteer job and wouldn't be too soul-destroying. I'd just put a pizza and a garlic bread in the oven, and the kitchen TV was playing the football highlights in the background. Everything was chilled, and it was the first time I'd sat in the kitchen and not felt like a stranger.

An advert caught my eye on one of the job sites—a bartender at Revolve, a gay club in Soho. Part-time shifts of varying hours, and the pay wasn't bad either. That could work. I quickly filled in the application form, attached my CV, and hit the Submit button.

The TV abruptly turned off, and my head shot up to see Huxley standing across the other side of the island, his arms folded across his chest and his usual glare on his face.

I jerked my head in the direction of the TV, glaring straight back at him. "I was watching that."

"Yeah? Do you have an extra set of eyes? Because it looked like you were looking at your laptop screen to me."

Dick. Pushing back my stool, I rounded the island to stand in front of him, mimicking his posture by folding my arms across my chest. "Did you want something, or did you just come in here to piss me off?"

His lip curled as he turned his body to face me head-on and took a step closer to me. "You're way too easy to piss off."

"Am I?" I stepped forwards. There was no way I was going to back down. "Are you sure it isn't the other way around? Let's think back to when we first met...oh yeah, I remember. You pushed me into a wall and told me that I shouldn't make the mistake of thinking I belonged here. Seemed to me like you were *very* pissed off then. I hadn't even spoken to you at that point."

"Fuck you. I didn't need to know you to know you were an asshole. You and your gold-digging, home-wrecker mum—"

He hadn't even finished talking before my fist swung at him, connecting with his face. "Don't you fucking dare talk about my mum that way, you absolute fucking wanker!"

His fist shot out in retaliation, and I tried to duck, but he still struck a glancing blow off my jaw that made my teeth smash down on my tongue. A burst of blood filled my mouth, and I did what any sensible person would do—I lunged for him. He was gripping his nose, which I now saw had blood pouring from it, but he caught me around my waist, twisting us and sending the back of my head smacking into one of the cupboards.

I shoved him away from me and ran for the sink, where I ran the tap, holding my mouth open under the stream of icy water until the bleeding had slowed right down. When I wasn't in danger of looking like a vampire who'd just fed when I opened my mouth, I withdrew my head and glanced over at Huxley. He was still standing there, blood dripping down his face.

For fuck's sake. I grabbed a handful of kitchen towels and stalked over to him, shoving the towels into his free hand. "Apply pressure to your nose with these, idiot. Tip your bloody head back too."

Surprisingly, he actually followed my instructions, probably because he was in shock or something. Pulling out a chair at the island, he climbed onto it and rested his elbows on the counter, keeping his head tilted back as he pressed the wad of kitchen towels to his nose.

I crossed back over to the other side of the island. The oven beeped, letting me know my pizza was ready, but I'd lost my appetite. After switching it off, I looked back over at Huxley. "Let me say one thing. You have a stupid fucking flimsy reason for disliking me on sight. But guess what—first impressions count, and yours was the worst first impression I'd ever had. And the second. And the third. And you know what I realised? I really fucking dislike you. You're a complete wanker, and I want you to stay the fuck out of my way."

He lowered the bloodstained tissue, and his eyes fixed on mine, hate burning in them. "I fucking hate you. I don't give a shit that our parents are getting married; you're not and never will be my brother. Stay the fuck away from me."

Gathering up my laptop and schoolbooks, I gave him a cold smile. "It looks like we both want the same thing, doesn't it? You stay out of my way, and I'll stay out of yours." I headed straight for the door and paused. "There's a pizza and garlic bread in the oven. Have it, or don't, I don't give a fuck. Your face has made me lose my appetite."

With that, I walked away.

THREE

COLE

Stretching out on my bed, reclining on my back, I clicked the link on my laptop. My mum and stepdad were out, and my step-wanker's door was shut and locked, so now it was time for me to spend some quality time with my right hand.

The video had only just started playing when there was a loud screech, and then it felt like the whole house shook as blaring guitar music blasted through the hallway.

Fucking Huxley. He wanted to disturb my alone time? I'd make him regret it.

Shooting upright, I hit a few buttons on my laptop, scrolling through the list of Wi-Fi speakers to the one labelled Huxley's Room. Lucky for me, he had a video speaker, and unlucky for him, he was about to get a sudden surprise. I connected my laptop to his speaker, making his music cut out, and then skipped to the part of the video where the two guys were really going at it.

Then I turned his speaker volume all the way up.

It took less than two seconds for his roar of outrage to sound from down the hallway, followed by pounding footsteps. I launched myself off my bed, reaching the door just as he flung it open.

"What the fuck are you playing at?" He barrelled into me, snarling in my face. I shoved back, sending his body flying into the door, but he grabbed a handful of my T-shirt and yanked me into him. "I said, what the fuck are you playing at?"

"I was getting ready to have some quality alone time until you started playing your shitty emo music."

"Shitty emo music? That's Oasis, you ignorant bastard. Classic Britpop. Get an education and listen to some of the classics instead of your soulless, brain-dead pop shit." Breathing hard, he curled his fingers into my T-shirt, holding me in place. I could've broken free of his hold, but the enjoyment of pissing him off made me stay in place. His chest rose and fell against mine rapidly, eyes narrowed and flashing with fire, and my lips curved into a grin.

"Brit*pop*? The clue is in the name," I said, just to rile him up even more. I knew it was some kind of indie rock genre, but I was happy to play ignorant if it meant I was getting under his skin. "Whatever it was, it was disturbing my alone time. You gonna turn your music down in future, or do I need to cast my porn to your speakers again to get my point across?"

His gaze flicked behind me to where my laptop was open on the bed, still playing the video I'd cast to his speaker. "Turn it off," he ground out, shoving me away from him. Unprepared, I staggered backwards but recovered quickly.

He'd better not be thinking what I thought he was thinking. "Why? Have you got a problem with two men together?" My voice was low and dangerous. "Because I'm gay, and if you think we have a problem now, we're gonna—"

"I'm bi, so no, I don't have a problem with that," he bit out, and I exhaled heavily. *Good.* Then my brain caught up with the information he'd just disclosed. He was bi? *Fuck.* No. That didn't matter. He was my future step-wanker, and that's all he ever would be. Plus, we hated each other, and yeah, hate sex was a thing, but not for me. I preferred my partners to actually like me.

Huxley derailed my unwelcome train of thought with a vicious glare. "You don't get to cast your shit to my video speaker. You do it again and you can say goodbye to your laptop."

Stepping back up to him, I pushed at his chest. "You play your music that loud again and you can say goodbye to yours."

We stared at each other, neither of us willing to be the first to back down. Finally, though, my stubbornness won out when he dropped his gaze with another angry snarl, turning on his heel and yanking the door open. His parting words hung in the air as the door slammed shut behind him.

"I wish you'd never come here. Just fucking leave me alone."

FOUR

HUXLEY

THREE MONTHS LATER

Happy fucking wedding day. No, scratch out the *happy* and add several more *fucking*s and it would be closer to the truth. I gritted my teeth, lifting my joint to my lips in an attempt to mellow my mood enough to get through this farce of a ceremony. My bastard of a father was marrying his *personal assistant* after what was, for all intents and purposes, an affair. Just how cliché could you get? And after throwing a world-record tantrum, complete with a screaming match and enough broken crockery to fill a mid-sized skip, my mother had waltzed out of my life. Hadn't even given me the option of going with her. If I was lucky, she remembered to call every few weeks, but the last I'd heard, she was taking a trip to "find her inner child" or some shit, and apparently, she couldn't use a phone while she was...wherever she was.

I angrily exhaled a cloud of smoke, not giving a shit

that the smell was permeating my suit. My dad was far more interested in his new wife-to-be than me, and after witnessing eight weeks of their nauseating behaviour, I'd got the fuck out of there. All summer, I'd rotated between friends' sofas and spare beds, all the while resenting the fact that Cole fucking Clarke was tucked up all cosy in *my* house, playing happy families with his mum and my dad.

Why had my dad taken such an interest in him? He said it was because Cole hadn't had the same privileges I'd had, and with his own dad a mostly absent figure, he wanted to try to make it up to him. Yeah, great, but he forgot about his own fucking son in the process, other than gifting me a car as my graduation present, which was basically him trying to buy me off. And I didn't see June Clarke making the same effort with me. In fact, she'd hardly even spoken to me...okay, maybe because I'd made it clear that I didn't want her near me, but whatever.

Bringing the joint back to my lips, I inhaled deeply. At the same time, the door to the hotel room opened, and I steeled myself, biting back words I wanted to let fly. This wasn't my room, after all. It was just a room that the groomsmen were using to dress in, where the photographer kept popping in and out and making us pose like imbeciles.

The words I wanted to bite out died in my throat anyway as I saw who was standing in the doorway.

Cole.

I hadn't seen him since early summer when I'd moved out, taking care to only come back home when I knew he was busy with his volunteering or his new job in a club in Soho. Now, I couldn't help taking him in as I exhaled slowly, the smoke curling in front of me and making it look like I was seeing him through a mist.

The first thing I noticed was his dark hair, usually on the messy side but now parted to the left in a boring-as-fuck, neat style that had the unfortunate effect of accentuating his features. His deep brown, thickly lashed eyes blinked the haze of smoke away as his lips curved into a lazy smirk. His tailored dark grey morning suit matched mine, which fucking sucked because I couldn't deny he wore it better than me.

"My favourite brother."

"Not your brother," I ground out, giving him the finger.

"Black nails. Cute." He made a show of pulling his phone from his pocket and glancing at the time. "I'll wait to call you brother in about...hmm...forty-five minutes."

Before I could reply with a cutting remark, he strode over to me, plucked the joint from my fingertips, and took a deep drag, blowing the smoke into my face.

"Give that back. I need it. It's medicinal."

"Yeah?" He raised a brow. "In that case, I think I'll finish it up."

Then he was gone, and the only thing I could do was punch the wall, wishing it was his smug fucking face.

The ceremony and speeches were finally over, and I was able to escape outside, avoiding well-wishers and other relatives I had no desire to ever see again in my life. Heading out onto the stone patio with a JD and Coke, I stopped dead when I saw Cole over to the left, deep in conversation with a cute guy with light brown, wavy hair. A weird feeling went through me. Almost like jealousy, except I knew better. I'd known for a long time that I was bi, but that guy wasn't my type at all. Still, it wouldn't hurt to find out who he was. Just in case.

"Cole." I lifted my hand in greeting, wincing internally as my bruised knuckles made themselves known as I stretched them out. It probably hadn't been the best idea to punch the wall, but it was the next best thing to punching Cole's face—which I still wanted to do. Strolling over to them, I watched the shock flare in his eyes as he took me in. Maybe I should've tried this sooner—faked playing nice to see what I could get out of it.

He recovered quickly, his gaze flicking to his companion before returning to me. "Ahhh. Huxley. My favourite brother."

I was temporarily struck dumb. He'd never said my name before, but the way it had rolled off his tongue so easily, like he'd been saying it for years... Gritting my teeth, I focused on the task at hand. The fucker was just trying to get a rise out of me by calling me his brother. "*Step*brother, you mean." I turned to his companion, holding out my hand. "I'm Huxley Granger. And you are...?"

"Elliot Clarke. Cole's cousin." He gave me a small, polite smile, and it was clear that he had zero interest in me, which was good, because now I was here, I wasn't feeling any kind of way about him. My gaze returned to

Cole, who was watching us both with his brows pulled together, his bottom lip trapped between his teeth.

There was something in his expression... I stepped closer to Elliot, and the second I did so, I saw Cole tense up out of the corner of my eye. "Elliot. I haven't seen you around before. Do you live locally?"

Shaking his head, he gave me another small smile. "Nope. Bournemouth. But I go to London Southwark Uni—do you know it?"

"LSU? Yeah." This time, there was nothing fake in my interest. "What's it like? I'm starting there this September."

His smile turned genuine. "It's good. There's a lot of clubs and societies and things if you like all that. Great facilities. Yeah. I like it."

"Good," I replied, briefly glancing back over at Cole, who was now giving me a look that was definitely a glare. Returning his glare, I took a step back. "Nice to meet you, Elliot. Maybe I'll see you around campus."

As I turned to head in any direction that led me away from them, I made sure to brush past Cole, dipping my head to his ear. "Call me *brother* again and I'll knock your fucking teeth out."

He huffed out a laugh, although it sounded a little shaky.

I didn't bother to look at the expression on his face as I walked away.

FIVE

COLE

"Cole!"

I glanced away from Elliot to see my mum beckoning me towards her. I sighed, turning back to my cousin. "Duty calls. Sorry."

He swiftly shook his head with a smile. "Don't worry about it. I should probably text Ander back anyway—he's been blowing up my phone asking how everything's going."

"Sounds like him." Ander was Elliot's best friend, and from what I'd seen of their friendship over the years, they were both obsessed with each other to a level that I could never contemplate with anyone. Then again, I'd never had a best friend, so who knew how best friends actually behaved?

Shoving my hands into my suit pockets, I headed over to my mum. I plastered on a smile when I drew closer, because despite my reservations about this whole thing—mainly Huxley, if I was being completely honest—it was clear to anyone with eyes that she was happy.

"Ah, Cole. The photographer would like to take a few photos down by the lakeside. David and I thought it might be nice to include a few family photos."

Great. More time with Huxley. I kept the smile on my face and hoped it didn't look too fake. "Okay, yeah. Where do you want me?"

"Just down here." She pointed down the gravel path that wound down the side of the lawn in the direction of the lake. I took her arm as we walked down, helping her to balance. Those heels were not made for gravel. As we drew closer to the lake, I saw Huxley and his dad waiting with the photographer, who was kneeling down and changing his camera lens. He had a bit of a pained expression on his face, and it soon became obvious why when I heard David's voice.

"...Smoking weed on our wedding day! All I ask is for you to behave for *one* day, and you can't even manage that."

I smirked, and at the same time, Huxley glanced over at me, shooting me a savage glare. "Cole smoked too."

My smirk dropped, and I shook my head. "Smoke? Me? You must be mistaken." Okay, I'd made a point of taking a drag in front of him, but after that, I'd put the joint out to save for later, and then I'd done my best to air out my suit and spray on some aftershave and made good use of the toothpaste and mouthwash in the facilities the hotel had provided. Huxley hadn't even bothered to try and hide it.

"You know you did, you lying wanker," he hissed.

"Huxley! Please." David pinched his brow, huffing

out a breath. "Can you just pretend to get on with Cole for one day? You're brothers now, so act like it."

"Yeah, come on, *brother*," I said, just to piss him off, and he gave me the middle finger, not even caring that his dad, my mum, and the photographer were all watching him.

"Huxley. That's enough." There was a finality in David's tone, and Huxley dropped his arm, his jaw clenched. David cleared his throat, giving Huxley a pointed look before turning to the photographer. "My apologies. We're ready for the photos now. Where would you like us?"

The photographer slipped into professional mode, directing us to stand in front of the lake. I was on the end next to my mum, and Huxley was thankfully at the other end, next to his dad. My smile came easily as I imagined various ways that I could piss off Huxley. It would be amazing if I could just trip him up so he fell into the lake.

After the photographer had snapped a few photos, he studied them on the screen on the back of his camera, then gave us a thumbs up. I exhaled, relaxing. Now I could get away from Huxley and drink myself into oblivion.

Except my mum stopped me in my tracks just as I was getting ready to make a break for it. "Cole? Huxley? Before you go, we'd like a photo of the two of you together. Just one."

I knew that the horror I saw in Huxley's eyes was reflected in my own. For once, we were both in complete agreement.

"Cole. Please." My mum's voice wobbled a bit, and

shit, I was going to have to do this, wasn't I? I wasn't going to be the heartless bastard that ruined her day. I could suck it up and suffer Huxley for one photo.

I nodded at my mum and then walked over to where Huxley was standing. He watched me coming, his teeth gritted and his fists clenched like he was getting ready to punch me. Fuck, why did he have to look so sexy, all wrapped up in that suit with murder in his eyes? There was clearly something wrong with me if I was finding my step-wanker hot. I kept some distance between us, or I tried to until the photographer instructed us to move closer. There was a slight smirk on his face, and I was betting that this was payback for Huxley's behaviour earlier.

"Closer." He was gesturing with his hand, and with a sigh, I stepped right up next to Huxley, so close that the sleeve of his suit jacket brushed against mine.

"You're a lying bastard," he spat between gritted teeth.

"Remove the word 'lying' and you've just described yourself." I smiled widely. "It's so easy to smile around you."

He legit growled, like he was an animal. "Fuck off."

"Wanna know why? It's because I'm imagining pushing you into the lake." The photographer raised his camera, and I added, "Smile."

Huxley pasted a smile on his face that actually looked genuine, and I wondered whether he was trying my trick of imagining that he was pushing me into the lake. Why did other people get nice stepsiblings, and I ended up with the one from hell? I knew I wasn't doing anything to improve relations between us, but he was a complete and utter wanker...and okay, so was I, as it turned out.

"Miracles can happen." David stared at us, shaking his head. "Maybe now you two can be civil to each other, you can consider moving back home, Hux."

Huxley snorted. "Unless he's moving out, I'm not moving in."

David muttered something about "immature behaviour," which I agreed with, although I had to say it was quite nice to not have to be on edge every time I left my bedroom. There was a tiny part of me that felt bad that Huxley had left his home because of me, though. A tiny part that I was able to ignore now that I had him standing next to me, all tense and bristling with hostility.

"Arms around each other," the photographer called, and I groaned. But I slid my arm around Huxley's waist, knowing that it would just make him more pissed off.

He stiffly placed his own arm around me. "Touch me again after this photo and I'll break your arm."

"Do you think you could? How much effort would it take to actually break someone's arm?" I took a breath to smile at the photographer, imagining Huxley flailing in the lake and coated in algae, before I continued. "I think we'd be pretty evenly matched, probably. Maybe I'd break your arm first."

Thankfully, the photographer called out to say we were done with the photos, interrupting whatever comeback Huxley had. He shrugged my arm off and stormed away, and I watched him go with the tiniest feeling of guilt. I shouldn't have pushed him, not at my mum's wedding.

With a sigh, I turned to my mum and David. "I'm sorry. I wasn't helping the situation. He, uh, I shouldn't let him get under my skin."

David shook his head. "Huxley's always been hot-headed and a troublemaker. I don't blame you at all."

Yeah, but I wasn't faultless, was I? "Even so, I apologise."

They both smiled at me, and David stepped forwards to clasp my shoulder. "Thank you, Cole. You have nothing to apologise for. I only wish Huxley could be a little more like you. I wish he'd give you a chance."

Those words made me feel even worse. I had to leave. "Congratulations again. I'll, uh, leave you to the rest of the photos."

David hugged me, taking me by surprise. As he circled his arms around me, he leaned in, lowering his voice. "I don't expect you to start calling me Dad, and I would never dream of taking the place of your own father. But I'd like for you to see me as a father figure. As far as I'm concerned, I've gained another son, and I only want what's best for you."

When he released me, my mum was dabbing at her eyes, smiling. "Thank you for being so good about everything, Cole. I'm very proud of you," she whispered as I hugged her and kissed her cheek.

I needed a drink because, somehow, I'd managed to make myself look like a saint and for Huxley to look even worse, and that guilt was rearing its head again. Making my way back up to the hotel, I headed straight for the bar, ordering a shot of vodka and tipping it back.

"Another," I said to the bartender, just as someone came up beside me.

"I'll have the same."

Huxley stood next to me. In the short time since I'd last seen him, he'd ditched his jacket, waistcoat, and tie, and now he was in his white shirt with the sleeves rolled up, exposing the tattoos that ran down one arm. His bleached hair was mussed up like he'd been running his hands through it, and as he drummed his fingertips on the bar, his black nails glinted at me. Fuck me, he looked hot.

It still didn't negate the fact that I loathed him, though. It just meant that I had working eyes.

He shot me a sideways look that warned me not to mess with him, and I wasn't planning to. All I wanted was to drink until I forgot he even existed.

When the bartender set our shots in front of us, I raised my glass. "Wanna toast?"

His dark brows lifted as his eyes met mine. "To what?"

I shrugged. "To forgetting that this day ever happened?"

"I'll fucking drink to that," he muttered, raising his own glass. At the same time, we tipped our shots to our lips and drank.

It was the second time today that we'd been in agreement, but I doubted it would ever happen again.

SIX

COLE

Yawning, I stepped outside the side entrance of Revolve, the club where I did shifts as a bartender. It was three in the morning now, and the only thing I wanted was to get back to the house and pass out in my bed.

I shoved my hand in my pocket, reaching for my phone to book an Uber. My fingers brushed against a crumpled slip of paper, and I pulled it out. The name "John" was printed in block capitals above a phone number. Maybe I should text him. He'd been cute, from what I could remember. As a gay man working in a gay club, I never had any shortage of numbers, plus I had quite a few suggestive messages on HookdLDN, the local hook-up app, but I took advantage of my opportunities less often than I suspected most people would.

As I pushed the paper back into my pocket, my phone began vibrating, making me jump a mile. My heart racing, I lifted it to see an unknown number, and I hit Answer straight away. As a rule, I didn't answer unknown numbers, but if someone was calling me at three in the morning, I couldn't take the risk of not answering in case it was important. My mum and David were off on their honeymoon, a month-long cruise, and if anything had happened to my mum...

"Hello?"

"Is that..." The voice at the other end paused. "Do you know a Mr. Huxley Granger?"

A shocked sound burst from my throat. *Huxley*? "Yes. He's...he's my stepbrother. What's this—"

"Right. Are you able to get to Honeybourne Road? Mr. Granger has been in an accident. He's quite shaken up. He asked us to call his stepbrother."

Attempting to calm my accelerating heart rate, I tried to keep my voice steady. "Yeah. That would be me. Cole Clarke. Is he okay? What happened?"

"It's probably best if you get here as quickly as possible. Someone will be available to answer your questions."

When I'd been given the exact location and hung up the phone, I wasted no time in flagging down a black cab. Fuck waiting for an Uber; I needed to get to Huxley right now.

My hands were fucking shaking the whole way there, my heart beating out of my chest. The same question ran through my mind over and over as the cab rumbled through the quiet streets.

Why had Huxley asked them to call me?

Flashing lights cut through the darkness, and my whole body tensed, my gaze scanning the road up ahead. Blue flashes lit up the night sky in front of me, throwing the buildings on either side of the road into sharp relief.

"This is as far as I can go. Police cordon." The driver thumbed at the distinctive tape stretching across the road, flapping in the night breeze.

"That's fine. Thank you." Slipping out of the cab, I ducked under the tape and stopped dead, taking in the scene in front of me.

Two cars. One flipped on its side, the underside of the car facing me. The other... My breath caught in my throat, my hand flying to my mouth as I took in the wreck that had once been Huxley's black Audi S5 Coupe. From the back, before I'd crossed the cordon, it had looked okay, but as I forced my feet to move, I saw that the front was now a heap of twisted metal. The car's hazards were flashing, combining with the blue lights of the police car parked at an angle across the end of the road, and reflecting off the shattered glass littering the tarmac.

"Sir! Excuse me! You need to stay behind the cordon!"

I was spun around by a hand to my shoulder. A policewoman was there, shaking her head at me.

"I—they—someone called me. Huxley. My, uh, brother. Th-that's his car. The A-Audi." I was stumbling over the words, my voice shaking like fucking crazy. As much as Huxley and I detested each other, I wouldn't wish this on him. Fuck. He wasn't *dead*, was he?

Through the glassy haze that had suddenly obscured my vision, I registered the policewoman's face turn from stern to sympathetic, and the hand on my shoulder squeezed gently before she removed it. "Come with me."

After saying something into the little radio she had clipped to her chest, she led me over to an ambulance a little way from the scene of the wreckage. The back of the ambulance was open, and a paramedic was there waiting for me. He gave me a smile, all friendly professionalism, and my pounding heart rate slowed just a little because surely he wouldn't smile if Huxley was in a bad state?

"Mr. Clarke. Is it okay if I call you Cole?" he addressed me gently, and I nodded, swiping the back of my hand across my face to clear my vision. "Good. Your brother was in an accident tonight. He's going to be okay —it's a miracle, to tell the truth, that he's only escaped with some minor cuts from the glass and bruising from the airbags. Someone up above must have been watching out for him."

"What happened?" I scanned the paramedic's face, relief coursing through my body at the knowledge he was okay.

The paramedic glanced over at the policewoman, who coughed discreetly. "Drunk driver on a joyride hit his car."

"Shit. Okay. So he's okay?" I asked, just to make sure.

"He will be. He's been thoroughly checked out in the ambulance. He's very shaken up, and he needs someone to be with him to make sure that he gets some rest and takes the pain relievers he's been prescribed. If the safety features in his car hadn't been so good, and if he hadn't had luck on his side...we might be looking at a completely different story."

"Okay," I said again. Fuck. My brain didn't seem to be working properly.

"Would you like to see him? He's been cleared to go

home." Leaning closer, the paramedic murmured, "Between you and me, I'd prefer to take him into the hospital and keep him overnight for observation, but with the cuts they've made recently...unless we have a strong case for taking him in, he has to go home."

"No, it's okay. I'll, uh, sit with him. Make sure he's okay."

The paramedic smiled. "If only more of our patients had siblings as caring as you." *If only he knew the truth.* "If you're concerned at all, call NHS Direct, and if his condition worsens, dial 999."

"Uh..."

"He's going to be okay. It won't come to that, I'm sure of it." With those words, he directed me to climb inside the ambulance, where I got my first look at Huxley.

He was sitting up on the metal bed, a blanket wrapped around his shoulders, sipping from a bottle of water through a straw. His face was drained of colour, with a grey pallor, and there were several small cuts on his face. The skin around his eyes was a little swollen and red, and I knew that it would darken to a purple by tomorrow.

I stood there for a moment, not knowing what to say or how to approach him. Eventually, I went for the simplest option. "Hi."

His eyes finally met mine, and I was taken aback by the hostility in them, far stronger than the relief that had flickered in his gaze for the tiniest moment. I shouldn't have been because that was standard from him, but I thought that maybe the accident would have made him a little less... He'd chosen to call me, hadn't he? Instead of one of his friends?

"You came?" he said, and there was a definite question in his words.

Had he thought I wouldn't? Okay, we couldn't stand each other, but there hadn't ever been a question in my mind. The second I'd picked up the phone and heard what had happened, all I could think about was getting to him as fast as I could. "Yeah."

"Just to make it clear," he rasped, "you're the last fucking person I wanted to call." Dragging his hand across his face, he visibly winced, gritting his teeth as his palm skimmed over the cuts and bruises. He let out a frustrated huff of breath. "I didn't have any other option. It was you or no one."

I stored that information for later, when I could interrogate him about it when he wasn't looking like he was about to pass out any minute. "Are you ready to go? I'll get us a taxi."

"Fuck's sake. Where's your car?"

"I was at work. I don't drive there; there's nowhere to park."

He grunted, and that was a good enough reply for me. As much as I enjoyed mentally sparring with him, even I wasn't enough of a dick to do it when he'd just had a traumatic experience. Pulling out my phone, I booked an Uber to take us both home.

"How's my car?" was the only question he asked me during the otherwise silent ride back to the house. We'd gone out of the other end of the road because I hadn't wanted him to see the scene of the accident.

I swallowed hard, staring out of the window. "It was... it was wrecked."

There was no reply, but when I turned my head and a streetlamp illuminated the inside of the car, I saw him bite down on his trembling lip as a tear fell from his closed eyes.

SEVEN

COLE

Huxley had spent most of the next couple of days sleeping. I figured it was good for his recovery. I'd called one of the bosses of my volunteer admin assistant job—hopefully, eventually to be a paid job—explaining the situation, and lucky for me, he'd given me the okay to work from home. Even luckier, I wasn't scheduled to work any shifts at Revolve until the following week, so I was able to keep an eye on Huxley.

Not that either of us was particularly happy about the situation. Huxley was resentful as fuck that I was the one to be looking out for him, and I was resentful as fuck for the same reason. I hadn't yet asked him why he'd chosen me to call, despite his network of friends, and I hadn't even wanted to broach the subject because every time I'd seen him, our conversations consisted of me reminding him to take his medication and him giving me the finger or telling me to fuck off out of his bedroom.

Today, though, enough was enough. Two and a half

days of rest was plenty. He was going to answer my questions, whether he wanted to or not.

First of all, I needed to sweeten him up. From what I remembered of the two-ish months we'd lived under the same roof, he was *not* a morning person. So I used the expensive coffee machine in the kitchen to make him a latte loaded with caramel syrup. By the way, his sweet tooth was the only thing sweet about him.

Entering his bedroom without knocking, I placed the mug of coffee on his bedside table. He hadn't bothered locking the door after the first night, when I'd threatened to smash his door in if he tried to lock me out. Just because I didn't like him didn't mean that I was going to leave him alone. As it stood, I had responsibility for him, no matter how unhappy we both were about it, and I wasn't going to be the one to explain to his dad why something bad had happened to him under my care.

"Big brother." Standing at the edge of his bed, I smirked as I gently shook his shoulder.

He reacted instantly, springing up in bed with a growl, then grimacing in pain. "Why the fuck is your ugly face the first thing I'm seeing this morning?"

"I made you coffee." I gave him a fake smile. "My face is not ugly, by the way. Are you jealous of the way I look?" I nearly said "not as ugly as yours," but that would be a lie. Despite his perma-scowl and over-reliance on hair bleach, he was unfortunately blessed in the looks department. Not that I'd ever tell him that.

Instead of replying, he did his usual gesture of giving me the finger before grabbing his coffee. I noticed that his black nail polish was all chipped and messed up, and something about it seemed wrong. I ignored that thought, moving on to more important things. Taking a seat at the side of his bed, I opened my mouth to speak, but before I could say anything, I found myself being shoved hard, crashing to the floor and knocking my head.

"You asshole. What was that for?" It took everything in me not to drag him out of the bed and pound him into the floor—with my fists. The only reason I held off was because he was injured.

"Stay off my bed."

"Fine." I picked myself up from the floor, shooting him a glare, and took a seat in his desk chair, wheeling it over so it was next to his bed. In the meantime, he'd arranged himself against the headboard, knocking back painkillers with the glass of water I'd left out for him last night.

"Why the fuck are you here? I know it's not to bring me coffee. I'm fine now. I don't need you playing nurse anymore."

That was debatable. But on with the reason why I was here. He was right; it wasn't to bring him coffee.

"I want to know why you chose me, out of everyone."

His shoulders stiffened. Lowering the glass of water, he switched it for his coffee mug, drawing his knees up under the covers and hiding his face behind the mug. "You were my last and only choice," he said flatly.

"But how? Surely you had friends you could call? Someone you like slightly more than me?"

"I like everyone more than you," he growled. "Dad's on his honeymoon, my mum's off on her spiritual journey, and my friends...look. I was in a lot of fucking pain, okay?

I didn't want to be crashing on someone's sofa or in a shitty spare bed with springs poking into my back. I wanted to be here, in my comfortable bed, with my own private bathroom. You being forced to be at my beck and call was an unexpected bonus."

He'd spent most of the time sleeping, so I hadn't exactly been at his beck and call, but whatever. I thought back over what he'd said. It made sense. "Okay." With a shrug, I got to my feet. "Thanks for letting me know. If you're feeling fine now, then you obviously don't need me here anymore."

I'd made it all the way to the open door when his voice sounded again. "Wait."

I turned back to him, and our eyes met. He moved, and the duvet slid down and...fuck. Across that lean, lightly toned chest was a myriad of bruises in blues and purples.

"I'm not...I'm not fucking fine, okay," he bit out. "But I don't want or need you looking after me anymore. Go and do whatever shit you want, I don't care."

"Huxley." I took a step towards him, trying to remember the instructions the paramedic had given us just before we'd left. "You need to apply pressure to that bruising."

Glancing down, he swore under his breath. He'd clearly been keeping those bruises hidden from me because every time I'd seen him, he'd either been under the covers or wearing a T-shirt, but now there was no hiding them.

"I fucking tried, but—"

I held up my hand. "Here's what's going to happen.

I'm gonna help you wrap your chest, and then I'm going to leave you. It goes both ways. You don't want me here; I don't want to be here. But, Hux, if you dare to fucking injure yourself any worse and your dad finds out, I *will* kill you."

It was only after he'd been staring weirdly at me for the best part of a minute that I rewound what I'd said to him. Shit. I'd called him Hux—where had that come from? We weren't, and would probably never be, at the name-shortening stage.

Ripping my gaze away from his, I crossed his bedroom and entered his en-suite bathroom to find the roll of stretch bandages. When I returned and took a seat on the edge of his bed, I steeled myself for him to kick me off again, but he remained where he was, staring at a fixed point on the wall rather than looking at me. Which was good.

I unrolled the bandages and then cleared my throat. His jaw clenched, but he lifted his arms enough that I could get the bandages around his torso. Shifting closer on the bed, I reached out. At the first contact of my fingertips on his skin, we both inhaled sharply. My heart rate was speeding up. I'd never been this close to him, never touched him without there being any hostility behind it. Carefully, I began to wind the bandages around him, doing my best not to touch him without a layer of fabric between my skin and his. When I'd finished, I wasted no time in shifting away from the heat of his body. Both of us were breathing more heavily than normal, and all I wanted to do was to get away from him. To purge my mind of the knowledge of his warm, smooth skin, the way his heart had beat under my palm as I was wrapping the stretchy fabric around him, the way his thigh had been pressed against mine, his breath hitting my hair as I lowered my head to fasten the bandages.

"Now, get the fuck out of my room." His voice was hoarse.

"Fuck you. I'm going." Without another word, I launched myself off his bed and out of his door, making sure to slam it as I left. The crash reverberated around me as I stalked down the hallway, down the stairs, and then out through the front door. I had no idea where I was going—all I knew was that I needed to put as much distance between me and Huxley as possible.

EIGHT

HUXLEY

Cole stayed out of my way for the next few days. Sometimes I'd hear the shower in the upstairs bathroom, and occasionally, I'd hear his footsteps padding past my room, but that was it. Every day, though, without fail, I'd come downstairs on a hunt for food and find takeaway food in the fridge that I just needed to reheat. The fucker somehow seemed to know the foods I liked, or maybe it was just that the accident had left me with a new appreciation for all the necessities required to keep me alive. Things like food, drink, sleeping in a decent bed, weed, my guitar…

My guitar. It sat there accusingly, staring at me from the corner of my bedroom. I'd left it here when I'd decided to move out, knowing that I couldn't keep lugging it from house to house, expecting it to still be in one piece when I eventually decided to settle. I loved playing it—it was a form of stress relief for me, along with the weed—a way to balance the thoughts inside my head, to give me some peace.

With a sigh, I forced myself to pick it up. As soon as my hand closed around the neck, a lump appeared in my throat out of nowhere. Maybe this accident had fucked with my head more than I'd thought. Cole hadn't let me see the scene, but there had been pictures taken by the police, pictures my insurance company needed also, and yeah...I hadn't realised until the moment I'd seen them just how fucking lucky I was. How had I managed to walk away from that accident unscathed, apart from some shallow cuts and bruising that was painful as fuck but would soon be nothing but a memory?

The thoughts were too much to process. I sank to the floor, cradling my guitar on my lap, and lost myself in the soothing process of tuning it. When it was ready to play, I ran my thumb over my guitar pick, gently strumming the strings as I tried to decide what to play. My fingers made the decision for me, playing the opening bars of "Somewhere Only We Know" by Keane.

As I strummed the guitar, I began to sing softly.

I was in the zone, and it took me a while to realise that there was another presence in the room. Raising my head, I saw Cole, frozen in the doorway, his gaze fixated on my guitar.

My fingers slipped on the strings, causing a screeching sound, and I growled in frustration. "What the fuck do you want? Get out." Why had I left the door open?

He blinked, his eyes flying to mine before he staggered backwards. "Sorry," he whispered, and then he disappeared.

I tried to get back into the zone after that, but of

course, fucking Cole had ruined it. Giving it up as a lost cause, I made my way down to the kitchen for a snack, and my least favourite person was there, perched on a stool at the kitchen island, rubbing the spot between his brows as he stared at his laptop screen.

"Why do you have to do that here? There's a desk in your room. Use it," I said irritably, heading over to the fridge.

"I'm used to working in the kitchen. There wasn't enough room for a desk in my old house, so I did all my homework at the kitchen table. I guess it's a habit that's stuck with me, even now with this charity work."

His voice was musing, and I wasn't prepared to have a civil conversation with him right now. Or ever. "Maybe you should make a new habit," I suggested as I reached into the fridge for a punnet of ripe red strawberries. Fucking delicious.

There was no reply, so I took that to mean our conversation was over. Good. I portioned out some of the strawberries to eat and grabbed a bowl from the cupboard. As I was running the tap to rinse them, I felt Cole's eyes on me.

"What are you doing?"

I rolled my eyes. "Washing strawberries. What does it look like I'm doing?"

"Those are my strawberries," he said, as if he actually had ownership of any of the products in my dad's fridge.

"I don't see your name on them." Picking up the biggest, ripest strawberry, I turned to face him and bit down into the soft, juicy flesh. "Mmmm."

He stared at me...at my mouth, and when my tongue

came out to swipe across my lips, making sure I caught all of that delicious strawberry juice, I could've sworn his eyes darkened.

"They're mine. I bought them this morning." His voice suddenly had a rasp to it, and why the hell did my dick decide that it would be a perfect moment to perk up?

"Too bad." I lifted the remaining half of the strawberry to my lips, and then everything seemed to happen in slow motion. There was the scrape of a chair, and then Cole appeared in front of me, biting down on the strawberry that I was about to place in my mouth.

I forgot what words were because *his fucking tongue was touching my fingers.*

He swallowed hard, lifting his head, and our eyes met for a long, charged moment.

"*Mine,*" he growled.

Then he swiped the bowl of strawberries from the sink and stalked out of the room, leaving me there fucking speechless and with a raging boner.

I had a text later on, after I'd had the most frustrating wank of my life. Frustrating, because my dick did not seem to get the message that Cole was not only a fucking asshole, but he was also my stepbrother and therefore completely off limits, even if hell had frozen over and he had been someone I was remotely into.

COLE:

FYI my mum texted to say she's going to call at 8pm our time. Do you want me to tell them about the accident?

ME:

No. NO. Do not say anything. The insurance is paying for repairs and it's in my name. If my dad finds out, I will fucking make you pay

COLE:

Chill. Smoke some weed. Seems like you need it

I fucking hated this asshole. Just to piss him off, I sent him a picture of my empty bowl with strawberry leaves inside. After he'd left the kitchen, I'd taken the rest of the strawberries from the fridge and eaten them up in my room. They were my favourite fruit, and yet the experience of eating them had been ruined for me because every time I bit into one, I had a vivid image of him taking the strawberry from my hand with his teeth, his tongue swiping across my fingers. That had led to my dick getting even harder, hence my frustrating wank.

A minute later, my phone buzzed again, and I opened it up to find a picture that immediately made my blood boil. Cole was reclining on his bed, a joint between his lips, smoke curling lazily through the air. He'd included the caption, "Thanks for rolling this. Saved me a job".

The bastard had stolen that from my stash.

So what if I'd had a few of his strawberries? He had

to have gone digging through my drawers to find that, and that was a total violation of my privacy.

That was fucking it. Ignoring the pain from my bruises as I launched myself to my feet, I stormed out of my bedroom and down the hallway to confront my step-brother.

NINE

COLE

I smiled to myself as I sent the selfie to Huxley. He was going to lose his shit, and I couldn't wait. It was time to get us back on an even footing after that weird-as-fuck moment in the kitchen. What had I been thinking? I fucking ate that strawberry from his fingers like we were in a bad porno. I wasn't sure who had been the most shocked—him or me.

Predictably, there was a crash—Huxley's bedroom door opening—followed by footsteps stomping down the hall.

"Hi, brother." I waved the joint at him, and I could almost see the steam pouring from his ears. His blue eyes were dark with rage, highlighted by the purple bruising around them, and his fists were clenched as he stormed over to my bed. Before he could take a swing at me, which I just *knew* he was about to do, I quickly set the joint down on the small plate I was using as a temporary ashtray-slash-burn protection thing and then threw my body into a lying position. With the mattress there to absorb the blow, the punch wouldn't hurt as much. And as much as I wanted to fight back, I had to remind myself that Huxley had only very recently been in a car accident, so for now, I had to bide my time. Incapacitate him in the least painful way possible. Then give him the beating of his life when he was all recovered.

My sudden move caught him off guard, and his fist encountered empty air as he swung at where I had been just a second earlier, making him fall forwards. He threw his hands out to brace his fall, one landing on the bed right next to my head, the other landing on my shoulder, twisting it down into the mattress.

Despite being caught off guard, he recovered quickly, moving to straddle me on his knees and delivering a solid punch to my stomach, making me gasp. "You fucking bastard," he snarled. "I wish you'd just—"

All my good intentions went out of the window as I reacted on instinct, even though the air had been knocked out of me. Bucking my body up to throw him off, I shoved hard at his bruised torso. He howled in pain, falling to the side and smacking his back into the wall. I felt bad for about two seconds until he threw himself back onto me, his nails scraping down my bare arms. A sound came from my throat that was halfway between a cry and a shout. His nails were short, but it still fucking stung as they dug into my skin. I grabbed a handful of his stupid bleached hair and yanked as I got my legs around him.

"Fucking get...off," he panted as I held him in place, throwing my arms around him to stop him from using his own arms as leverage...or to hurt me again.

"Why the fuck would I do that?" I kicked the back of

his shin with my heel as hard as I could. It probably didn't hurt much, but if it pissed him off more, it was a bonus.

His head raised, and for a moment, that bruised, hate-filled stare was so close to mine I could have counted his individual eyelashes. If I'd wanted to. His breath ghosted across my skin, and I smelled strawberries.

My strawberries. "You ate my fucking strawberries. I can smell them on your breath."

"Well, I can smell my fucking weed all over you," he ground out, and then he dropped his head.

Then, teeth were sinking into my throat, and the sudden shock and pain meant my grip on him loosened. He scrambled upright, breathing hard, and I just stared up at him, open-mouthed. "You *bit* me? What the fuck?"

"I'll do a lot worse than that when I've recovered." His tone was a low, threatening rasp, and fucking hell, the sound went straight to my dick.

No. That was not happening, not now, not ever.

I quickly pulled myself upright, which unfortunately meant that he was suddenly sitting in my lap, and our bodies were way too close for comfort. Even so, I got in his face, my hair brushing against his forehead as I spoke through gritted teeth.

"Yeah? I'd like to see you try. I can and will fight back, and you'll look and feel a whole lot worse than you do now. *Asshole.*"

He scrambled off me, wincing as he did so, his hand gingerly clasping his chest. The remorse that I'd pushed away hit me again, along with a vivid flashback of his wrecked car and his face, pale and shaken, no matter how much he'd tried to hide it.

"Huxley. I—"

"Save it," he hissed, making his way to the door. "Just stay the fuck away from me, and don't take any more shit from my room."

The door slammed closed behind him, and I fell back onto my bed with a groan, throwing my arm across my face.

I had the feeling that I'd just made things between us a hundred times worse. How did I fix it? Did I even want to fix it?

It was no surprise that I hadn't been able to fall asleep. At two in the morning, I decided to go down to the kitchen and make a hot chocolate, something I vaguely remembered my mum doing for me a few times when I was little and had trouble falling asleep.

I flipped on the cooker hood light rather than blinding myself with the overhead lights and screamed at a pitch I hadn't achieved since my balls had dropped.

Huxley burst out laughing as I collapsed back against the oven, my heart beating out of my chest at the sudden shock of seeing him standing there. I could only stare at him. I'd never even seen him smile, let alone laugh. It transformed his entire face. With the animosity temporarily gone, he looked... He looked like someone I'd — *No.*

"Why the fuck were you standing there in the dark? I'm too young to die of a heart attack."

He didn't answer, his laughter dying away as he appeared to realise that all I was wearing was a pair of tight black boxer briefs. His gaze tracked across my body as he shifted on his feet, and there was something in his eyes that made my breath catch in my throat. *Fuck*. I cast around for something that would get his eyes off me. As my gaze lowered, I noticed two things. Firstly, he was also in a similar state of undress. That would have been epically bad, for various dick-related reasons, but the second thing I noticed put everything else out of my mind. He was holding a bag of ice to his chest, and above the top of the bag, I could see the bruising, so dark against his pale skin.

I'd hurt him there, when he'd already been hurt. I'd made it worse.

"Huxley. I'm sorry." My voice was scratchy and loud in the silence. "I didn't mean—"

"Forget about it," he snapped. "I did just as much to you. And I'm not gonna apologise for it."

"Yeah, but you were already—"

"I said, forget it." His jaw clenched. Turning around, he opened the freezer and placed the ice back inside. I didn't miss his wince.

I *really* hated feeling bad. Even though he'd been a complete and utter wanker, I shouldn't have done what I did. "Uh." Rubbing the back of my neck, I directed my gaze at the floor. "I'm making a hot chocolate. Want one?"

There was a long pause. His head was still buried in the freezer, but finally, I heard his quietly muttered reply.

"Okay."

Okay? To be honest, I hadn't actually expected him to agree. But he had, and now I...now I needed to get my fucking feet to unfreeze from the kitchen floor so I could get to the coffee machine to use the milk frother.

"Can't sleep either?" Why were words still coming out of my mouth?

He grunted in reply as he closed the freezer door. It was then that I noticed the glass of water and his painkillers on the kitchen island in front of him, along with a tube of bruise relief cream. His pointed glare warned me not to offer to help him with the cream, so instead, I got my legs to unstick from the floor and headed over to the coffee machine to make a start on our hot chocolates. I pulled two slightly chipped blue-and-white-striped mugs from the cupboard, both of which belonged to my mum and had come from our flat, and then got to work.

While I'd been occupied making the hot chocolates, Huxley had disappeared. Carefully carrying the mugs, I padded down the hallway to find him, heading for the lounge. The games room door was open, and there was a sliver of light spilling into the hallway, so I changed direction.

According to David, when he'd shown me around, the games room had once been Huxley's playroom when he was a kid. As he grew older, they'd adapted it, and now the small room contained a huge TV and two games consoles, along with a large squashy black leather sofa and a bookcase full of a combination of books and PlayStation and Nintendo games. The walls had posters of bands and concerts, and a guitar stand was placed in one corner, along with a stool, microphone, and small amp.

When I entered, I saw Huxley curled up on the sofa with the TV remote in his hand, illuminated by a tiny lamp on top of the bookcase. He didn't look up, but I knew that he knew I was there because his shoulders tensed.

"Here's your hot chocolate," I said, stating the obvious as I carefully placed his mug on the small coffee table in front of the sofa. When I straightened up to leave, he cleared his throat.

"I'm gonna watch something. You...you can stay...if you want."

Was this him accepting a tentative truce? I'd made him hot chocolate, and now he was inviting me to watch a film with him? Images of his bruised torso played through my mind, and I knew there was only one acceptable way to reply.

"Uh. Yeah. Okay."

His legs and most of his chest were covered by a dark blue chunky knit blanket, hiding his bruises. Even though I wasn't cold, I grabbed my own blanket from the basket next to the bookcase. Because sitting there in just my underwear was too much. It left me feeling too exposed—and not only literally. We needed the comfort of a barrier between us. Taking a seat on the sofa, I sipped my hot chocolate while he scrolled through the film selections. He didn't ask me what I wanted to watch, and I didn't offer my opinion. It was enough that we were here in the same room and not at each other's throats for once.

He selected *Bohemian Rhapsody*, muttering something about having seen it before and it not mattering whether he fell asleep, which I took to mean that I wasn't the only one having trouble sleeping tonight. When he hit Play, I stretched out my feet to rest on the coffee table as I cradled my mug of hot chocolate in my palms. I watched out of the corner of my eye as he took his first tentative sip of his hot chocolate, his lips curving into a tiny smile as he swallowed before taking another, larger sip.

He liked it.

I forced myself to concentrate on the TV.

I blinked.

Then blinked again.

And again. This time, it lasted longer.

I placed my mug on the table.

My eyes fell shut.

This time, they didn't open again.

TEN

HUXLEY

My eyelids peeled open in time for me to see the end credits of the movie scrolling on the TV. Shit, I hadn't meant to fall asleep, and wait—what the fuck? I stared down at the head resting on my shoulder, soft, tousled dark hair catching on the stubble on my jaw. An arm was slung across my stomach, a warm weight that didn't hurt my bruises. How had Cole ended up over on my side of the sofa? It wasn't a big sofa, but even so...I would've thought that even subconsciously, we'd be trying to get as far away from each other as we could. Yet here he was, cuddling me without a care in the world, as if we hadn't been at each other's throats earlier today—yesterday? Whichever. The point was... I lost track of my thoughts as I stared down at him, his face relaxed in sleep, his dark eyelashes fanning out over his cheekbones, his chest rising and falling softly in his sleep. Speaking of his chest...the blanket he'd been under had slipped down to pool around his waist, and acres of smooth, tanned skin were exposed. Guess he'd been spending a lot of time sunbathing in the garden this summer. Why the fuck had I decided to move out again? I wanted to trace the ridges of his abs with my tongue, to—

Fuck. No. I jerked, physically trying to shake the thoughts out of me, but too late, I realised that I'd woken Cole. He grunted something unintelligible, his eyes slowly blinking open, his gaze soft and unguarded.

Then his whole body stiffened, his eyes widening. "Shit," he muttered, scrambling away from me and yanking his blanket up around his shoulders. His gaze bounced around the room, eventually settling on the TV. "Uh. Sorry."

That sleepy rasp in his tone did things to me that I really did not fucking appreciate, and I was glad that I was covered by a blanket. "S'okay." I didn't look at him either.

We fell silent, and I'd never stared so hard at film credits in my life. But now I took a sudden interest in the names of the gaffers and the grips and even the catering companies that were used. Anything was better than the alternative, which was to break this uncomfortable silence.

In the end, though, Cole was the one who broke it.

"Hux?"

"Yeah?" My voice came out way too raspy. The way he'd said my name, though...

"What happened that night?" His words were hesitant. "I...uh, you don't have to talk about it. I just keep picturing your crashed car, and I just..." He huffed out a breath, shaking his head. "Never mind."

My mind flashed back to that night, and I shivered,

suddenly cold. "It was a blur. One minute I was driving back from a mate's house. Then this car came out of nowhere—I didn't even have a chance to avoid it. We collided. My car kind of jerked back really hard, and I remember the pop of the airbags and all this screaming metal. Then, I dunno, I must've blacked out for a minute because the next thing I remembered was flashing lights and being checked by the paramedics."

I could feel Cole's gaze on me, but I kept staring at the TV like my life depended on it.

"Were you scared?" His whispered question came from much closer than I'd been expecting, and I realised that I could feel his body heat against my side.

Licking my suddenly dry lips, I nodded.

"Y-yeah." There was no disguising the crack in my voice.

He exhaled harshly. "Shit." I heard him move on the sofa, his arm brushing against mine for a second, and then he said, "I'm...uh...I'm glad you thought to call me."

There was no point in lying anymore, to him or to myself. "Me too."

As soon as I'd managed to get the words out, his hand was lifting, his fingers curving around the edges of my blanket. "Can I see?"

All I could do was nod, and then he carefully lowered the thick fabric until my torso was bared to him. Even in this dim lighting, I knew that the bruises were stark against my pale skin. My hands fell to my sides, letting him look his fill.

A finger traced across my sternum, Cole's touch so light and careful. "Does this hurt?"

I shook my head, swallowing around the lump in my throat. "No. The bruises...they don't hurt like they did to begin with. I'm so lucky that's the only thing I have to worry about. I was...I was so fucking scared. I thought...I thought for a second that I was going to die."

The finger tracing across my chest trembled. "Huxley. Look at me."

Slowly, I turned my head to meet his gaze, even though it was the last thing I wanted to do. His brown eyes were wide and so fucking pretty, brimming with emotion. When he leaned forwards, I couldn't breathe, but he just wrapped one arm around my shoulder and gently pulled me into him. I buried my face in the crook of his neck and let him hold me.

"I'm so fucking glad you're okay," he said thickly.

I lifted my arm and tentatively placed it around his waist. My words were whispered into his warm skin, and they were the sincerest words I'd ever spoken. "So am I."

I felt him sigh against me, and then his hand stroked up and down my back until I let myself relax into his body, realising that for the first time in a very long time, even if it was only for this stolen moment in the early hours of the morning, I had someone to lean on.

When I made it downstairs the next morning at around 11:00 a.m., Cole had gone into the office to pick up some samples of promotional material that his boss wanted him to work on—or so the note propped up against the coffee machine said. It also said that my lunch was in the fridge, which, yeah, it kind of pissed me off that he was treating me like an invalid, but it also made me feel things that I wasn't used to feeling in relation to my stepbrother. *Nice* things.

Fucking hell. I needed to smoke some weed to take the edge off my thoughts. Probably wasn't a good idea with the painkillers, though, was it?

Fuck it. I'd have a small one, just to take the edge off.

I rolled a quick and messy joint, then headed out into the garden with my guitar and a notebook and pen. The sun was shining, but I sat in the shade of the oak tree at the bottom of my garden. First off, I smoked the joint, propped up against the tree, music playing from my phone speaker. When the turmoil in my head was finally quiet, I began strumming my guitar and stopping to jot down lyrics whenever they came into my head. I wasn't trying to write a song, but I found the process helped my brain to settle down. I had notebooks full of lyrics, unfinished songs, and music compositions. Maybe one day I'd find a good songwriter who I could collaborate with, make some music of our own.

I huffed out a quiet laugh. That wasn't likely to happen. I couldn't even pass my A levels, which was why I was about to begin a foundation year course at LSU in order to be able to even get onto a degree course the following year. How could I expect anyone else to want to work with me? Not that songwriting was work per se—unless you were getting paid for it—but a collaborative relationship required commitment, and my track record was shit.

Cole's record was almost as bad as mine. He'd barely

scraped through his A levels, but a degree wasn't on his radar, so he had no reason to care about his results. As it was, he'd been able to charm his way into not one but two jobs. Meanwhile, I was still living off the allowance my dad was providing me with on the proviso that I took my upcoming course seriously and managed to pass all my modules.

I glanced down at my notebook, my eyes widening as I realised what I'd written.

Pretty brown eyes and a charming smile.

Fuck's sake. I drew a line through the words, then another two for good measure. Closing my notebook with a snap, I climbed to my feet. No more songwriting. Time for lunch because that weed had made me really fucking hungry.

When I inspected the fridge, I found that Cole had made me a ham-and-mustard sandwich, and he'd also left me a small pot of strawberries. My face felt weird for a minute before I realised why. I was *smiling.* A small smile, but it was there.

As I was plating up my food, adding a bag of crisps from the cupboard, my phone buzzed with a text. It was Rav, one of my mates that I'd crashed with over the summer.

RAV:

Philips came through with the good shit. Fuckin buzzin. Wanna come over tonight to sample the goods?

I paused. Normally, I'd be all over that. But I wasn't in the mood for it right now.

ME:

Not tonight but thanks for the offer

RAV:

Tom's bringing Michelle & her fit AF friends. Pills & pussy! You can't say no to that

Tempting, and I knew how the night would go down because half of my summer had passed in a haze of pills and pussy and the occasional dick. But...something had changed, and I couldn't put my finger on what it was.

ME:

Thanks but I can't. Had a car accident & prob shouldn't take pills with the shit they prescribed me. Don't want to end up having my stomach pumped

It probably wasn't true because the pain relief was mostly extra-strong ibuprofen, but I was no doctor. Who the fuck knew how it would react? It was a good enough excuse as far as I was concerned, and it was the reason I needed to stay in.

RAV:

Shit. You OK?

ME:

Yeah all good. Nothing major

That was a half-truth, but Rav...actually, none of my friends were particularly close to me. Our relationships were mostly superficial, and therefore, I wasn't interested in going into details. Rav was a good guy, although my dad would disagree because he was a "layabout with no prospects." But yeah, not close. He wasn't someone who'd hug me when I was feeling scared, not like—

I groaned loudly, thumping my head against the cabinet in front of me. My phone buzzed again, and I glanced down at the screen to see Rav had written, *Next time*. Not bothering to reply, I carried my food upstairs.

As soon as I was set up on my bed, I picked up my phone again and typed out a message, hitting Send before I could talk myself out of it.

ME:

Thinking of cooking tonight. Do you have plans or do you want me to make extra?

The reply came through a few minutes later as I was biting into the soft, crusty bread of my sandwich, the chunky pieces of ham topped with the perfect amount of wholegrain mustard.

COLE:

Depends on two things. Can you cook and what are you cooking?

ME:

Scared I'll poison you?

COLE:

It's a distinct possibility

ME:

If I was going to poison you I'd do it in a way you weren't expecting

COLE:

Like what?

ME:

If I told you then you'd be expecting it

COLE:

That's reassuring

OK if you promise not to poison me I could be tempted... What are you cooking?

I read back over the messages. It was hard to tell a person's tone over text, but it kind of felt like he was flirting with me. Something inside me fucking flipped at the thought, and another part of me filled with panic. What was the protocol when your former or maybe still current adversary—who was also your stepbrother—flirted with you?

My fingers were already flying across the screen, ignoring the panicked part of my thoughts. I'd never had a text conversation like this before, but that didn't mean I was going to stop.

ME:

It's a surprise

COLE:

I'm not convinced

ME:

You won't regret it. I'm good

COLE:

Big words. OK let's see if you've got what it takes to impress me. Should be back by 6

I glanced down at my tray of food, another small, unexpected smile tugging at my lips.

ME:

Bring strawberries

COLE:

I KNEW you'd eat the rest. OK see you later. Be ready to impress me

ME:

Oh I will

Throwing my phone down next to me, I leaned back against the headboard. Now I just had to work out how to cook an edible meal in the four hours before Cole came home.

Yeah. I was fucked.

ELEVEN

COLE

When I entered the house, I could hear what sounded like the banging of pots and pans coming from the kitchen, accompanied by several swear words, all set to the soundtrack of "Bad Place" by The Hunna. Huh, Huxley liked that song too, did he? I guess I shouldn't have been surprised. It fit his general vibe, if that made sense. I took the opportunity to head upstairs to shower off the grime of the Tube and London's streets, dropping my bag in my bedroom.

When I was changed into comfortable grey joggers and an old, faded green T-shirt, I made my way back downstairs. Huxley was next to the oven with his back to me, plating up something that smelled good into two pasta dishes. I forced myself to look away from him, instead focusing on the kitchen table, where he'd laid out two place mats and cutlery.

I cleared my throat to give him some warning that I was there. "Hi."

I wasn't looking at him, but I still heard his sharp

intake of breath over the music, telling me that he'd been surprised by my presence. "Hi."

"Want a drink?" Keeping my voice casual was more difficult than I thought it would be, mostly because I hadn't seen him since our unexpected hug in the early hours of the morning. I was one hundred percent certain that something between us had changed, but I was also very fucking sure that I wasn't going to bring anything up unless he did first.

There was a minute of strained silence, and then he said, "Yeah, okay. There's beer in the fridge."

Glad of something to do, I went to the fridge and grabbed two cans of beer, decanting them into pint glasses. I carried them over to the kitchen table and took a seat, fiddling with my phone while I waited for him to finish dishing up the food. Something told me that he wouldn't appreciate me offering my help.

He eventually approached the table with the two pasta dishes, steam wafting from the top of them. When he set them down, I got my first good look at what was in them. Egg noodles, chicken, what looked like green peppers, and maybe spring onions? I was impressed. This was already much better than I'd been expecting.

"It's a stir-fry. Uh. Sesame and soy sauce and honey. With chicken and peppers. And, uh, spring onions."

A smile tugged at my lips. "It looks good. Smells good too."

When I darted a glance at Huxley, he was clearly tense, his mouth twisting as he eyed his own bowl of food. "It looks good," I said again softly. His gaze flew to mine, and he breathed out heavily before nodding.

"Yeah. I make this all the time. One of my, uh, favourites."

My tiny smile widened into a grin. He was clearly nervous and exaggerating about how often he made this, and it was so fucking cute—

No. It wasn't cute.

I picked up the chopsticks that had been laid next to the fork beside my dish and shoved them into the noodles. The chicken and sauce clung to the noodles as I brought them to my mouth, and I made sure to force my face into a neutral expression as I chewed.

It wasn't even necessary. Okay, if I'd been in a criticising mood, I might have said that the egg noodles were overcooked, as was the chicken. But honestly, it tasted good despite that. And even better for the fact that he'd voluntarily invited me to share his meal with him.

"This is good," I said around a mouthful of food and instantly wished that I'd waited to swallow before speaking.

He smirked at me, seeming to relax. "Yeah? I told you I was good."

"You did."

His eyes met mine, deep blue depths that I knew I could easily get lost in, and I couldn't look away. Something was happening here, and it was making a combination of dread and excitement pool low in my belly.

"Have you, uh, spoken to your dad?" Changing the subject was probably the best idea.

He shrugged, his smirk falling away to be replaced with a frown. "A couple of times. I haven't told him about the accident, and I don't want to."

"I won't say anything," I reassured him around another mouthful of noodles. "It's up to you if you want to tell him or not. You're okay, so it doesn't seem like there's any point in telling him."

"Thanks." His voice was quiet. He turned his attention to his food, and we ate, the silence between us tempered by the music that he'd left playing. I was pretty sure we were both glad of it because without it, the quietness would've seemed a lot more awkward.

Every now and then, he looked up or I looked up, and our eyes met. I knew my cheeks were flushed, but I was maybe eighty percent sure that his were too.

I had no clue what we were doing.

"Did you get the strawberries?"

It was an effort to lift my gaze from my pint glass, which I'd designated as a safe spot since I'd finished my meal. When my eyes met his again, I felt a jolt, like I'd been shocked with electricity. "I got them," I said, and my voice was too fucking raspy. "I'll get them now."

Because I'd taken them upstairs when I'd come home, not wanting to intrude on his food prep time in the kitchen.

When I made it back downstairs, Huxley had cleared the table, and he had two plates laid out on the counter. "Brownies with the strawberries?"

There was something in his voice... "Are those hash brownies?"

He nodded, the corners of his lips kicking up into another almost-smile. "Wanna eat these and watch something while we get high?"

How the fuck had we gone from hating each other to

this so quickly? However it had happened, I knew I didn't want to go back to how we were before. "Yeah, okay," I said. "I'll wash the strawberries and bring them in."

The rest of the evening felt like a dream, a hazy cloud, thanks to the weed and the beers. There was the sweetness of the strawberries and the delicious chocolatey goodness of the brownies, combined with the high of the weed. The bitterness of the beer. The noises and colours of the TV, a distant presence that my brain couldn't focus on. The warm, delicious press of Huxley's body all up against my side.

By the time I went to bed, all I knew was that I was feeling more relaxed than I had done in a really long time and that I was very fucking interested in the one person I knew I wasn't allowed to like.

TWELVE

HUXLEY

The club was busy, packed full of guys who looked like underwear models, and the music was horrendous. Fucking *Britney Spears* was playing, of all things. So why was I here? Because my stepbrother had texted me earlier, luring me here with promises of a surprise that I was apparently going to love. So far, all the signs were pointing to me not loving it. This place was not my scene. I preferred dark, dingy pubs, preferably with cold beers and a live band playing.

I pushed my way to the long, crowded bar, my gaze scanning the bartenders for Cole. When I spotted him, I sent a text.

ME:

At the end of the bar. My left, your right. Where's my surprise?

I saw his hand go to his pocket to pull out his phone, and my stomach flipped when his lips curved into a smile. This was not good. He raised his head, his eyes meeting mine, and his smile widened. I bit the inside of my cheek to stop myself from returning his smile. He was unfairly hot, but when he smiled, it did something to me that really shouldn't be happening.

When he reached me, I tapped my fingers on the bar top. "Where's my surprise? If the club is my surprise, I'm not impressed."

His gaze dropped to my hand. "You repainted them."

"Huh?" was all I managed to get out before he was lifting my hand and examining my nails in the dim club lighting. My skin was prickling with...something. I didn't want him to stop touching me, and that shocking thought was enough to have me snatching my hand back.

"What's with this music?"

He blinked a few times, still staring at my hand, and then lifted his gaze to mine. A smirk tugged at his lips. "Revolve theme night. Throwback Tuesday. Not to your taste, huh? There's different music on the other floors."

"This is not my scene. I'd rather—"

Leaning across the bar, he placed a finger to my lips, his eyes widening as if he couldn't believe he'd done it, which made two of us. I...I wanted to kiss his finger—no, to pull him across the bar and kiss the smirk right off his tempting mouth.

He exhaled hard, dropping his hand. "Sorry. I wanted to guess." His smirk reappeared. "Let me see if I get this right. You would rather be...in a pub. Probably one with a band with a sexy, growly guitarist playing alternative rock so loud you can't hear your mates talking over the music. And...beer." I stared at him, and he grinned. "Damn, I'm good. You're so easy to read."

My eyes narrowed, and I curled my fingers into a fist. I wasn't actually angry, more like irritated with his smugness. "Nope. You didn't get it all right. I like listening to a lot of different music, not just alt-rock, and I don't want it to be too loud to speak." Not *all* the time, anyway.

"But I got the rest right, didn't I?"

"Yeah," I admitted grudgingly. "Now, are you going to tell me what this supposedly amazing surprise is before my ears start bleeding from this music?" Britney had finished, but now we had some fucking awful dance remix of what might have been Katy Perry or Taylor Swift or some other singer I never wanted to hear again.

He rolled his eyes. "Stop being so sulky. Wait here." When he turned away from me, his attention on whatever drink he was making, I took the time to study him, unnoticed. He was wearing a tight black sleeveless T-shirt with the club logo splashed across the front, which nicely showcased the lines of his fit body. His dark hair was artfully tousled, and his deep brown eyes were sparkling with humour as he smiled to himself. I bet he got hit on all the time working here. He was so fucking sexy.

When he returned, he placed the drink down carefully in front of me. "I call this one the Huxley," he said.

"Excuse me?" I glanced down at the...whatever it was. A cocktail of some sort, in a tall glass with crushed ice. Smooth, jet black in colour, and garnished with a strawberry. When I looked back up at Cole, he was biting down on his lip in a really distracting way.

"Uh. The Huxley. It reminds me of you. Same colour as your nails. Bitter coffee for your broody personality, a bite of ginger for your, uh, fiery moments, and sweet blackberry and raspberry for your sweet tooth. And vodka, because of those vodka shots we did at the wedding. It's supposed to be garnished with blackberries or raspberries, but I used a strawberry because they're your favourite."

I wanted to kiss him more than I'd ever wanted to kiss anyone in my life.

I had to stop myself from wanting things I couldn't have. Tearing my gaze away from his, I studied my drink. "Is this...how did you get it to be so black?"

"It's a secret," he said, leaning across the bar. "But if you behave, I might tell you." Was he *flirting* with me?

Before I could come up with a response that didn't involve me grabbing his stupidly tight T-shirt and pulling him all the way across the bar to attack his mouth, he was flagged down by one of the club patrons, a hot, shirtless guy with an amazing body. I watched as Cole gave him an easy smile, which the guy returned, batting his lashes at Cole, and jealousy coiled low in my gut. Fuck this. I didn't want to be feeling this way.

To distract myself, I took a cautious sip of my drink.

Wow. It was good. Really, really good. After taking another, bigger sip, I pulled the strawberry from the rim of the glass and placed it in my mouth. Mmm. When I'd finished it, I glanced back over at Cole, still mixing drinks for the hot guy, and swiped my tongue across my lips. The bottle of white rum he was in the process of tipping up stilled in mid-air, his eyes darkening as his gaze went to my mouth.

Fuck. I discreetly adjusted myself in my jeans, glad I was pressed up against the bar so no one, especially him, could see the evidence of how he was affecting me. The hot guy said something to him, laughing, and he shook his head with a grin, returning his attention to the drink he was making. He swiped at the guy's arm jokingly, and I gritted my teeth.

When he came back over, I'd managed to get myself back under control. Mostly. "Do you flirt with everyone like that?" I immediately wanted to take the words back.

He raised his brows. "What's it to you?"

"Nothing," I muttered. I shouldn't have said anything.

"I flirt because it adds to the positive experience for the people that come here, and it comes naturally to me while I'm working. It's fun. But that just now...that was JJ. He comes here a lot, and he flirts with everyone, not just me specifically. He's one of my cousin Elliot's housemates."

"Have you two ever..." I trailed off, not sure I wanted to know.

"Nah. I always thought it would be a bit weird with him being my cousin's housemate."

Not as weird as wanting to fuck your stepbrother, though.

"He's not my type, anyway," Cole added after a couple of seconds.

I swallowed hard. "No? What's your type?"

We stared at each other for a moment, Cole gripping

onto the bar counter so tightly I could see his knuckles standing out in sharp relief.

He cleared his throat, his dark gaze fixed on me. Damn those pretty fucking eyes. "Did you like the drink?"

My dick was hardening again, and I pressed into the bar, unable to adjust myself with his full attention on me. It was so fucking unfair that he could affect me like this. "Yeah. What was the secret ingredient?"

He leaned right across the bar, his lips brushing against my ear, and I couldn't stop the shiver that went through my body. "Activated charcoal," he murmured, and those two words had never sounded sexy until he said them in that low, husky tone.

"I wouldn't have guessed that. What's your type, Cole?"

His lips touched my ear again, his breath hot on my skin. "I think you already know the answer to that."

"Hux, meet Tom." Cole's voice cut through the stupor I'd fallen into. I'd had three of those cocktails now, and to my horror, I'd found myself *humming along* with yet another Britney throwback. I couldn't believe it.

I was sitting on a stool at the very end of the bar, away from the crowds, and I'd spent ninety percent of the time I hadn't been drinking just watching Cole work. I saw several people try to get his number, but he turned all of them down, which shouldn't have made me feel as good as it did. Now I turned to the new guy. He was tall and pale-skinned, with jet-black hair that I was pretty sure was dyed. Tattoos snaked up his arms, and metal piercings glinted in his ears. He grinned at my perusal, sticking out his hand. "Huxley. Nice to meet ya. I'm Tom."

I darted a look at Cole, who was also grinning. "Tom's your surprise."

My *what*?

Tom chuckled at my confusion. "I work here. Last week, I mentioned to Cole that I was looking for a new singer and guitarist for my band. He told me that you were...what was it...a fucking amazing guitarist and you had an incredible voice and that I'd be crazy to even think of picking anyone else."

Glancing back at Cole, I noticed his cheeks were flushing, and I was helpless to stop my smile. He'd thought of me? Even though that would've been before we'd been on more civil terms? "Yeah. I'm alright. Can you tell me a bit about the band?"

He took the stool next to me, and we fell into a discussion about the band and the music they were into. The band was named the 2Bit Princes, and they played a mix of covers and original music. Apparently, they'd previously had a lead singer, but he'd moved to Manchester, and now they were looking for a replacement. As well as Tom, the other member of the band was a guy called Curtis, who was a drummer and was also going to be starting at LSU this September. Tom was the songwriter and lead guitarist, and when I mentioned that I dabbled in writing lyrics, he seemed genuinely interested in collaborating on some of the songs with me.

Honestly, it sounded too good to be true. I'd been

scraping by in life, and I'd never really given serious thought to doing anything like this because it had seemed so out of reach for me. No one had ever encouraged me to go for what I wanted before.

But Cole had seen something in me.

He'd believed in me, even when no one else had.

THIRTEEN

HUXLEY

The large garage was piled high with junk all along one wall, but the majority of the space was clear, with three guitars and amps, microphones, and a set of drums grouped around the centre.

Tom inclined his head at me when I entered, sliding my sunglasses onto the top of my head and blinking as my eyes adjusted to the dim interior after the brightness of the sun outside. "Alright?"

"Alright." I nodded back, wandering over to the guitars. I ran my fingers over the smooth wood of the closest guitar. "Gibson. Niiice. I wasn't expecting a setup like this."

He grinned. "Yeah. Perks of having a brother that runs a music shop. All the discounts." Sliding onto one of the stools in front of the guitars, he picked up a cherry-red Gibson Les Paul and strummed it experimentally. "We got lucky. Just after I left you last night, I heard from Curtis. He thinks he's found us a bass player. So we'll play these two guitars, and Rob—our new bass guitarist, if all goes well—gets the Fender all to himself. He can play keyboard too, so we'll try a mix and see what works."

"Sounds good." I picked up the second electric guitar, another Gibson, this one in a glossy jet black. While Tom fiddled with the amps, getting everything ready, I began tuning my guitar, getting a feel for it. I had an electric guitar back at home, although it wasn't anywhere near this quality, and I normally defaulted to my acoustic. This particular guitar was fucking gorgeous, and I already knew we were going to make magic if Tom and Curtis were half as talented as Tom had said they were.

"Tom!"

I glanced up from the guitar to see two guys entering the garage. One was lean, around the same height as me, with tousled brown hair and a smile on his face. The other was tall and hulking, with wild red curls and a thick red beard. Flesh tunnels adorned his ears, and he had piercings in his septum and the bridge of his nose.

"Curtis." Tom raised his hand at the brown-haired guy, who shot me a curious look before turning to his companion.

"This is Rob. Rob, meet Tom, our songwriter and lead guitarist, and that means...you must be Huxley. Our potential new guitarist and lead singer."

I nodded, and he stepped over to me, shaking my hand. We all exchanged greetings, and then Tom suggested that we get started with a song we all knew to see how well we could work together. After going through the 2Bit Princes' usual set list, we settled on "Dakota" by Stereophonics, a song we were all familiar with.

As I sang the opening lines, the metal of the microphone almost brushing my lips, I couldn't stop my smile. Glancing to my left, I caught Tom's wide grin, and as he joined me in the chorus, it was like we'd been singing together for years instead of this being our first time. Our guitars in sync with Rob's smooth bass lines, and Curtis controlling the rhythm with his drums...this was pure fucking magic.

Right then, I wished that Cole could be here to see me.

"It went well, then?" Cole raised a brow at me, a smug grin on his face as he leaned back against the kitchen counter. His smugness was irritating as fuck, but I couldn't bring myself to be annoyed with him. If it wasn't for him, today wouldn't have happened. Not only that, but even smug, he was still hot. Memories of the club assaulted me—how he'd leaned across the bar, his lips touching my ear, sending shivers through my entire body. How I'd wanted to kiss him.

"Yeah, it was alright." Attempting to bite back my smile, I shrugged, heading over to the fridge to grab a drink. Inside, though, I was buzzing, on a high that only music could provide. And maybe a bit of the buzz was from his proximity. Why did I have to be attracted to my fucking stepbrother, of all people?

"High praise from you." The voice was so close behind me that I jumped, and when I turned around,

Cole was right there, his grin widening as his eyes met mine.

Our gazes connected, and it was like time fucking stopped. My heart was pounding, and I could feel the heat of his body as he stepped even closer. When I let go of the fridge door handle and turned to face him fully, his smile dropped, his soft lips parting. The desire to kiss him was almost overwhelming, even more so than it had been at the club. My dick was already hardening in my jeans as his darkened eyes held my gaze, and I knew he was going to be aware of it at any second, but this time, I couldn't bring myself to care.

What were we doing?

"Cole," I whispered helplessly, and it broke the spell. He blinked, inhaling a shaky breath before taking a large step back. Then another, and another, until the kitchen island was between us.

"I'm glad." He cleared his throat, attempting to get rid of his sudden low rasp that went straight to my dick. "I'm glad it went well. Really glad. I've...I'm, uh... See you later."

Then he bolted from the kitchen, leaving me alone, but not before I caught sight of the outline of his erection in his joggers.

My dick throbbed as obscene thoughts played through my mind. I wanted my stepbrother, badly, and I couldn't deny it any longer.

I wanted Cole to fuck me.

Fucking hell.

Something had to give.

Soon.

FOURTEEN

COLE

Rubbing at my eyes, still half asleep even after a shower, I stumbled into the kitchen. Coffee. Need. Coffee.

My sleepiness was gone in an instant as I took in the person who currently had his back to me, swearing at the coffee machine while he slammed his fist down on the counter. "Fuck's sake! Why won't you just work, you stupid fucking piece of shit?"

I would've smirked at Huxley's frustration, but I was too busy running my gaze over the lines of his bare back, down over his tight ass, currently covered in a pair of loose cotton shorts, and down those lean, muscled legs. Sexy as fuck.

I gritted my teeth.

Thou shalt not ogle thy stepbrother.

As I drew nearer, with Huxley still unaware of my presence, occupied with wrestling with the machine, I noticed that there were two mugs set out on the counter.

The same blue-and-white chipped mugs that we'd used when I'd made him the hot chocolate.

My heart fucking skipped a beat, and it made me do something that I knew was completely, one million percent ill-advised, but I didn't stop. I came up behind Huxley, planting my hands on the counter on either side of his arms and caging him in. He jumped a little, a gasp falling from his lips as his swearing stopped mid-tirade.

"Having trouble with the coffee machine?" I kept my voice low, and Huxley shivered against me. Fucking hell, he was so tempting. I'd never allowed myself to entertain thoughts of just how hot he was before now...okay, that was a lie, but right here, today, all my defences were gone, and my cock was in charge, apparently. "Need a hand?"

"Uh, yeah." He cleared his throat, leaning into me just the tiniest bit, and I took the invitation for what it was, pressing against his back. As we'd established, my cock was more than interested, already hardening in my boxers, and he was going to become aware of it any second. But neither of us moved, and he cleared his throat again. "Yeah. Fucking machine."

I laughed softly, balancing on the balls of my feet and angling my head to peer over his shoulder. His hair, still a little damp from his own shower, tickled the side of my head as we both looked at the machine. "Okay. First thing. You need to turn it on." Flipping the switch to the On position, I heard the familiar beep, followed by a whirr as it came to life.

"I should've realised that," he muttered. "I've never... My dad bought this machine for your mum. We had one of those pod ones before."

"It took me a while to work it out," I reassured him, forcing myself to focus on the machine and not the way his back was so warm against my chest and the way my dick was pressing against the delicious curve of his ass. "Uh, so. Next thing. You attach the milk frother—" Good, he'd already filled it. "—and then you pick the coffee you want on this bit here."

His finger swiped over the LED display, his head rolling to the side just enough that my lips would be brushing against his skin if I angled my own head any further. "Latte?"

"Is that what you want?" My head turned, my lips connecting with his skin in the barest touch.

He shivered again, subtly pressing back against my dick. He moved his hips in a small, circular movement, and I had to bite down on my lip to stop a groan from escaping.

His words came out low and raspy as fuck. "It's what I was going to make for you."

I really, really wanted to kiss him. To do way more than kiss him, if I was completely honest. To yank down those shorts and bend him over the kitchen table, pounding his tight ass with my hard fucking cock—

Shit.

With the final remaining bit of blood that hadn't gone to my dick, I managed to get my brain to formulate a reply. "Mmm. Good choice." What the fuck were we doing here? Were we really doing this? My brain was scrambling to catch up with the way we'd apparently done a one-eighty over the past week, but my dick was having no such trouble.

I let my mouth press a little harder, my touch purposeful. "Put the mug here and press for the latte."

One of his hands slid slowly over mine. It trembled slightly. "That's it?"

What had he just asked me? Oh, yeah. "That's it."

"Thank fuck," he ground out, and then he was spinning in my arms, and suddenly, my very hard dick was pressing against another very hard dick and— Whoa. Huxley's mouth was on mine.

I stumbled, off balance, but he grabbed me and yanked me into him, his hands gripping my back, his fingers digging into my muscles. When I regained my balance, I gripped the counter on either side of him again and kissed him back.

Oh. My. Fucking. Days. Soft. Wet. Warm. Addictive. Fucking delicious.

"Cole," he groaned, and I had to kiss him again, pressing him into the counter, my tongue sliding against his as we kissed harder, deeper.

He broke the kiss, his mouth going to my throat, where he lightly bit down. His stubble scraped across my skin, and I fucking loved it.

I lifted a hand to the back of his neck, scratching my fingers across his nape, making him groan into my neck as he ground his hips against me. My dick was fucking leaking as it slid alongside his, I was so hot for him. "So this is mutual, huh?"

He raised his head, his gaze meeting mine, all heavy-lidded, blown pupils. "If by mutual you mean do I wanna get my dick inside you, then yeah, it's mutual."

I let go of the counter, finally getting my hands on his

ass. Fuck, it felt so good. "I was thinking more that I'd bend you over the kitchen table and fuck this sexy ass that I can't stop looking at."

He moaned low in his throat, but then his brows lifted. "You can't stop looking at it?"

It probably wasn't a good idea to give him too much ammunition. "You were sticking it out when you were trying to work out the coffee machine. I couldn't help noticing it."

"Hmmm." He eyed me, clearly unconvinced, so I kissed him again. It was better when he couldn't speak. Better when neither of us spoke, probably.

"I wanna fuck you so hard," I rasped against his mouth as I manoeuvred us over to the kitchen table, immediately failing at my plan of not speaking. It was like now the barrier of our animosity was gone, even if it only turned out to be temporary, there was nothing stopping this insane attraction that had flared to life between us. But even so, this was a bad idea for so many reasons. But weren't bad ideas the best ideas?

I ground against him, my cock fucking throbbing against his. "This is such a bad idea. But I don't want to stop. I have to have you."

His teeth clamped down on my lower lip, sending a bolt of pleasure-pain down my spine. My dick was so hard it almost hurt. "Then fucking do it. One and done. Get this out of our systems."

We were on the same page. I spun him around, pressing down on his back to make him bend over. His ass stuck out perfectly, his shorts draping over the curve in a way that made my cock jump and my breath catch in my throat. Yeah, that was what I was talking about.

"Stay there." There was a condom and a sachet of lube in my wallet, which I'd dumped on the console table in the hallway yesterday. I ran for the hallway, hoping that he wouldn't change his mind in the minute I was gone. But I didn't have to worry. When I returned, he was still in place, his hands on the edge of the table and his back curved.

As I drew closer, he turned his head to look at me, his gaze going straight to the huge tent in my shorts, and his eyes darkened. "Excited?"

"Yes. But so are you." Reaching him, I curved my arm around his body, lowering my hand to his dick and rubbing my palm across the head. Even through the cotton, I could feel how much he was leaking for me, and it made me even hotter.

"Fuuuck. I don't even think I like you, but I fucking love what you're doing to me," he panted, pushing into my hand while I rubbed my own straining erection against his ass.

Ripping the condom packet open with my teeth, I circled my thumb around his cockhead. "We don't have to like each other to fuck. And like you said, this is a one and done."

"You make a good point. Fucking get on with it, then."

Impatient. Then again, so was I. Wasting no more time, I released his dick and yanked my shorts down, glad I hadn't bothered with underwear.

"Shorts off," I instructed hoarsely, and he temporarily

let go of the table to tug them down. Rolling on the condom, I stepped out of my shorts. I opened the lube packet, getting it all over my fingers and a bit on the kitchen floor, before I kicked his legs farther apart. What a sight. All that creamy skin and that sexy, tight ass bared for me. It had clearly been way too long since I'd last fucked someone because all I could think about was ramming my cock inside him. I reminded myself that I had to prep him, and I needed to be careful of his bruises, but I'd never struggled so much to hold myself back.

He pressed back into my fingers as I opened him up way quicker than I usually would, his hands flexing against the wood of the table as he held on with a death grip. He lowered his head, exhaling hard. "Fuck me. Can't wait anymore."

"You read my mind." Withdrawing my fingers, I smeared the remaining lube over the condom and positioned the head of my cock at his hole. Then I pushed inside until I was balls-deep.

Fuuuck.

We both groaned. Huxley's ass was fucking heaven, so tight around my dick. I wrapped my fingers around his hip for purchase as I began thrusting in and out, the noises spilling from Huxley's mouth driving me crazy as I picked up the pace, pounding in and out of him. He arched his back a bit, changing the angle, and he gasped.

"Fuck. Yessss. So. Fucking. Good."

My thoughts exactly. How we'd managed to get to this point, I had no idea, but he was so fucking hot and tight around me, and the way he looked...the curve of his back, his fucking sexy ass that was taking my dick so well, his lean arm muscles tensed and holding himself up on the table as he took everything I gave him. Beautiful. The way the back of his neck already had beads of sweat, teasing the ends of his hair. It made me want to lick it. This man was fucking gorgeous.

My balls were already drawing up, and there was no way I was going to be the first to come. I wasn't the best at multitasking, but wanking a partner while keeping up a good rhythm fucking them was something I'd become quite proficient in.

Swiping my palm across the head of his cock, coating it with his precum, I wrapped my hand around Huxley's thick erection and began stroking him up and down.

"Good?" The word was punched out of me as I slammed back into him.

A low, desperate sound came from his throat that went straight to my cock, and I just *knew* he was close. I was too, balancing on a knife edge, but I really wanted him to come first.

"You're so fucking hot, Hux. Feel so good around my cock. Fucking love your dick in my hand." Words were falling from my mouth unchecked.

Huxley's whole body stiffened, and his dick pulsed in my grip. My eyes fell closed as I finally let go, following him over the edge as I unloaded into the condom, still gripping his dick, my hand slippery from his cum.

So. Good.

But there was no time to bask in the afterglow. He was already pulling away from me, awkwardness descending between us, reality hitting hard as we both took in the fact that we'd just fucked in our parents' kitchen at ten in the morning with the sun streaming through the windows and doors, and there was now lube and cum desecrating the polished herringbone flooring. And the table.

That wasn't even the worst part.

No.

The worst part happened about two seconds after I'd pulled out of Huxley. His phone started ringing.

It was on the kitchen table in front of us. Maybe he was cum-drunk or something, so he wasn't thinking straight, but he immediately swiped to answer it, and as he did so, I saw the name on the screen.

Dad.

Fucking fuck.

"Huxley." David's voice came clearly through the phone because I was standing so close to Huxley. With my softening dick in my hand and a condom full of cum. There were so many things wrong in this scenario.

Stepping away from where Huxley was slumped over the table, I dealt with the condom and then crossed back over to him so I could grab my shorts. I winced at the mess on the floor. I was *so* going to nominate Huxley for clean-up duty.

"Yeah...he...uh, he's here," Huxley was saying into the phone, shooting me a quick, panicked glance. My eyes widened, and I shook my head, backing away, but it was too late. He'd already put the phone on speaker.

"Cole! You're here!" My mum's voice came through the phone. I gritted my teeth. Out of the corner of my eye, I saw Huxley pull his shorts back on.

"Hi, Mum."

"I'm glad we caught both our sons together," she said, and I bit down on my tongue so I didn't give in to the sudden urge to laugh. Not that there was anything funny about the situation.

"This is an unexpected surprise. We have some news for you both." David came back on the line. "Unfortunately, we've had to cut our honeymoon short. There's been a rather urgent situation at the office that needs to be managed delicately, and I have no choice but to be there to oversee things. We'll be back tomorrow night."

They said some other things, but my brain could only focus on the words *we'll be back tomorrow night.*

When the call ended, Huxley dropped his head to the table with a groan, thumping it against the wooden surface once, then again for good measure.

"Great. They'll be back tomorrow, and I found out yesterday that my car was written off. They can't repair it. My dad's gonna lose his shit when he finds out that I've wrecked it."

"Do you think he'd react worse to that than finding out his stepson fucked his son?" I mused, trying to lighten the situation. Not that he'd find out—that would be a secret we took to the grave.

"Fuck off, you're not helping." Hux spun around, shoving at my chest, and I stumbled backwards, unprepared.

"Okay, okay. Calm down. Let's think about this for a second." My brain started whirring, finally recovering from the epic dicking I'd given him. "You had a private number plate, yeah?"

"Yeah..."

"Okay, so, we just need to find another car that's the same model and colour as your scrapped car and then transfer the plate to it. Your dad will never know."

"Yeah, but that will take time. I don't even have the insurance money yet, and who knows how long it'll take to find a new car that looks the same?"

I stepped closer to him, needing to calm the panic flaring in his eyes. Fuck, he was so tense. Ready to explode any second. Carefully, I stretched out my hand and placed it on his arm, my fingers curling around his bicep. "Hux. We'll sort it. We can make up a story about the car being in the garage for some reason to buy us time."

His eyes met mine, his expression wary. "We?"

I nodded, lightly squeezing his arm. "Yeah. *We*."

I still didn't know what was going on between us or whether we even liked each other or not, but the aftermath of his car crash would be forever imprinted on my mind. There was no way I would let him face this alone.

FIFTEEN

HUXLEY

Yesterday had been...weird. I'd woken up, only concerned with making coffee. The next thing I knew, my stepbrother had his dick in my ass. Then I'd come harder than I could ever remember coming in my life, and just as I was basking in the afterglow, my dad had called and dropped the bombshell that he was coming home tomorrow—or today, now.

I dragged my plectrum across my guitar strings, the twang of the notes sounding loud in the quiet. I wasn't even in the mood to play, but I wanted to try and relieve some of this tension that was giving me the headache of the century.

Cole had fucked me.

We'd agreed it was a one-time thing to get it out of our systems, but here was the problem: he'd fucked me so good I was already craving it again. What the fuck was that all about? I didn't even like him. Much. And more importantly, he was now officially my stepbrother, despite the fact that neither of us had a say in our new relationship status.

Cole Clarke.

He wasn't even my type. Aside from girls, I generally went for boys that were more like me. Fuck-ups. Outsiders. Not interested in toeing the line. Not that Cole was squeaky clean—in fact, I was more or less certain that he wasn't, based on the way that his A-level results had been subpar, his antagonistic temperament, and the fact that he smoked weed like a fucking stoner. But he was a good boy in my dad's eyes. A suck-up, as far as both our parents were concerned.

And...there was also the fact that he was fucking hot. There was no way that I could deny it. I'd had a hard-on for Cole Clarke for way too long, and I needed to get over it. Now. Yesterday would never, ever be repeated.

My fingers flew across my guitar strings, and it took me a moment to realise that I was playing "Strawberries & Cigarettes" by Troye Sivan. Fucking strawberries. I'd never be able to eat them again without thinking of Cole. It still didn't stop me from singing the lyrics, though, even if Cole was on my mind.

Maybe I had a problem.

There was a knock on my door, and I groaned, my head thumping against the wall behind me. The one person I wanted to avoid, and of course he was here.

I hadn't even answered Cole's knock, but he entered my room anyway. I did my best to ignore his sexy-as-fuck body, clad in loose black football shorts and yet another faded T-shirt that stretched across his pecs. This time it was a navy colour, and even though it made no sense to me, the colour managed to make his brown eyes pop beneath his thick lashes.

He hovered in my doorway. "Hi. Uh. Your dad and my mum will be back in a few hours. I tried to tidy the house, but do you want—"

His words were cut off by my mouth. Somehow, and I didn't know when, I'd laid down my guitar, climbed to my feet, and walked over to him. And now my hand was curved around the back of his neck, holding him in place while I kissed him.

I was kissing Cole.

His hands gripped my hips, his touch like a hot brand. I'd never had a kiss like this in my life, and never from someone I thought I hated up until very, very recently. Everything I'd ever done with anyone prior to Cole coming into my life had been based on mutual interest.

But now it was Cole, the person who'd been antagonising me for months.

And he was giving me the best fucking kiss I'd ever had.

"Fuck." He ripped his mouth away from mine, breathing hard. "I didn't mean to do that."

He didn't? *I* didn't. My brain scrambled for a reply, but I didn't have one.

Eventually, my brain rebooted. "Our parents come back today. We need to make sure that the house is sorted. Show me what's left, and I'll help you."

It was a good enough excuse to avoid talking about what had just happened, and I was horrified that I'd kissed him so easily. Focusing on something else was what we both needed.

Cleaning up the rest of the house took a couple of hours. After we'd done the inside, we ended up in the garden, where I helped Cole drag the patio furniture back into place. The early evening sun bathed the patio with its rays, and as I watched Cole's muscles flexing, the sunlight caught his eyes, and my breath caught in my throat.

"Cole. Come here." My voice was low.

His gaze shot to mine, and he immediately dropped the cushion he'd been holding, stalking over to me. When he drew close enough, I grabbed a handful of his T-shirt and yanked him into me, slamming my mouth down on his.

Yes. This was what I needed.

Backing me up against the side of the house, his arms came around me, his hands threading into my hair and tugging my head back to expose my neck. Then his mouth moved to my throat, his stubble scratching lightly across my skin. "Hux. We shouldn't be doing this," he panted, grinding his erection against my thigh.

"I know." I closed my eyes as I rocked my hips forwards. "Can't stop."

"Cole? Are you here?"

We sprang apart, both of us breathing heavily, panic filling the space between us. That was Cole's mum's voice. *Shitshitshit*. Our parents were back early.

"Quick." Cole grabbed my hand, tugging me forwards. "Games room. Side door."

We made it into the games room, and Cole got busy

flipping the TV on while I grabbed the Switch controllers and booted up Mario Kart. He dived onto the sofa next to me, taking one of the controllers.

My heart was still racing, but my dick had deflated, thanks to the sudden scare we'd had. I glanced over at Cole, who looked a bit dishevelled but otherwise normal. I just had to hope that neither of our parents noticed anything odd about us.

The handle of the games room door turned, and I gritted my teeth, steeling myself. Next to me, I could tell that Cole was holding his breath, his whole body poised on a knife edge.

"There you are." Cole's mum appeared in the doorway, closely followed by my dad. Both of them wore wide smiles as they took us in.

"We should go away more often if this is the result." My dad laughed loudly. Cole and I didn't join him. "Look at the two of you getting on so well."

"Two brothers having fun together," June added, and I saw Cole visibly wince out of the corner of my eye. I didn't dare to speak.

After giving us a comprehensive rundown of their cruise, they eventually left us alone. Thank fuck.

Cole collapsed back against the sofa with a groan, rubbing his hand across his face. "That was way too close."

"Way too close," I agreed. "That can never happen again."

He nodded. "Yeah, but what do we do about it? Because I keep wanting to kiss you. And more."

That was true for me too, but I wasn't at a point

where I could admit it out loud. I thought for a minute. "Maybe I should move back in. Properly. There's a thing called exposure therapy, isn't there?" My stomach churned as I made myself say, "This...thing between us, it's just a temporary attraction. You're not even my type."

Biting down on his lip, he lowered his gaze so I couldn't read his expression. "Yeah. It'll fade soon enough. It always does." A tiny smile tugged at his lips, although it seemed forced. "If you're staying, I'm pretty sure I'll stop finding you hot after a week or two when I've had to look at your face every morning over breakfast."

"Same." I nodded, ignoring the sick feeling in my stomach. "You're right. It'll fade away quickly, especially with our parents reminding us every five fucking seconds that we're a family."

Cole finally raised his head to meet my gaze, his expression unreadable. "I'm going to go now. Because I know you're right, but I also know that I really want to carry on what we started outside. My willpower isn't that strong yet."

He'd made it all the way to the door when I launched myself to my feet. My arm shot out, and I grabbed him around the back of his neck, spinning him around to face me. His eyes were wide and full of emotions he was trying to hide. It fucking hurt to look at him, but I couldn't look away. "One last kiss. And then we're done."

"Huxley," he whispered, and I kissed my name from his lips, soft and slow and so fucking sweet, like I'd never kissed anyone before.

And then he was gone, and I was left alone.

SIXTEEN

COLE

"Where are you going with that?"

Stopped in his tracks by his dad's voice as he passed the kitchen, Huxley turned, carefully placing down the guitar case he'd been carrying.

"I'm in a band, and I'm going to play." Leaning against the door frame, he stared at his dad, a stubborn slant to his mouth, his body tense as if he was waiting for—

"More time-wasting? You need to be preparing for university, not—"

"Wait a minute!" Somehow, I realised I'd climbed to my feet, and I was leaning across the kitchen island, my hands planted on the edges. "Huxley's really talented. This isn't time-wasting."

David's brows rose, his mouth opening and closing a couple of times before he cleared his throat. "Well. That's...well. Okay." He sighed, returning his attention to Huxley. "I'm sorry. Maybe I was a little harsh. I only want what's best for you, Hux. I'm happy you're living back here again and the two of you have managed to put your differences aside." Stepping closer to Huxley, he placed his hand on Huxley's shoulder. "I want you to promise me that it won't affect your upcoming studies, and I'll say no more about it."

"Yeah, okay," Huxley mumbled, shrugging off his dad's hand. "I've got to go. The others are waiting for me."

"Why don't you go along, Cole?" David's gaze flicked to mine, narrowing on me. "The two of you can tell us how it went over a late dinner, perhaps."

It probably meant that he wanted me to be his spy, to make sure that Huxley really was doing what he said he was doing. David trusted me, after all. But all I could think was that it gave me an excuse to spend time with Huxley. Time I wanted so fucking badly but I knew I wasn't supposed to want.

"Uh..."

"Tom's picking us all up in his van. There's room." Huxley stared down at the floor as he spoke quietly, his fists clenching and unclenching at his sides.

What did it say about me that I just wanted to wrap my arms around him? It wasn't even about my dick this time. I wanted to reassure him, to let him know I was on his side. "I'll come. Give me two minutes to grab my wallet and my phone."

His head rose, and as soon as his sapphire eyes connected with mine, my heart fucking jolted, and my breath caught in my throat. Outwardly, I did everything I could to play it cool, but the sudden heat that flared in his gaze before he quickly turned away told me that I hadn't been successful in hiding the way he affected me.

"I look forward to hearing about it later." David gave me a genuine smile as he headed over to the kettle.

I escaped the kitchen, hearing him saying something about making my mum a cup of tea, but my attention was diverted by Huxley.

When we were both outside the front door, waiting for Tom's van to show up, I found myself moving closer and closer—

"Cole. Don't." He took a pointed step back from me.

Fuck. "Shit. Sorry." I scrubbed my hand across my face, slumping back against the wall. How long would it take for this attraction to fade? How long would it take before I stopped wanting to kiss his soft fucking lips every time I saw him? No one had ever held my attention for this long before. Especially not when that someone had been the bane of my existence until recent events had irrevocably changed things between us.

"Tom's here," he said, and the relief was clear in his voice. I straightened up, watching as a battered black van came to a stop in front of the house.

Introductions were quick, and I soon found myself squeezing into the back of the van with all the band's equipment, along with a huge, red-haired guy named Rob. In the front, Huxley and the 2Bit Princes' other bandmate, Curtis, were seated with Tom. I'd immediately jumped in and insisted that Hux should take the space in the front seat because it would have been a bad idea for the two of us to be alone back here together. A very bad idea when I was having so much trouble staying away from him.

"You're Huxley's brother?" Rob had a bit of a Brummie twang to his accent, and it intrigued me.

"His stepbrother, yeah. We only met this year. What about you? You're not from around here, are you?"

He chuckled softly. "Did the accent give it away? I lived in Birmingham until I was ten, but my parents split up, and we ended up in Watford. I met Curtis a few weeks back when I moved to central London for my job—he's one of my housemates for now, until I find a more permanent place to live. I'm an accountant."

I stared at him, and he laughed again. "Yeah, yeah. I've heard it all before. I don't look like an accountant. But what's that even supposed to mean? How should an accountant look?"

"Boring? Yeah, that's probably a stereotype." A smirk pulled at my lips as I eyed him. "Is it true that accountants have the most boring jobs but the wildest lives?"

Rob tapped the side of his nose, amused. "I can't give away any secrets. It's part of the accountants' code."

I hadn't realised that the van had come to a stop until the doors were flung open, and I found Huxley there, glaring at both me and Rob. Rob snorted under his breath at Huxley's face, shaking his head, while I blinked at the sudden brightness, frowning at him. What was his problem now?

When we climbed out of the van, I sidled up to Huxley, dipping my head to his ear and speaking in a low voice. "What's the matter?"

"Nothing. Enjoyed your time with Rob, did you?" he

bit out and then stormed off, yanking equipment out of the back of the van with short, jerky movements. *For fuck's sake.* I rolled my eyes. *Jealous little fucker.*

Wait a minute. Huxley was jealous? A smile curved over my lips. I shouldn't have liked the idea so much, not when I was trying to get over him, but I did like it. A lot. After all, he'd told me that he didn't even think he liked me right before I'd fucked him, *and* he'd told me I wasn't his type. Yet here he was, getting jealous over a completely innocent conversation...so maybe this thing wasn't as one-sided as I thought.

"Stepbrothers, hey?" Rob smirked as he pulled his guitar from the van. "There's more to this story, isn't there?"

"We're not..." I trailed off with a shrug. There was no point in incriminating either myself or Huxley any further.

When we were inside the small, dingy pub where the 2Bit Princes were apparently doing a test gig, the band members quickly and efficiently set up their equipment, like they'd done it hundreds of times before. I knew they'd only been together for a short time, and both Hux and Rob were new, but to look at them now, it was like they'd been together for years. Wandering over to the bar, I ordered a pint while they tuned their instruments and did a sound check, or whatever it was bands did to get ready to perform. I couldn't tear my gaze away from Huxley, and I knew Rob had noticed because he kept shooting me winks and grinning every time Huxley caught his eye. It both amused and worried me. The worry was because if Rob could tell from one short interaction that Huxley was jealous and there was more to our relationship than just stepbrothers, then that meant we were being way too obvious. I had to dial it back even further, somehow. There was no way our parents could even get a hint that anything non-brotherly had happened between us.

I settled on a stool at the bar, leaning back against the worn wooden counter as the band took their places. From here, I had a great view of the small stage area. The pub was fairly quiet, but those who were there had turned their attention to the stage, and a feeling of pride rose inside me. Pride that Huxley was up onstage, getting to do something he loved. His angry look had disappeared, replaced by concentration, and there was a look of pure joy in his eyes that made something inside me go all fucking soft and melty.

Get a fucking grip, Cole.

Then Huxley began to sing, and every single thought flew from my mind.

This was what he was born to do. He was so fucking talented. His black-tipped fingers danced across the strings of the glossy ebony electric guitar, his sexy voice sending shivers down my spine as he rasped into the microphone. I shifted in my seat, sliding my arm across my lap to hide my growing erection. Fucking hell. He was giving me a boner from just seeing him up on the stage. I hadn't even known that was possible. How was I going to make it through the whole set list? My cock was going to be seriously unhappy with me.

I managed to calm my dick down by reminding myself of how unimpressed our parents would be if they knew that their sons had recently defiled their pristine kitchen and concentrated on taking slow pulls from my pint as the band began their next track. Forcing myself to focus on things other than my hot-as-fuck stepbrother, I let my gaze drift across the rest of the band. Tom was playing a red electric guitar and harmonising with Huxley on the choruses. I knew he was good because he'd shown me video clips at work before, but it was different seeing him in action in person. Curtis was on the drums, pounding out a rhythm, and Rob played what I thought was a bass guitar, crooning into his mic every now and then.

They were amazing. Maybe I was biased because I knew Hux and Tom, but based on the reception from the pub patrons, probably not. Quite a few people were up on their feet, pressing closer to the stage, cheering and singing along with the lyrics, and more were piling into the pub. I didn't recognise the song, but I made a mental note to get the set list from Huxley or Tom afterwards so I could learn the songs they were singing. I wanted to be involved. Yeah, I could only just about carry a tune, but that wasn't the point, was it? The point was to be supportive. Fuck it. Maybe I could help them out some other way. I knew a lot of people. There had to be a way to get them some more publicity. Maybe I could look into them doing a charity gig for my work, perhaps. Or maybe... My mind raced. I was sure one of my bosses had mentioned that he had a brother or cousin or someone that owned a music studio somewhere south of the river...

Tugging my phone from my pocket, I fired off a quick email and then settled back to watch the band. The next song came to an end, and Huxley glanced over at Tom, mouthing something to him that I couldn't see. Tom grinned and nodded, and then Huxley motioned for one of the pub staff members to come forwards. They conferred quietly for a minute before the guy disappeared into the crowd. He returned with a stool, which he passed up to Huxley. When Huxley had placed the stool behind his mic and adjusted the stand height, he headed over to the side of the stage and brought out his acoustic guitar.

Rob took his place behind the keyboard to the left of the stage as Huxley lowered himself onto the stool and began to strum his guitar softly. He searched the crowd, his eyes finding mine, a tiny smile curving over his lips as he sang the opening lines of "Strawberries & Cigarettes" by Troye Sivan, holding my gaze the entire time.

It gave me fucking goosebumps. There was nothing that could've made me take my eyes off him, not even if our parents had suddenly appeared.

This was so much worse than I'd thought. I was in way too deep.

When the song had finished and Tom announced that the band was taking a break, I found myself sliding from my bar stool and making my way through the crowd, all the way up to the stage.

"You're fucking amazing," I whispered when I reached Huxley, staring into his gorgeous blue eyes rimmed with smudged black liner.

"Yeah?" He gave me a small, almost shy smile as he placed his acoustic guitar back on its stand. His chest was rising and falling rapidly, probably with the adrenaline from performing onstage. Then again, my heart was racing, and that had everything to do with the boy standing right in front of me, close enough to reach out and touch.

"Yeah. I mean it. You're amazing," I repeated, and his smile widened. Picking up the black electric guitar, he bit down on his lip, lowering his gaze to the scuffed wooden boards beneath our feet.

"I sang that song for you. It...reminds me of uh, you, I guess. You know. The whole strawberries thing."

My eyes widened as I replayed his words in my mind. It reminded him of me? And he chose to sing it...for me? "Fuck. *Fuck*. I wish—"

"I know." His lashes swept up as his eyes met mine again, his smile disappearing. "But we can't."

Reaching out, I placed my hand on his wrist, lightly squeezing, the only touch I'd allow myself to give him, even though I wanted so much more. My other hand landed on his mic stand, my fingers curling around the cool metal as I held his gaze. I couldn't disguise the crack in my voice when I replied. "I know. I'll try...I'm trying...I can't—"

"Cole."

I shook my head, taking a step back and letting my hand fall from his arm. "Don't. It's okay. I'm, uh, gonna get another pint. I'll see you after, yeah?"

He gave me a sad smile as I walked away.

SEVENTEEN

COLE

Do you know what doesn't work? Exposure therapy. Not when it came to me and Huxley. Ever since he'd officially moved back in, it had been an exercise in torture. Not only that, but it also felt like I'd had to almost constantly listen to our parents talking about us being brothers and being a family, and I honestly couldn't stand it. I wanted Huxley more than ever, and it was taking everything in me to hide it.

Hux himself didn't seem so affected, but he'd spent a lot of time in his room with his guitar, so I couldn't help wondering if he was trying to avoid spending time with me for the same reasons I was trying to keep my distance from him. Every time I looked at him, I wanted to kiss him again, so I was doing my best to stay away unless our parents were there as a buffer.

The attraction would fade soon. It always did.

"Cole." My mum interrupted my thoughts, peering around my bedroom door. "You're not working tonight, are you?" I shook my head, and she smiled. "Could you come downstairs? David and I have something we'd like to speak to you about."

"Uh, yeah. Alright." Climbing to my feet, I followed her downstairs. David was in the living room, standing in front of the fireplace, but all my attention was taken by Huxley, who was sprawled on one of the sofas. I hadn't seen him since yesterday, and my eyes drank him in like he was a glass of water in the desert. Fuck, why did I want him so badly?

His gaze met mine, and a flush appeared on his cheeks. He bit down on his lip, quickly looking away. Yeah...I was pretty sure he was at least a bit interested in me, whether he wanted to admit it or not.

"Ah, Cole. Take a seat." David pointed to the sofa Huxley was sitting on. I took a seat as far from Hux as I could, while my mum crossed the room to stand with David.

When David had tugged my mum into his arms, he cleared his throat. "Now that the two of you are finally being civil and acting like the brothers you now are, we would like to try something new." He paused dramatically. "Family bonding nights."

Huxley and I exchanged wary glances. *What the fuck?*

David continued speaking, "I know that you're both adults in the eyes of the law, but you live under our roof, and until either of you move out permanently, you're under our house rules." His mouth twisted. "It's not about us trying to control the two of you. It's...I know that we haven't been the most attentive of parents when you were both growing up—" He paused, his gaze darting to my mum, who squeezed his arm. "—but now we're all together, we both want to rectify that. To build a family. While we were on our honeymoon, we discussed things, and the idea came to us."

My mum picked up where he'd left off. "You're brothers now, and we want to strengthen our relationship with the two of you." Her gaze turned to me, wide and pleading. "We know that it's difficult for you both, but we want to be a real family."

Fucking hell.

I managed a small nod, keeping my jaw clenched tightly so I wouldn't accidentally say anything because I had no sentences that didn't begin with "what the" and end with "fuck."

She gave me a smile. "We're not trying to put rules in place, but we'd like to at least try this. Once every two or three weeks, we'd like an evening with us all together. Catching up on your lives. Perhaps playing board games or watching a film."

Me and Huxley in the same room was bad enough, but a whole evening with our parents forcing us to act like the siblings we absolutely fucking weren't?

Fuck my life.

David matched my mum's smile as he removed his hand from around her waist. "On that note, I have a couple of surprises for you all."

Blowing out a heavy breath, I tipped my head back against the sofa, waiting for whatever was coming next. It wasn't like this evening could get much worse.

David disappeared, but before I had time to wonder where he was, he reappeared with a bag in his hands. He held it out to my mum, who gave him a questioning look as she dipped her hand into the bag.

She pulled out a framed photo. It was of her and my stepdad on their wedding day, posing in front of the lake. They both looked so happy. Okay. That was kind of sweet.

"Oh, David. This is beautiful." Her fingers caressed the glass. "The pictures were ready much sooner than I thought they'd be."

David shook his head. "Only a couple. I spoke to the photographer because I wanted you to have at least one when we returned from our honeymoon." He reached into the bag. "This was the other one."

Huxley and I both stiffened in our seats because it was a photo of the two of us in front of the lake. We were both smiling, and even though I knew that moment had been filled with loathing, we looked so good together I couldn't look away.

"Our two boys. You both gained a brother that day." David appeared in front of us, slapping Huxley and me on the shoulder, and I did my best to hide my grimace. Luckily, he didn't seem to notice because he took a step back, clapping his hands together.

"Now. Who wants to play Trivial Pursuit?"

EIGHTEEN

COLE

The next morning, as I covertly eyed Huxley across the kitchen table because I couldn't help myself, I took in the way his black eyeliner was smudged across his skin, like he'd rubbed his hands over his eyes. His face was paler than usual too, and when I lowered my gaze, I noticed his fingers drumming on the surface of the table, which made it seem like he was on edge. Shit, what was wrong with him?

David and my mum were eating breakfast with us, so I couldn't ask him outright. Instead, I sent him a text.

ME:

What's wrong? Don't tell me nothing because I don't believe it

He shifted in his seat, his hand going to his pocket, and his gaze dropped to his lap. He didn't look at me, but a few seconds later, I got a reply.

HUX:

Nosy fucker. Nothing's wrong. Got to pick up my replacement car this morning

Oh. *Oh*. I immediately put two and two together. He was going to pick up a new car, and it would be the first time he'd driven since the crash. No wonder he was on edge. I was relieved that his dad had bought the story about his car being in the garage so he wasn't going to have any added stress, but at the same time, it meant that I was the only one who could know what had him so worried.

ME:

I'm coming with you. No arguments

I saw Huxley bite down on his lip when he read my reply, and there was a tremor in his fingers as he tapped out a response. Fuck. He played the part of being strong and untouchable, but there was a vulnerability deep inside him that just made me want to be his fucking protector or something. He didn't really need protecting, but he did need someone to lean on. And today, that person would be me. If he let me.

My phone buzzed.

HUX:

OK

I exhaled, relieved that he'd agreed without a fight. Quickly finishing up my breakfast and skilfully avoiding my mum's questions about my weekend plans, I dumped my plate in the dishwasher and made my way to my bedroom.

ME:

What time?

HUX:

20 mins. I booked an Uber

After a quick shower, I grabbed my phone, wallet, and sunglasses. I thought for a minute and then headed to his room. There was no answer when I knocked, so I pushed inside, immediately spotting the item I wanted. *Huxley's guitar*. I carefully placed it inside its case and zipped it up and then tugged the straps over my shoulders. There was an idea prodding at the back of my mind, and I didn't know if it would help, but I had a feeling it might.

When I saw Huxley waiting by the front door, he frowned at me, his mouth opening, but I shook my head at him. "Trust me."

He huffed out an irritated breath but remained silent.

Good.

Huxley's hands trembled as he placed them on the steering wheel.

"Hux." I curled my fingers around his thigh. "It's going to be okay. I'm here."

He gave a short, jerky nod, staring straight ahead with his jaw clenched.

"We'll go slow, okay? Follow the satnav instructions,"

I said, trying to keep my voice low and calm. It worked with spooked animals apparently, and Huxley may not have been an animal, but he was definitely spooked. I couldn't blame him. Scenes of that crash flashed through my mind for the hundredth time, and I unconsciously tightened my grip on his thigh. If I were his boyfriend, I'd—

Never mind that, because being his boyfriend wasn't an option. All I was here to do was to support him through this.

His breaths came hard and fast, and I rubbed my thumb up and down over the seam of his jeans. "We'll go slow," I repeated as he turned the engine on.

A choked noise fell from his lips that he instantly stifled by clamping his mouth shut, his fingers gripping the wheel like it was a lifebelt. When we began moving, I loosened my hold on his thigh but kept my hand where it was. It felt like he needed it, needed to know I was right here with him.

His brow furrowed in concentration as he slowly and carefully navigated towards the outskirts of London, until we were seeing more greenery than built-up areas. I'd set the satnav destination to Coulsdon Common, almost on the border of Surrey, a huge green area far enough south of the centre that we could avoid the heavy traffic and far enough that it would give him a chance to get used to driving again.

When we reached our destination and Huxley had cautiously parked, I took his guitar from the boot and set off so he had no choice but to follow me.

"Where are we going?" His voice was a little breathless, and I really, really wanted to hold him in my arms and tell him that I was so fucking proud of him for not only facing his fears but for driving all the way here. This infatuation I had for him wasn't going away anytime soon. And I couldn't really even call it an infatuation because it was turning out to be so much more than that.

Why did he have to be my fucking stepbrother?

Sliding my sunglasses on to hide my eyes, I glanced over at him, hoping my voice wouldn't give away the feelings I was doing my best to push aside. "I thought we needed some fresh air." He nodded, accepting my response easily.

When we'd been walking for a little while, I stopped. We were in an open space filled with tall grasses and flowers—maybe a meadow or something, I didn't know. The sun was shining, and there was no one else around other than a few dog walkers in the distance.

"Here," I said, removing Huxley's guitar case from my shoulders and setting it down. "Let's sit."

He flopped down onto the grass next to me, leaning back on his elbows. "Cole."

I turned my head to meet his gaze behind the barrier of my sunglasses. "Yeah?"

"Thanks." His smile was small but genuine. "I don't know how you knew when I wasn't even aware of it myself, but I really needed this."

The way he was looking at me...

I wanted him. Way more than I'd realised. And I couldn't have him.

"I'm glad it helped." My voice betrayed me, cracking at the sudden rush of emotions. Quickly turning away from him, I cleared my throat, attempting to regain my composure.

"Cole," he said again, so fucking soft. "Look at me."

"I can't."

"Why?"

Moving into a seated position, I shoved my sunglasses up to the top of my head, my arms encircling my drawn-up knees. Burying my face in my arms, I let the truth fall from my lips, even though it was the last thing I wanted to do. I couldn't keep it in anymore.

"I'm so fucking tired of pretending that I don't want you."

I wasn't looking at him, but I heard his sharp intake of breath. He fell silent, and I pressed my lips together, my fingertips tightening around my knees, digging into my skin.

Sudden, small sounds filled my senses. A guitar case opening. A plectrum against strings. A heavy breath.

"Come here," Huxley murmured, and I lifted my head. He had his legs stretched out in front of him, his guitar resting on his upper thighs, and his hand held out in invitation.

Swallowing hard, I unfolded my body, taking the invitation for what it was. I shifted into a lying position with my head on his legs.

His fingers came down to carefully stroke through my hair. "Don't think this is one-sided," he said quietly. "I want you too. It's been killing me to stay away from you."

"What does that mean?" It felt like my voice was so loud, breaking the peaceful silence of this open space.

His soulful blue eyes met mine, and for once, he

wasn't hiding anything. "It means that I don't think I can stay away anymore. I thought I could get over this, but you're in my head, and I can't stop thinking about you."

I reached a hand up, tracing it across his jaw, feeling the rasp of his stubble beneath my fingertips. "Me neither. I know we're stepbrothers, but we didn't ask for this. I want you, and you want me. Why shouldn't we be together?"

We both knew why. Our parents. I was sure they wouldn't understand, and with us all living under the same roof, things were bound to blow up. How could we tell them when we didn't even know ourselves if this was something between us that would burn hot but could burn out just as quickly? Was it worth the inevitable fall-out? What would it mean for family relations if and when this whole thing crashed and burned? Our family was so new...how could we risk what our parents were doing their best to build, with a relationship that had no guarantee of lasting?

Then there were other people. We weren't even related, but there would be those who wouldn't understand. Not that I cared what they thought, but our parents would.

Huxley sighed, his fingers curling around mine. He pressed my fingers to his lips and then lowered my hand to rest on my stomach. "Let's stay here and forget about the rest of the world for a while."

Our eyes met again, and I exhaled slowly, pushing aside my worries. I kept on breathing in and out, slow and steady, until everything else faded away. There was just the two of us, here in this grassy meadow, with the rays of the late-summer sun playing across our bodies. Me and the boy that I somehow, against all odds, wanted to be mine.

A smile curved over Huxley's lips, and he looked so fucking beautiful. "That's it. Breathe. It's just you and me, Cole."

"You and me," I repeated, holding his gaze.

His fingers moved across the strings of his guitar.

The familiar opening notes of "Somewhere Only We Know" sounded, and then he began to sing.

NINETEEN

HUXLEY

Curtis studied the campus map that we'd been given in our induction packs when we'd arrived at the LSU student union. Tapping the piece of paper, he glanced over at me. "My induction meeting's over there at three. Want to meet back here afterwards to look around the campus?"

"Yeah, okay." My business course was a foundation one, whereas Curtis was going straight into the first year of his computing degree, so we wouldn't be sharing any classes together. "See you in a couple of hours."

I made my way over to the building indicated in my own induction pack and entered a small lecture hall with tiered seating. The hall was already half full, and I scanned the room, spotting a free seat towards the back. Good. Sitting at the front was never a good idea, in my opinion. I'd been in trouble enough times as it was; I didn't need to draw any extra attention to myself by being in full view of the lecturer. I'd promised my dad I'd behave, but based on my track record, it might not be smooth sailing. So the back of the hall it was. I just needed to keep my head down and pass this year so I could start my full degree.

My phone buzzed in my pocket, and I pulled it out, setting it to silent before I checked the message.

COLE:

Good luck today. You're gonna smash it *fist bump emoji*

A smile tugged at my lips as I replied. This was so fucking weird—I'd never smiled much before Cole came around. Never really had a reason to, I guessed.

ME:

I'm officially a poor student

COLE:

Poor?!!! The bank of dad is funding your student life

Yeah, he was. For as long as I toed the line. And I would because I was tired of being a fuck-up, and I was lucky to have even been given a place on this course.

But that wasn't even the real reason. I wanted *Cole* to have a reason to be proud of me, to see those soft brown eyes light up the way they had when he'd seen me play with the 2Bit Princes for the first time.

His eyes were so expressive. When he looked at me, it felt like he could see into my soul. Like he could see the real me behind the superficial mask that everyone else saw. And he let me see him too. The way he'd looked up at me last weekend, lying in that field, his gaze so full of emotion that he hadn't even tried to hide... I hadn't been able to get that look out of my head ever since.

At first, avoiding him had been easier than I'd thought. Physically. I'd just reverted to my hermit ways, either shutting myself in my room with my guitar or hanging out with the band. Getting him out of my head had been a different question. My right hand had been getting a constant workout, thinking about him fucking me so good, and the worst thing of all? I hadn't even had a chance to touch his dick. Now I never would.

And why did he have to be so fucking sweet? I preferred it when he was just my unwanted asshole step-brother, not the guy who'd looked out for me and taken care of me ever since the car accident. He'd hooked me up with the band, he'd seen how fucked up I was over driving again for the first time since the collision, and he'd somehow managed to penetrate the hard shell I kept around me. He'd given me exactly what I hadn't even known I needed, every step of the way.

With a defeated sigh, I gave in to the impulse to text back words that I shouldn't be saying to him.

ME:

What are you doing later? Want to meet after work? I'll be on campus until 6ish

COLE:

I'm going straight from the office to the club. Got called in for a shift to cover sick leave

My heart shouldn't have sunk at his words, but it did.

ME:

OK

COLE:

You can come and keep me company if you're not busy? Widen your musical taste with some pop *tongue out emoji*

ME:

You want my ears to bleed again? You're the one who needs to expand your musical taste

COLE:

Yeah I know. That reminds me. Send me your set list. I wanna learn all the songs you guys play so I can be your first official groupie

My stomach fucking flipped. He wanted to learn our songs?

ME:

OK. I'll bring it to the club tonight. If my ears bleed you owe me

COLE:

Drinks on the house. I'll make you your signature cocktail. When you've had a few you'll be singing along with Britney again. Or was it Kylie?

I sent him a row of middle finger emojis, to which he sent back the word "haha" repeated over and over a million times. *Dickhead.*

ME:

What time do you want me?

Shit.

ME:

AT THE CLUB

COLE:

I want you all the fucking time Hux

Sorry shouldn't have said that

Doesn't matter what time. Come straight from uni if you want. The club won't be open but I can let you in

Cole. Why did he have to be so fucking tempting?

ME:

It's not like it's a secret and you know I want you too

I'll text when I'm on my way

COLE:

Wait are you texting me from uni? Aren't you supposed to be in an induction???

ME:

Yeah but it hasn't started yet

I turned my phone camera on, and after glancing around me to make sure no one was paying attention, I took a quick selfie, sticking my middle finger up at the camera, although the effect was ruined by the grin I couldn't manage to bite back.

ME:

Proof

COLE:

Ruuuude giving me the middle finger but hot AF. That's gonna be my wallpaper

Have I mentioned I'm weirdly obsessed with your nails? I like that you paint them black

ME:

Bit weird but I won't judge. I'm weirdly obsessed with your eyes

As soon as I'd sent it, I wished I could take it back. Fuck it, who was I kidding? I'd crossed a line the second I'd decided to send the first text.

COLE:

That's not weird. My eyes are amazing. Yours aren't too bad either

ME:

Wow what a compliment

COLE:

I'm not the songwriter poet person here but ok I'll give it a try. Hold please

I waited, staring down at my phone in anticipation. What the fuck was he doing?

COLE:

Strawberries are red, your eyes are blue. They're also sweet, and so are you

Oh yeah!!! I've got skillllssss

ME:

That was horrible. Never do that again

COLE:

Jokes aside your eyes are fucking gorgeous and you know it. Everything about you is hot

ME:

Stop it. You're making things harder

COLE:

Mmm yeah I know what's getting harder

SORRY. I keep forgetting I'm not supposed to do that. Pretend I didn't just make a comment about you making my dick hard

Despite myself, I laughed. We shouldn't be doing this, but it was getting harder to remember why we shouldn't. But I had to get us back on track. The lecture hall was almost full now, and it was very much *not* the place for sexting. Time to change the subject.

ME:

Aren't you at work? Send me a photo as proof

COLE:

You could just say you want a selfie of me

Before I could reply, he sent through a picture of himself, sitting at his desk and grinning at the camera. Even under the unflattering office strip lights, he was gorgeous.

COLE:

Is that going to be your wallpaper?

ME:

No. My wallpaper is a close-up photo I took at the one-night-only Oasis reunion at Glastonbury after two hours of me pushing through the crowds to get to the front. That was EPIC. Best night of my life. I'm NEVER replacing that photo

COLE:

Fair enough

A tall man in a suit entered the room and took his place behind the lectern at the front. He was closely followed by another man and a woman, who also took their places at the front of the room. Conversations began to stop as people's attention were caught by the new arrivals, and I shot another quick text to Cole.

ME:

Induction's starting. See you later

COLE:

Good luck

Before I put my phone away, I scrolled back to Cole's photo. Biting down on my lip, I studied it for a minute, my gaze tracing the lines of his face. With a sigh, I gave in to the inevitable and set his photo as my wallpaper. *Goodbye, Oasis. It was nice knowing you, but now I have a new obsession.*

I'd just have to remember not to leave my phone lying around where our parents could see it. Or Cole.

TWENTY

COLE

I finally heard from Huxley just after 8:00 p.m., saying he'd eaten at the campus with Curtis and was on his way to Revolve. A wide grin stretched over my face when his text appeared on my screen, and I sent back a casual "see you soon" as if I hadn't been obsessively checking my phone for the past two hours.

I couldn't fucking wait to see him. I wanted to hear about his day, to find out how he got on at uni, and to—

"Someone's extra happy today. Got laid last night?" Smirking, Tom nudged me with his elbow as he passed me with a crate of beers, which he lowered to the floor.

"Can't I just be in a good mood?" Kneeling next to him, I began loading the fridge in front of us with bottles from the crate.

"You're normally in a good mood when you're here. But not like this. What's up with the Cheshire cat grin?"

"Mind your own fucking business," I told him without any heat in my voice. He just grinned and shook his head as he placed bottles on the shelf below the one I was refilling.

"Sooner or later, you'll crack."

"Will I?" So he didn't question me any further, I directed the conversation to something else I'd been meaning to speak to him about. I purposely wanted to mention my proposal to him rather than Huxley because I didn't want Huxley to feel like I was doing him favours all the time, in case he got weird about it. "I spoke to one of my bosses about you."

"Me?" Pausing in the middle of lifting a bottle from the crate, he swung his gaze to mine. "Why? They're not thinking of letting me go, are they?"

"Uh, sorry, no. I should've clarified. Not the bosses here. Kaito, my boss at my volunteer job. I remembered he had a relative who owns a music studio, so I mentioned your band to him. He put me in touch with his brother, and I sent him the link to the band's YouTube and social media links."

"And..." Tom's full focus was on me, and it looked like he was holding his breath.

"And. Long story short, he was impressed with you guys. He said you could use the studio once a month on a Sunday morning if you want, free of charge. Something about him paying it forward and helping out local up-and-coming talent. It's somewhere south of the river, can't remember where exactly, but I don't think it's too far from Waterloo station."

"Are you serious right now?"

"Yeah. If you want it, he told me to pass on his number, and he'll sort it."

Tom suddenly lunged at me, sending us both off balance as he threw his arms around my waist. "I fucking love you!" he shouted, way too close to my ear. The crate of beers wobbled as his thigh crashed into it, the bottles inside clinking ominously.

"Whoa, watch the beers." As we straightened up, I shot him a grin. "No need to proclaim your love for me. It was nothing, really. All I did was mention you to my boss in passing." Or harassed him both by email and in person and then sent his brother a long email detailing exactly why he should care about this band that he'd never heard anything about before in his life. But whatever. No one else needed to know that part.

Climbing to my feet, I grabbed my phone from the shelf under the bar, where I'd temporarily stashed it, and then sent the details to Tom.

"Bro, I'm buzzing. We need to celebrate this, and you're coming with us. A proper studio. Fucking hell, look at these mixing decks! This amp!" Tom had already pulled up the studio website and was scrolling through the images at a rapid pace.

"Glad you approve. But you guys should celebrate as a band. And you should make the most of the studio. Put out some of your own songs, as well as the covers."

He hugged me again, mumbling his thanks, and when he let me go, Huxley was there, standing on the opposite side of the bar with a dark look on his face.

My stomach churned. Bad fucking timing. I hoped he'd listen to my explanation because everything he'd just seen was completely innocent. "Hux. I, uh, didn't know you were here already."

"Yeah, I can see that." He folded his arms across his chest. He looked so fucking gorgeous, even pissed off, and I had a sudden flashback to the way we used to be with each other, when we were on opposite sides. Had I wanted to fuck him even then? Yeah. Yeah, I had. Looking back, I could admit that much to myself now.

Tom stared between us, and I could see the moment comprehension dawned. His brows rose, and his mouth fell open before he forcibly snapped his jaw shut with an audible click. "I'm going to get another crate of drinks. I'll leave you two to it."

"No, you won't." I reached out and gripped his arm, holding him in place. "*I'm* going to get another crate of drinks, and I'm taking Huxley with me." I shot him a hard look, and the second he'd recovered from his surprise, he smirked at me.

"Don't do anything I wouldn't do."

I raised my middle finger at him as I backed away, ducking out from behind the bar and coming round to where Huxley was still standing, bristling with tension.

"Hux." Curling my fingers around his wrist, I gently pulled his arm away from his chest. "Come with me."

Thankfully for all three of us, he came with me without protest, following me back behind the bar and through the door that led into the staff-only area. I led him through the corridors to the set of doors that led to the outside area where our supplies were delivered. There were a few large storerooms here and a smaller room where we kept the cleaning supplies. That was the room I aimed for, and as soon as the door closed behind us, I spun around, pushing his back up against the door and pressing my chest into his.

"Want to tell me what that was all about?"

His eyes met mine, glittering in the dim overhead lighting. "Why don't you tell me?"

"I might have given Tom some good news. But I'll let him share it with you." Dipping my head, I exhaled softly, feeling him shiver against me. Yeah. He'd definitely been jealous. "You know he's straight, right? Even though he works here?"

"He had his hands on you," he growled, pushing at me, catching me off guard. Before I knew what was happening, he'd spun us so now my back was up against the door, and he was holding me in place.

"Hux. Calm down. It was just a hug."

"I know." The fight went out of him, and he let his head fall forwards against my shoulder. "I don't like that he gets to do that and I don't."

My arms wound around his waist, pulling him closer. "We can hug. It's just, uh, inadvisable with the way I, y'know."

"The way you what?" Raising his head, he stared at me, a challenge glimmering in his eyes. He wanted me to say it out loud, did he? Fine. If he wanted me to torture myself, I'd torture him too.

My hands slid lower, palming his sexy ass through his jeans. His breath hitched, but he stayed completely still. He wasn't going to make this easy on me, was he? I pulled him in, letting him feel the line of my hardening erection. His tongue slid across his lips, and he almost imperceptibly angled his hips, enough that his own hard length pressed against mine. He planted his hands on the door on either side of my head, his gaze never leaving mine.

"The way you what?"

I angled my head forwards, speaking against his lips. "The way I want you. Do you want me too, Hux? Is that why you were acting all jealous out there?"

"Yes, I fucking want you," he growled, and as soon as the words had been torn from his lips, his mouth was on mine.

Hot, hard, and so fucking hungry. We went at each other like we were fighting, instantly riled up. The tension between us had been building ever since the last time we'd kissed, when we said we weren't going to do this anymore, and I needed more. Needed Huxley. To make him mine.

It was like he was reading my mind. "You're fucking mine," he said harshly, his hand working at my jeans, freeing my throbbing cock and getting his hand wrapped around it.

"You're mine, and you know it," I panted between mind-melting kisses, opening his jeans so I could get my hand on his dick and make him feel as good as he was making me feel. "I haven't even looked at anyone else." Maybe I shouldn't have said those words, but they were true. From the minute we'd first collided, he'd consumed me, and now I only wanted him. Even when I'd been trying to get over him, the thought of being with someone else hadn't crossed my mind. Huxley was the one I wanted.

I slid my thumb across the head of his leaking cock, and he groaned against my throat, his teeth scraping over my skin. His body shivered against me, and I fucking loved how responsive he was. It got me going like nothing else. That, and the feel of his hand around my cock... there was no way I was going to last. But neither was he. I wrapped my hand around his length and began to stroke him, matching his pace. He groaned again, thrusting up into my hand, his knuckles knocking against mine as we both chased our release.

"Harder. Fuck. So close," I gasped. Two more strokes and I was coming, my cock throbbing in his grip, cum soaking his hand and my work shirt. So. Fucking. Good.

"Fuck," he whimpered, shuddering as he came, his body falling into mine. We were a fucking mess, but I didn't even care. I wiped my hand off on my shirt since it had already been ruined and wrapped my arms around him again, stroking up and down his back. His arms came up and tentatively wound around my waist, and I smiled, pressing a soft kiss to the side of his head.

Eventually, we pulled apart. We needed to clean up, and I still had work to do before the club officially opened for the evening. I eyed Huxley warily as I pulled wipes and tissues from the supply cupboard next to the door and handed them to him. Was he going to tell me he regretted this or it shouldn't have happened?

He remained silent, cleaning up and straightening his clothes. My stomach flipped, but this time, it wasn't from his proximity. I stripped off my work shirt, balling it up, and exhaled. Okay. I just needed to ask him—

The words I'd prepared died in my throat when I turned around to find his heated gaze devouring my body. I couldn't have stopped my grin if I'd tried.

His eyes flicked up to meet mine. "No need to look so smug."

"All those press-ups I've been doing to work off our sexual tension have paid off," I said, flexing my muscles.

He shook his head, attempting an eye roll, but his hands were already reaching out, sliding over my chest, and lower still, his fingertips tracing over the ridges of my abs. "I thought this might have broken the tension, but I want more."

"Mmm, me too." Letting my shirt fall to the floor, I cupped the back of his neck, bringing him close enough to kiss. "But as much as I want to stay, I have work to do."

"Kiss me first," he murmured against my lips, and so I did.

Later that evening, when there was a lull in the action, I slid a Huxley across the bar—the cocktail, that was, not human Huxley—and watched as those black-tipped fingers curved around the glass with care. Huxley smiled at me, and it was warm and open, and it gave me fucking butterflies.

"Hux?" Leaning across the bar, I beckoned him closer with a crook of my finger. "Do you still want more?"

His gaze searched mine, his bottom lip pulled between his teeth as a million different emotions played across his face. Finally, he nodded.

I exhaled, trying to hide my relief. "Me too. Are we gonna do this?"

He sighed, reaching out his hand and hooking his little finger around mine. "I want to try. But if we do this...I don't want to tell anyone. Definitely not our parents, and not our friends, either. It's... The pressure will be too much. I can't afford to fuck anything else up."

Could we do this? Have a secret relationship, just friends and brothers on the surface, and we'd be the only two people in the world who knew the truth?

There was no other answer but yes. My feelings weren't going away, and if both of us were prepared to give this thing between us a try, then I'd be a fool to say no. I wanted him too badly for that.

"Okay. We'll give this a try. And it'll stay between us."

He gave me another smile again, releasing my finger, and leaned back, away from me. His lashes lowered as he slowly, deliberately licked his lips. Little fucking tease. "Get back to work. Once your shift is over, you're mine."

Two could play that game. I held his gaze as I planted my arms on the bar, my muscles flexing as I moved closer to him again. "Stay there. Drink your cocktails and enjoy the music. Watch me working. I'll be serving all those other guys, but I'll be thinking of you, knowing that your cock's hard in your jeans watching them flirt with me, wishing it was you instead. Your jealousy turns me on, and it turns you on too, doesn't it, baby? Makes you want to pin me up against the nearest surface so you can remind me who I belong to. So you can drive away all those jealous thoughts with your hand on my—"

"Cole." His eyes darkened, his jaw clenching as he

ground out the words between gritted teeth. "Shut the fuck up. Now."

My grin widened. "I'm going back to work. Think of me." Blowing him a kiss, I stepped away.

The end of the night couldn't come soon enough.

TWENTY-ONE

COLE

It was official. I had blue balls. For the past few days, either Huxley's dad or my mum seemed to be around *all the fucking time.* And not only during the day. When we'd arrived back from the club sometime after four in the morning, Huxley's dad had been up, working in his dressing gown. He said something about a big project that they needed to prepare for, but it had meant that neither Huxley nor I had felt comfortable giving in to our baser urges. The most we'd managed was a few stolen kisses, not helped by the fact that Huxley was now properly starting his degree course alongside his band practice, and with my paid job begging me to do extra hours as well as the hours I was already committed to in my volunteer job, we were struggling to carve out some time to see each other. Yeah, it had only been a few days, but now that we'd agreed to try to be together, it was frustrating as fuck.

I paused my playlist—or, more accurately, the 2Bit Princes' set list—pulling out my earbuds. I had to make time to see Huxley. I wanted him to understand that this was important to me. That *he* was important to me. And today was a big day because it was his first proper day at uni. I'd left this morning before he was up, so I'd only been able to send him a good-luck text, but I was sure he was feeling a bit nervous about it, even if he kept it hidden.

My bosses were pretty easy-going—as long as I completed the tasks they'd set, they were usually flexible with hours. Still, it was best to check before I made any plans. I headed over to Meera's desk in the corner of our office and took a seat in the empty chair to her left.

She glanced up with a smile. "Cole. Perfect timing. I was hoping to speak to you. I was going to wait until Kaito was here, but he's going to be out of the office most of the week, so it'll just be me delivering the news."

News? Meera and Kaito were my bosses...fuck, I hoped they weren't going to tell me they were letting me go.

Shaking her head with a light laugh, she patted my arm reassuringly. "No need to look so worried. It's good, or at least we hope it is. We talked about this a little before, but I hadn't mentioned much about it until I knew we had something solid to offer you. How would you feel about being a paid member of staff? I can tell you now, it's not much more than minimum wage—as you know, we run this charity on a shoestring—but we'd love to have you on board as a permanent member of staff. Still part-time because we don't have the budget for full-time yet, but if you're interested, the job's yours."

I grinned, the tension instantly leaving my body. A

proper job, in a place I loved working in? "Yes! I'd love that. Thank you for the opportunity."

She beamed at me. "We all love you here, Cole. You're quick and efficient, and you get on with everyone. We don't want to lose you."

"In that case, I'd be honoured. Would it still be the same hours?"

Reaching into her desk drawer, she withdrew a small sheaf of papers. "Flexible days, sixteen hours a week to begin with, and we should be able to increase that to twenty hours in the new year. It's up to you how you want to spread your hours out, and you can still work from home as necessary too. Here's the contract, plus all the boring details about sick pay and annual leave and pensions. Have a read through, and bring the signed contract next time you're in."

"Thanks," I said, taking the paperwork with a smile. "I really appreciate it, and I'm glad you took a chance on me. Uh, I hope this doesn't sound like I'm pushing my luck, but do you need me here until a set time today? That's actually what I came over here to ask about in the first place."

"Leave whenever you're finished with those emails. Got any exciting plans?"

There must have been something in my expression that gave me away because she arched her brows, amused. "From that look on your face...I'm guessing... new love interest?"

Yes. Obviously, I wasn't going to tell her that. Shifting in my seat, I shuffled the papers, attempting to act normal and not like a person with something to hide. "Uh, I wanted to meet my stepbrother. He's just started at uni over at LSU."

"Oh, he did? I went to that uni. Do they still do the annual kiss an eel day?"

"The what day?"

A laugh burst from her throat. "Silly tradition dating back years. It's not as exciting as it sounds."

It didn't sound even slightly exciting, to be honest. "Oh. I don't know. I haven't had a chance to speak to him yet. This is his first proper day."

"His first day? Leave now, finish up the emails at home. There's nothing that can't wait. Family comes first, and this sounds like a big day for your brother."

"Cheers, Meera. I'll text him and find out when he finishes." She waved me away, and I left her in peace, tugging my phone from my pocket. Huxley's selfie stared up at me from my screen, a mischievous look in his eyes that, for some reason, made me fucking ache. This must be how it felt to have Huxley withdrawal symptoms.

ME:

What time do you finish uni today? Want to meet? Need to celebrate your first day at uni

I received a reply almost straight away.

HUX:

I'll be finished in an hour. Yeah we can meet. Where?

ME:

I'll come to your campus. Is there somewhere easy to find?

HUX:

Meet outside the library. It's the big building in the middle of campus and it's signposted

ME:

OK see you soon

A grin spread across my face. Only an hour to go until I got to see Huxley again.

Leaning against the library wall, I glanced around the parts of the campus I could see. It seemed to be mostly modern glass and steel buildings, although there were a couple of older buildings in the distance. In front of me was a quad with benches and trees, with more buildings beyond, and behind it all, I could see the top of the Shard glinting in the sun. It almost made me wish for a second that I'd chosen to go to uni, but it was an option I'd never seriously entertained.

"Cole? What are you doing here? I didn't know you went to LSU."

I spun around to see my cousin Elliot's best friend, Ander, exiting the library with a good-looking guy with mid-brown hair and aqua-blue eyes. He was the kind of guy I might have taken a second look at if I'd met him in Revolve or on my HookdLDN app, but now I only had eyes for one person, even if this guy wasn't straight.

"Alright?" I nodded at Ander. "I don't go here. I'm meeting my stepbrother, Huxley. It's his first day today."

"Oh, yeah. Your mum remarried." He paused, frowning, before he clicked his fingers. "I remember now. Elliot said that it seemed like there was a bit of tension between you and your new stepbrother at the wedding. You're getting on better now, then?"

Yeah. There had been tension. A lot of it. "Uh, yeah. We managed to put aside our differences. We're friends now."

"Ha, maybe you should try putting aside your differences and becoming friends with that Noah guy. He's gonna be your housemate, after all." Ander nudged the guy standing to his left.

"Fuck that," the guy muttered with an irritated huff of breath. "That wanker fucked up my car and wouldn't even accept that he was the one to blame. He'd better be good at football. That's the only thing that could even partly redeem him in my eyes."

"Chill, Liam." Ander was clearly amused, smirking at his friend, who was now getting more irate.

"If you saw the damage he did to my paint job, you wouldn't be telling me to fucking chill."

I coughed pointedly, and both of their gazes swung to me. A flush appeared on Liam's cheeks, and he grimaced. Yeah, this moment was awkward. Lucky for me, I spotted a flash of bleached blond hair, and my heart skipped a beat as my full attention turned to Huxley.

"See you later. Say hi to Elliot from me," I threw over my shoulder as I made my way towards Huxley. Drawing closer, I watched his eyes light up and his lips curve into a smile, and I fucking loved that he smiled so easily around me now. I wanted to kiss him so fucking badly, but I couldn't, not here.

He glanced behind me to where Ander and Liam were still standing, and his smile disappeared. "Who were those guys you were speaking to?" he asked when he reached me.

"That was my cousin Elliot's best friend. Remember Elliot? You met him at the wedding. I guess that was one of his other friends with him. And before you start getting all jealous again, let me remind you, *again*, that I'm not interested in anyone else."

He groaned. "Yeah, I know. I know. It's harder than I thought it would be, being with you in a public place and not being able to let anyone know you're with me."

"You're right, because you know what I thought when I saw you?" Stepping closer, I lowered my voice. "I thought that I wanted to kiss you. I want to kiss the fuck out of you, in fact."

His smile reappeared, deepening into a knowing smirk. "I had strawberries just now in the student union."

"Fuck, Huxley. You'd taste like my favourite thing." This fucker knew exactly what the thought did to me. "We need to be alone *right now*. Come with me."

I had no idea where I was going, but I picked a random direction and started walking. As we headed away from the LSU campus, I forced myself to think with my brain instead of my dick and turned to Huxley. "How was your first day?"

"Good. Different. I think I'll like it."

That was high praise coming from Hux.

"Good. I worried about you."

His gaze shot to mine. "Seriously?"

"Of course I did. It's your first day in a new place." Reaching out, I gripped his arm, spotting a tiny, narrow road just behind him. "Down here."

"I'm not used to anyone worrying about me," he said, his tone soft and almost wondering, and I inwardly cursed his parents for making him feel like they didn't care. And I knew for a fact that his dad cared.

"Get used to it." I came to a stop in a doorway. We were down a little cobbled side road with the backs of buildings around us. I could just about catch a glimpse of the Thames between the buildings in front of us, and that meant there'd be people right up ahead, but for now, we were alone. Wasting no time, I wrapped my arms around Huxley, tugging him into me. But before I could kiss him, he placed his fingers to my lips.

"Wait. I brought you something." He dipped into his messenger bag, pulling out what looked like a lumpy napkin. Lifting his hand, he stared at me expectantly. Holding on to him with one hand, I unwrapped the napkin with the other, until his gift was completely unwrapped, the napkin falling to the floor.

"Oh. Hux."

"I saved the best one for you." On the palm of his hand was a large red strawberry, glistening in the early evening sunlight, plump and juicy and delicious. The way to a man's heart was definitely through his stomach, if it involved my favourite fruit.

He lifted it to my lips, and I bit into it, moaning as the delicious flavour burst on my tongue. Juice ran down my bottom lip, but before I could swipe it away, Huxley angled his head forwards and licked across the curve of my lip, capturing the droplets. "Fucking delicious," I rasped, my fingers curling around his, pushing the fruit towards his mouth. He bit into it, swallowing the remaining piece, and then all that remained for me to do was to suck the remaining juice from his fingers.

"Fuuuck," he groaned as I dipped my head and took his finger into my mouth, swirling my tongue to get every drop of juice. By the time I'd finished fellating his fingers to my satisfaction, both of us were breathing heavily and hard, his hips rocking against mine in a teasing, dirty grind that made me want to do very bad things to him, right there on a public street.

Curving my hand around his nape, I pulled his head to mine. "Need to kiss you."

The second our lips met, it was like fireworks went off in my belly. I couldn't get enough of him. I wasn't sure I'd ever get enough.

We kissed and kissed and kissed, slow and deep and so fucking good. When we finally broke apart, it was only because the sound of heels on the cobblestones alerted us to someone else's presence. I cupped Huxley's jaw, brushing one last soft kiss over his lips, and then let my forehead fall against his.

"Cole." His chest rose and fell against mine in unsteady breaths. "We should...uh...walk."

"Yeah." He was right. We couldn't do anything else here, even though my entire body ached for him. Clearing my throat, I forced myself to release him and take a step back. "Do you wanna walk by the river or something?"

He nodded, falling into step next to me as we made our way towards the Thames. Every now and then, his knuckles grazed over mine, and it was almost enough. *Almost.*

I broke the silence to stop my thoughts from turning melancholy. "I got a new job today. Well, not a new job, but they offered me a paid version of the volunteer admin job I've been doing at the charity."

Huxley glanced over at me with a smile as we turned left onto the Thames river path, the sunlight shimmering on the water. "Yeah? That's great." He paused for a second and then added, all in a rush, "I'm proud of you."

"You'll be even more proud when you hear the next bit." Pulling my phone from my pocket, I scrolled to my playlists. "I memorised all your songs. All the cover versions, anyway. The only ones I don't know are the new ones you're writing with Tom."

"You did?"

"Yeah. I'm allowed to listen to music at work, so I've been using my time wisely."

Coming to a stop, he rested his arms on the wall that ran alongside the river. His throat worked, his gaze fixed on the water as he blinked rapidly. The urge to tug him into my arms was so strong that I had to clench my fists and fold my arms across my chest to stop myself from reaching for him.

"I used to come down here with my dad when I was little." There was a tremor to his voice, but I didn't call him out on it, accepting the subject change for what it was. I hadn't wanted to make him feel uncomfortable or too emotional, and I'd somehow managed to with my words.

Casually leaning back against the wall, I gave a short nod. "Me too. Imagine—we could've passed each other hundreds of times and never even known it."

"We probably did." His voice was steadier, and I exhaled, relieved, as he continued. "Where else did you go?"

I turned to face the same direction as him, our arms almost touching. "My favourite place was Camden Lock. Did you ever go there?"

He nodded. "Yeah. Sometimes we used to get the waterbus from Little Venice to Camden. When we got to the part of the canal that goes through London Zoo, I used to imagine that the animals were escaping and I could escape with them."

Oh, Huxley.

"Do you still want to escape?"

He was silent for a long time. When he finally spoke again, he turned his head to mine, his gaze soft. "Not really. Not anymore."

TWENTY-TWO

HUXLEY

Something had changed. Cole was...paying more attention to me than he ever had before. Things like constantly checking up on me, little touches out of sight of anyone else, making quiet comments when he thought our parents couldn't hear. Don't get me wrong, I fucking loved having his attention, but he needed to dial it back before our parents got suspicious.

I pushed those concerns out of my mind for now. Our parents weren't at home, and I was enjoying this uninterrupted time to be with him. Neither of us had anywhere else to be.

Sprawled out on his back on my bed, head and shoulders propped up against the headboard, he was switching between smoking one of my joints and demolishing a bowl of strawberries. I would've personally had one or the other, but he didn't seem to care. His fingers tapped against the ceramic side of the bowl as I strummed my guitar from my position on the floor with my back resting against the side of the bed. I was going over the chords for one of the songs the band was going to perform tomorrow. We'd been invited back to the same pub where we'd done our test gig, and this time, we were getting paid for it. By the time we split the funds between the four of us, there wouldn't be much of it, but that didn't matter. What mattered was the fact that someone liked us enough to pay us to play. We were going to stick with covers at our gigs for now and gradually add our own songs to our social media before eventually performing them live.

"Your nails," Cole suddenly said, his fingers coming out to slide through my hair.

"Huh?" My hand paused on the strings, my plectrum slipping from my grip as I shivered at his touch.

"The polish stuff is all chipped on your thumb. Can I have a go?"

Straightening up, I twisted around to stare up at him. "Have a go?"

"Yeah. Can I try painting your nails?" He grinned down at me, all happy and relaxed from the weed, his eyes wide and slightly unfocused as he blinked at me from behind his long lashes. Who was I to say no to him when he looked so cute?

Cute. With a mental shudder, I reminded myself that I didn't think of people as "cute."

"If you want." Placing my guitar carefully on the floor, I shifted onto my knees, opening the drawer where I kept the small bottle of nail polish.

"I didn't mean now. I didn't want to interrupt your playing." His smile disappeared, and we couldn't have that. With a sigh, I closed the drawer again and climbed to my feet. I reached for the joint, inhaling deeply before crawling onto the bed, straddling his thighs. Lowering my head to his, I sealed my mouth over his, exhaling. His hands came up to settle on my thighs, his palms sliding up my legs, under the hem of my shorts.

When I raised my head again, I stared down at him, happy that the smile had returned to his face. I stroked my fingers through his soft hair as I took another hit of the joint. "You didn't interrupt me. I was mostly messing around, anyway. I know all the chords." Placing the joint back in the ashtray, I leaned over to my drawer again, trusting him to hold me in place while I pulled it open and felt around for the polish bottle and remover pads. "You can paint my nails now if you want."

"Really?" His eyes lit up as he moved into a seated position, somehow managing to keep me balanced on his thighs, and it made something inside me grow warm. It was fucking crazy that this guy had so much of an effect on me. "I might not be very good. I've never done it before."

I shrugged. "Doesn't matter to me. Look at me. You know I don't care about looking perfect or any of that shit."

"But you still look fucking hot, even when you put in zero effort." He shook his head, pretending to be annoyed. "Some people have all the luck."

"Is this your sneaky way of trying to get me to compliment you?" Handing him the bottle, I opened the pot of remover pads and began to remove my chipped polish. I was looking down at my hands, concentrating on what I was doing, when I felt a soft kiss being placed on the tip of my nose. My eyes flew up to see Cole studying me intently.

"I wouldn't coerce you into complimenting me. It wouldn't be genuine. But I know you like the way I look."

"Arrogant," I muttered, not even slightly annoyed, climbing off him to go and wash my hands in my en-suite bathroom. When I returned, the bastard had taken his fucking top off and was relaxing on my bed in just a pair of red football shorts, all his golden, tanned skin on display. He wasn't playing fair.

Accepting the inevitable, I lifted my own T-shirt, tugging it over my head, enjoying the way his eyes darkened as he stared at me. I knew I wasn't anywhere near as defined as he was, and I'd never achieve a tan like his—not that I wanted to or cared about getting ultra-ripped muscles, for that matter. But I knew that Cole thought I was hot, and that was the biggest ego boost I could ask for.

"See something you like?" His lips kicked up, his gaze sweeping over me. I dragged my bottom lip between my teeth before releasing it, prowling closer to him, watching the way he spread his legs a little, just enough for me to get a good view of the tent he had going on in his shorts.

"See something *you* like?" I climbed onto the bed again, taking my place back on his thighs. My dick was in a similar state to his, but I kind of liked the thought of torturing us both by holding off, and I was fairly certain he did too. "Are you gonna paint my nails, then?"

He smirked, his gaze flashing down to my erection, but he nodded. Uncapping the small bottle, he wiped the excess polish off the brush before he brought it up to my hand, which was splayed across my thigh. Slowly and carefully, he painted a stripe down the middle of my thumbnail, then another on either side. His brows were pulled together, his hand steady as my nail was covered in jet-black polish. "How's that?"

"You said you haven't done this before?"

"Yeah. I watched a quick YouTube tutorial while you were in the bathroom, though. I didn't want to fuck up too badly."

He was so fucking *sweet*. I couldn't find the words to say back to him, so I sat there in silence like a dickhead while he carefully painted the rest of my nails. When he was finished, I inspected his handiwork. It was messier than if I'd done it, and there were splodges of black on the skin around some of my nails, but overall, he'd done a good job. To be honest, even if he had completely fucked it up, I wouldn't have cared.

"Nice." I shot him a small smile. "They won't take long to dry."

"You can't touch anything while they're wet, can you?"

"Not with my nails."

"Good." He placed his hands on my waist, and somehow, I found myself lying on my back on my bed with Cole hovering over me. "Just keep your hands to yourself. I'm gonna make you feel good."

His lips trailed across my jaw, avoiding my mouth. Before I could protest, he began to move lower, over my collarbone and lower still until his mouth was right above my left nipple.

"When I first saw you, I wondered if you had these

pierced." He dragged the flat of his tongue across my nipple, making me gasp. Fucking hell. How was this the first time I'd realised I had sensitive nipples? Glancing up at me, he continued. "You had your pierced ears and septum, and those tattoos. I thought you might have piercings in other places."

"No other places," I gasped again as he repeated his tongue action with my right nipple, while his body pressed between my legs, creating a teasing friction on my dick that had my hips thrusting up, desperate for more contact. "I—I've thought about it. Getting my nipples...*fuck*, Cole...done. Just haven't...done it yet."

"It would be so fucking hot." He took my nipple between his teeth, gently biting down, and I moaned as sparks raced down my body. It was like he had a direct line to my dick. I wanted to touch him more than anything, and the *only* thing that stopped me was the thought that I might mess up the job he'd done on my nails.

When he kissed lower, his lips and tongue running over my stomach and down to the waistband of my shorts, my breath hitched audibly, and it caused him to raise his head, his eyes gleaming with a combination of mischief and lust that shouldn't have been as heady as it was.

His tongue slid across his lips as he slowly and deliberately ran his palm over my erection. "Want me to suck your cock, Huxley?"

"Do it." I forgot about the fact that I was supposed to be keeping my hands off him, my fingers gripping his head and pushing down. He laughed against my skin, easing my shorts down at a glacial pace, as if he was determined to torture me. *Asshole.* But when I was finally bared to him, my dick glistening with the clear evidence of my arousal, his humour died away.

"Fuuuuck," he moaned, and then he lowered his head and made me see fucking stars.

His lips closed around the head of my cock, his tongue licking across my leaking slit, humming as he tasted me. It was clear that he knew exactly what he was doing, but for once, I didn't let my jealousy get the better of me because he was here with *me*, and this was better than anything I'd imagined all those times I'd wanked myself off thinking of him giving me a blow job.

Lowering his head further, he alternated between taking me to the back of his throat and paying attention to the sensitive tip, his hand adding to the sensations, stroking my length, rubbing on my perineum and caressing my balls. It was the best blow job I'd ever had in my life, and I was fucking helpless as my orgasm crashed over me, leaving me panting and gasping for breath.

"Come here," I rasped when I finally managed to form words, and fucking finally, his lips met mine. I tasted strawberries, and weed, and myself, and somehow, it was an intoxicating combination that made me moan into his mouth, my hands kneading his ass, dragging his erection against me as he kissed me harder, his hips thrusting against my body. He shuddered against me with a muttered "fuck," his movements stilling.

I felt his lips move against my throat, where he'd buried his head. "I can't believe you made me come like that."

"Maybe we need a round two." The second I'd

suggested it, I knew it would be happening. I couldn't get enough of Cole Clarke.

"Yeah. I need to redeem myself."

Kissing the top of his head, I ran my freshly painted nails up and down his back. "There's nothing to redeem. It was so fucking good, Cole."

He huffed out a breath, tickling my skin. "Yeah. It was. Okay, round two it is. Gimme a bit of time for my dick to recover. And to, uh, clean up."

"Cole?"

"Yeah?" When he lifted his head, his fucking pretty eyes bored into mine. My stomach flipped, and I almost forgot what I'd been planning to say.

I swallowed hard. "I'm fucking you this time. I want you in my bed, and I want my dick in you. Got any problems with that?"

A smile curved over his lips. "No problems. Ready when you are, baby."

TWENTY-THREE

COLE

Curled up in Huxley's bed, his body a warm weight against my side, I almost forgot that we'd planned to fuck. I was so relaxed, so happy, and I had the person I wanted resting his head on my chest, his fingers idly tracing over my abs.

I said, I *almost* forgot. Almost.

My stepbrother wanted his cock inside me, and I was more than happy with that. I'd topped more than I'd bottomed, but I actually didn't have a preference one way or the other. As long as we both had a good time—and I had no doubt that with Huxley, we would.

While I recovered, though, I wanted to talk to him. We'd gradually been opening up to each other ever since our hostilities had dimmed, but I wanted more. I wanted everything. For him to share all his secrets and know I'd keep them hidden. And in return, I'd trust him with mine.

"Hux? What was it like for you when you found out about your dad and my mum?"

He stiffened against me, and I quickly placed a kiss to the side of his head, my arm tightening around him.

"You don't have to answer," I reassured him. While I wanted him to open up to me, I didn't want to make him uncomfortable.

"It's okay. I...I wasn't expecting it." His exhaled breath skittered across my throat. "It wasn't good. My parents...they fought about it a lot. I didn't get on with my dad at the time, and my mum...I thought she was on my side." His voice dropped to a whisper. "Then she left, and she didn't even ask if I wanted to go with her. I guess she didn't love me enough."

"*Huxley*." I wrapped both of my arms around him, holding him as close as I could. I wanted to take all of his hurt away.

We spoke a little more about how we'd felt, finding out about their relationship, and although most of what Huxley said had been what my mum had told me, I hadn't realised the depth of his pain and sadness until that point.

Something he'd said the other day clicked into place. He'd told me he wasn't used to anyone worrying about him. How could his parents go through life not letting their only son know just how fucking special he was? Yeah, from what he, his dad, and my mum had told me, he'd fucked up in the past and had a tendency to go off the rails, but that didn't mean he was *unlovable*. He was fucking amazing. Talented, gorgeous, interesting, and he cared deeply—which anyone could see if they took the time to actually get to know the real him.

"Hux?" I stroked down his arm. Touching him like this wasn't enough. I needed more.

"Yeah?"

I want to show you that you mean so much to me. "I want you to fuck me."

His sapphire eyes darkened, lust filling his gaze. Without another word, he rolled over so he was on top of me. We were both naked, and the second his hard length made contact with the inside of my thigh, my own cock responded, my body taking over, with everything in me wanting to feel him inside me, to bare myself to him in the most intimate way I could imagine.

"Cole," he murmured against my lips, capturing my mouth in a kiss that quickly turned hot and hungry. I pulled him closer, my hands all over his body as he deepened the kiss, grinding his hips down against mine.

Breaking away from his tempting mouth, I ran my hands down his back to grip his gorgeous ass, loving the feel of him beneath my palms. I wanted him so fucking badly. "I wanted this to be slower than when we fucked in the kitchen, but I can't wait, Hux. Need you inside me now."

Pushing himself upright to straddle my thighs, he ran a hand through his hair, making it stick out in crazy directions as he stared down at me. My gaze tracked over his body, all the way down to the evidence that he wanted this as much as I did. His gaze followed mine, and he bit down on his lip, his hand coming out to wrap around both of our cocks.

"Fuuuck." My head fell back, my eyes closing as he stroked us both together, slow and so fucking maddening.

"Stop teasing," I gasped when he dragged his thumb across the head of my dick, making me tremble with pure fucking need. He laughed, and it was so bright and genuine it gave me butterflies. *That* was something I'd never experienced during sex.

I was in so deep with this boy.

Thankfully for us both, Huxley decided not to torture us any longer, grabbing lube and a condom, and then lowered himself down onto the bed. I widened my legs, angling my hips to give him better access, so fucking happy and excited that we were getting to do this. If I only had one word to describe how I felt, it was *euphoric*.

His head dipped, his mouth opening to take the tip of my cock inside his warm, wet heat while his lubed fingers gently circled my hole. As one of his fingers slowly pressed inside me, his other hand curved around the base of my dick, all while his tongue slid across the underside of my cockhead in a slow drag that made my eyes roll back in my head. I made a mental note to give him a reward later for multitasking. Maybe it was all that singing and playing guitar at—oh, *fuck*. Why was I even *thinking* about that when he was pressing another finger inside me along with the first, so fucking carefully, brushing across my prostate and sending sparks right through my body?

"More." I gripped his shoulder, digging my fingers into the muscle. "Please."

His head came up, and he gave me a slow, dirty smile, scissoring his fingers and stretching me out before he worked a third finger inside. "I like hearing you beg."

Yeah, I bet he did, the fucker. "Get on with it. I want your dick inside me. I'm ready."

Our gazes held, and his pupils were so fucking blown, his eyes hazy and heavy-lidded with lust, but there was a curve to his lips as he looked down at me that I could only describe as fond. Was he feeling something more to this connection too? Maybe it was in my head because I was already dick-drunk, and he wasn't even inside me yet.

Withdrawing his fingers, he gently pushed my legs up towards me, smearing lube over my shin in the process, and then began to press inside. It had been a little while since I'd bottomed, so I concentrated on staying relaxed and breathing evenly, bearing down slightly as he entered me.

When he was all the way in, we both took a minute to adjust. I was so full, his cock throbbing inside me, hot and hard and perfect. "How do you feel so fucking good?" he muttered, his brows pulled together in a way that seemed almost angry. It made me smile.

"Because I'm just that good," I told him, earning me a glare and a sharp thrust of his hips that made me gasp. With a smirk, he did it again, and again, and then suddenly, we were in perfect rhythm with each other, moving like we'd done this a hundred times before, and it was everything.

"*Cole.*" He pushed at my leg with one hand, and I took the hint, letting my legs fall apart and then when he lowered his body to mine, I wrapped them around him. Now, the lengths of our bodies were pressed together, my dick finally getting some friction.

His face was so close to mine. Arching up, I took his

mouth in a hard kiss. He moaned, thrusting harder, his dick somehow at the perfect angle to send those fucking sparks through my prostate. That, the feel of his body against mine, and the grind on my cock sent me flying over the edge. I was unprepared, and I came so fucking hard, my dick pulsing cum between our bodies, my lungs gasping for breath. "Huxley...you fucking..."

I never finished my sentence because he shuddered against me and then stilled, his head falling forwards to the crook of my neck. I could feel his heart pounding against my chest, little tremors rocking us both as the aftershocks raced through our bodies.

Neither of us moved for a long time, other than Hux easing out of me. We didn't break the silence with words. We just held each other, alone together in our own world, sated and spent.

It was the best, most intimate sex I'd ever had in my life. Fucking him in the kitchen had been amazing—quick, dirty, and sexy as fuck—but there was something mind-blowing about what had happened here. And the one thing I knew was that I never wanted to give this up.

I never wanted to give Huxley up. He was in my bloodstream, and I was addicted.

When Huxley eventually peeled himself from me, both of us wincing at the mess of cum and lube that seemed to have smeared everywhere, he finally spoke. "Shower?"

"I'm too tired to move." I gave him my best pout, complete with wide eyes.

"My bed's a mess, and we need to change it before our parents get back. You have to move. Come on." With a hard tug, he yanked at the duvet underneath me, sending both me and the duvet to the floor, my limbs flailing as I slid off the side of the bed.

"You fucker!" Staggering to my feet, I lunged at him, but he was too quick. He darted for his bathroom door, closing it in my face. "Huxley Granger! Open this fucking door, now!" I shouted, hammering my fists on the wood. The door suddenly flew open, and there was Huxley's smirking face. His arm shot out, and he gripped my wrist, tugging me inside his bathroom. I found myself pressed up against my hot, very naked stepbrother, and I forgot all about being annoyed with him, wrapping my arms around him and backing him up against the tiled wall.

Needless to say, it took us a while to get around to showering.

When we were finally back in Huxley's bed—with clean sheets—he threw his arm across my chest, burying his head in the crook of my neck. This intimacy between us was sudden, but it felt so comfortable and so easy. I guessed we'd been working towards it for a while now.

"Hux?" I trailed my hand down his bare back.

"Yeah?"

"When you were a kid, what did you want to be when you grew up?"

He huffed out a soft breath against my throat and then brushed his lips across my skin, so lightly that I half wondered if I'd imagined it. "A few different things. I remember that I wanted to be an astronaut. Then my parents took me to the Natural History Museum, and I wanted to be a palaeontologist. After that, I wanted to be in a band, hitting it big, playing to sold-out stadiums."

"Do you still want that?"

There was silence for a while, and then he shrugged against me. "I know it won't happen. For a long time, I didn't even think of any future. I didn't think I'd be anything." He drew in a long, shuddering breath. "I fucked up, Cole. A lot. You know it. Everyone knows it. Even now, I have no fucking clue what's coming next or if I can even keep up with this uni course and the band."

"Yeah, you can. You can, and you will. Don't even fucking think that you won't."

"Thanks." His voice was a whisper. "If I can just get this degree, get a job that pays enough for me to live on, and carry on playing in the band, my life will be made."

I twisted on the bed so I could look at him. Leaning forwards, I kissed his lips. "You'll do it, and I'll be there for you."

A shy smile appeared on his face, and he ducked his head. "Thanks. Uh, what about you? What did you want to be?"

It was clear he wanted to change the subject, and I was happy to oblige him. I didn't want him to feel uncomfortable. "Hmmm. I wanted to be a palaeontologist for a while too. Then I wanted to be a footballer. For Arsenal, of course. What else...oh, a pro YouTuber making loads of money doing pranks and shit. You know Unspeakable? He was my idol for a while."

Huxley laughed against me. "There's still time to achieve your dreams."

"Nah, being a YouTuber seems like too much work, thinking up all the content and making it interesting. I'll be happy with what you said. A job that pays enough for me to live on. I like what I do, and although I'll probably stop working at Revolve when Mind You increases my hours enough to live on, I'm happy I had the experience working there. The only other thing I want is to have people I love around me. We don't all need huge career ambitions to be happy. If I have a job I like and people I love, then I'm good."

"Me too."

My response to that was to kiss him, and fucking hell, had anyone's mouth ever been as addictive as Huxley Granger's was?

"Cole? Huxley?"

"Oh, shit!" My exclamation was drowned out by Huxley swearing up a storm and jackknifing up off the bed. We stared at each other, wide-eyed, as the footsteps drew closer. I finally managed to get my body to work, diving for the bathroom and pulling the door almost all the way closed, my heart hammering in my chest.

When the knock came on Huxley's door, he was on his stomach, his duvet thrown over the lower part of his body, and one of his uni textbooks in front of him opened to a random page.

"Huxley." From the small gap in the bathroom door, I watched as Huxley's dad peered into the room. "Have you seen Cole?" His eyes darted to the left, taking in Huxley's guitar where he'd left it on the floor hours earlier. "What are you up to?"

I couldn't see Huxley's face from where I was, but I

could hear him easily enough. "I—I think Cole's at work. I'm doing coursework."

David stepped into the room. "Just coursework?" The suspicion in his tone made me wince. He still didn't trust Huxley, that much was clear.

"No. Uh, I've been practising for a gig. The band's playing tomorrow."

"Oh, really?" David's eyes narrowed on Hux. "June and I will come to watch you. Let me know the details."

It was a threat, and it made me want to run to Huxley's side and defend him.

But all I could do was hide there in the bathroom with my heart pounding out of my chest until his dad went away.

Tomorrow, though.

I'd make sure that our parents both realised just how fucking talented Huxley was.

TWENTY-FOUR

COLE

Unlike last time, the small pub was packed. I tried to tamp down my smile in case my mum and David wondered why I was so happy, but I was buzzing for Huxley and the rest of the band. While David went to the bar for drinks, my mum and I found a spot off to the side where we could see the band easily but wouldn't be caught up in the middle of the crowd. I commandeered a chair from a nearby table so my mum could sit down and then leaned back against the wall, watching the band setting up.

"Here." David appeared next to me, handing me an overflowing pint. "I wasn't expecting it to be this busy."

"This is where they played the last time I watched them. That one was a test gig, and this time it's their very first paid gig." It was impossible to keep the pride from my tone as I pointed out the band members. "That's Tom—he works at Revolve with me. He's studying for a PhD as well. Curtis is their drummer, and he's a student at

LSU. That guy behind the keyboard is Rob, and he's an accountant in his day job."

When I glanced over at David, he looked completely taken aback. "I...I have to confess, I had a different image in my head. I assumed they would be...well, layabouts, I suppose." He muttered something about how he should stop making assumptions, shaking his head.

My mum reached up, patting his arm. "Don't be so hard on yourself, love." Turning her gaze back to the stage, she said, "Do you think Huxley has noticed us yet?" Before I could stop her, she started waving madly in his direction. Huxley glanced up from his guitar, probably catching the movement out of the corner of his eye, and he tentatively lifted his hand in a brief acknowledgement, a flush on his cheeks.

Hi, I mouthed, and his flush deepened. So fucking cute.

I could feel David's gaze on me, so I quickly looked away from Huxley, tipping my pint to my lips and letting the chilled liquid slide down my throat. Mmm, cloudy lemon cider. So good.

Tom tapped the mic, doing his soundcheck, and thankfully, it took David's attention away from me. Huxley had more or less ignored me earlier today when we'd been around our parents, and I knew it was because he was shaken up by almost getting caught last night. I was too, but Hux seemed to be struggling. He'd kicked me out of his room as soon as David had gone, his eyes wide and his face even paler than usual, and the last thing he'd whispered to me was that I should lay low. Then this morning, he hadn't even spoken to me at breakfast other than to inform me of the time of the gig after I'd asked him directly, tired of being ignored. I could tell that our parents were wondering what had happened because they kept shooting us both concerned glances and then having wordless conversations with each other.

I needed to speak to Huxley to make sure we were on the same page. We'd agreed to give this thing between us a go, and I just hoped that what had happened last night hadn't given him second thoughts. I wasn't ready to give him up. Not at all.

Lost in my thoughts, I jumped when there was a loud screech of a guitar, and the band launched straight into their first song. The crowd roared its approval, the noise filling the small pub, and thanks to the time I'd spent learning the lyrics, I was one of the many people singing along with Huxley's sexy-as-fuck voice. My mum rose to her feet, gripping David's arm as she leaned over to shout in my ear, "They're brilliant, Cole! I'm so glad we came."

I nodded at her with a grin, glancing over to David to see if I could gauge his thoughts on his son's performance. His face was mostly impassive, but a small smile was ticking at the corner of his lips.

He liked it. Exhaling heavily, I slumped back against the wall. I hadn't realised just how important it was to me to know that Huxley's dad appreciated what he was doing with the band, that it wasn't just an exercise in time-wasting, that he was so fucking talented. His voice gave me chills, and the way he played a guitar... I knew nothing about guitars, and guitar bands had never really been my thing until Hux came along, but one thing I could say for sure was that watching him play was so fucking hot.

The time seemed to fly by, and before I knew it, the first chords of "Stop Crying Your Heart Out" by Oasis sounded. I knew that the band had a more upbeat song prepared for a potential encore, but this was the final one on their main set list. It was the first time I'd heard Huxley sing it in person, and from the first note, I was completely fucking mesmerised by him. I couldn't look away, my heart swelling and overflowing with feelings that were so powerful it made my breath catch in my throat, and I had to blink rapidly to stop anyone from noticing the sudden moisture that had appeared in my eyes.

"Oh, Cole."

I turned, wide-eyed, to find my mum's gaze fixed on me, a sad smile on her lips. Spinning away from her, I scrubbed my hand across my face, forcing myself to take deep, even breaths in an attempt to regain some semblance of composure. Fuck. What had she seen on my face?

I could have kissed David for breaking the tension right then when he spoke up over the applause of the crowd and the shouts for more. I was so fucking proud of Huxley, but now I didn't even dare to look at him.

"Oasis. I introduced Huxley to their music. Did he ever tell you that? When he wouldn't settle as a baby, I used to put their album on, and it worked like magic to stop him crying." He laughed softly. "Takes me back. He was so tiny then. My little boy." His gaze returned to the stage, and I saw something in his eyes. *Pride.* "My little boy's all grown up now. He's a man."

The way he said it, it was like it was the first time he'd realised that Huxley was a man and not the fuck-up child that he'd been seeing him as. It filled me with another wave of emotion, and I really needed to get a fucking grip because everyone was going to know there was something going on with me.

"Need air," I choked out, pushing off the wall and into the crowd, heading for the door. Behind me, the band began their encore, and I was sad I was missing it, but there was no way I could stay in the pub with my mum and David without them learning my secret.

I wasn't just in lust with my stepbrother. I was completely, one hundred percent head over heels in love with the bleached blond fucker.

Fuck.

TWENTY-FIVE

HUXLEY

Still on a high after the end of the gig, I made my way through the crowd towards my dad and June, accepting back slaps and words of congratulations. It felt amazing, but there was one person's approval I cared about above all others.

"Where's Cole?"

My dad shrugged. "He said he needed air. Listen, Hux, I want to tell you something."

The seriousness of his tone caught my attention, and my stomach churned. That was the voice he always put on when he was disappointed in me. *Shit*. He'd found my stash of weed—no, *fuck*, he'd found out that Cole had been in my room last night.

"I'm proud of you."

It took a moment for my dad's words to penetrate my spiralling thoughts, but when they did, I was sure I'd heard him wrong. "Huh?"

A small smile pulled at his lips as he took in my shock

at his words. "I said, I'm proud of you. You were great out there today. Honestly, Hux, I had no idea you were so talented, and I'm sorry that I haven't been more supportive of this musical venture of yours."

Was I in a parallel universe?

"I'm proud of you too," June added, lightly squeezing my arm. "The whole group of you did so well."

This was surreal. I had no idea how to respond—my brain was having trouble processing the things they'd just said to me. "Uh. Thanks. Uh." Waving my hand in the direction of the stage, I mumbled, "Got to go and pack up," before I escaped back the way I'd come from.

I wanted to find Cole, but I'd have to wait because I couldn't risk our parents getting suspicious. Last night had been way too close, and it had made me pull back from him this morning. I didn't know what to do—this was a fucked-up, weird situation that we'd found ourselves in, and neither of us knew how to navigate it.

By the time we'd got all the equipment packed up and into the van, my dad and June had left, and the pub was much quieter than it had been earlier. It was getting close to closing time, so Tom got us a round of pints before the bartender rang the bell for last orders. We needed to toast a successful night.

"I thought I saw Cole somewhere—oh, there he is." Tom nodded to my left, and I followed his gaze. Cole was leaning against the wall, looking fucking hot with his tousled hair, Oasis T-shirt—that he'd liberated from my room—and frayed, worn jeans that I knew from memory hugged his ass in a very distracting way. His attention was on a guy talking to him. A dark-haired guy who was also objectively hot, with artfully ripped jeans and a Ramones T-shirt.

Jumping off the low stage, I stalked through the pub to the bar, undetected by Cole. I stopped when I drew close enough to hear the conversation, a table in between me and the bar.

"...hot as fuck. Do you know if he's single and if he's into men?"

Cole crossed his arms over his chest, straightening up as he stared at the guy. "He's not single, so you're wasting your time."

"Okaaay...no need to get aggro, bro. I was only asking. You don't ask, you don't get." The guy's gaze flicked to the side, and he did a double take when he saw me standing there. He recovered quickly, turning to face me fully and holding out his hand. "Hey. I watched you play tonight. You've got some serious talent."

I couldn't help smiling as I stepped around the table to shake his hand. "Thanks." Being complimented on my skills by random strangers was new and weird and exciting. They had nothing to gain by telling me that they liked what I did, and even though I didn't know what to say in return other than "thanks," it felt good to be recognised. It was validating, I guessed.

"Can I buy you a drink to celebrate? I heard this is your first proper gig." As the guy spoke, the bell for last orders rang, and Cole stiffened next to him.

"Uh." There was still most of a pint left in my glass, but I nodded. Why not? Who would turn down free drinks? "Okay. Thanks."

The guy grinned. "I should probably introduce

myself before I buy you a drink. I'm Scott. You're Huxley, right? I've been following the band on social media."

A noise came from Cole. Did he just *growl*? What the fuck was his problem? "Yeah. I can introduce you to the others if you want."

"That would be amazing. What are you drinking? IPA?"

I eyed my pint. I had no idea what it was. It was cold and wet, and that was all I cared about. "Whatever's on tap. I'm not fussy."

"A man after my own heart." He winked at me, which was a bit weird, but he was a fan, so I brushed it aside. The 2Bit Princes had a *fan*. Fans.

Another noise came from Cole. When Scott turned to the bar to order our drinks, I glanced over at Cole, my brows raised in question. He glared at me. *What?* I mouthed, and he huffed out an angry breath, stomping around to my other side and dipping his head to my ear.

"That wanker's into you, and you're encouraging it. I already told him you weren't single." His words were hissed out through gritted teeth.

What? I stared at my fuming stepbrother. "Into me? No, he's not. He's just being friendly."

"He's fucking not. Before you got here, he told me you were hot as fuck and asked if you were single."

"You're joking."

"I'm fucking not," Cole bit out. "Now you're letting him buy you a drink, and I can't tell him that I'm your boyfr—the guy you're seeing because Tom and the rest of the band might find out."

"Shit." I grimaced, seeing the interactions in a new light. It had never crossed my mind that the guy might be flirting with me. People didn't flirt with me. Or did they, and I'd just been ignorant? Either way, I had to let Scott down and, more importantly, calm Cole down. I angled my head towards Cole's. He was so close that my nose brushed against his cheek. "I didn't realise he was into me. I'll set him straight, but you need to calm down. People are gonna get suspicious if you go around acting like a jealous boyfriend."

"Now you know how I feel when you act all jealous." Some of the anger bled from his tone. "I'm not even pissed off with you. It's him that's the problem." He sighed, taking a step back from me, turning to the bar and resting his elbows on the scratched and dented wooden surface. "I'm sorry. I wanted to tell you how good you were tonight. I was so proud of you, Hux."

"My dad told me he was proud of me. Can you believe it?"

"He did?" Cole's eyes lit up, and he finally smiled. "You really were amazing tonight. That—"

"Amazing. Epic. Unbelievable. Your voice!"

Cole's expression darkened again as Scott rejoined us, handing me a full pint glass dripping with condensation. "He's right. You really are amazing. I must've watched your cover of "Stop Crying Your Heart Out" a hundred times. If your YouTube views suddenly jump, it's because I've been watching you on repeat." He winked again, angling his body towards mine and getting up in my personal space, and okay, Cole had probably been right. It did seem like this guy was interested in me.

My gaze shot to the stage. There was no sign of the rest of the band, but I still didn't want to risk anything. I did the only thing I could think of. "Some of those views are probably from my boyfriend. He likes to watch us play on repeat too."

Scott deflated, his smile disappearing, but I was more interested in Cole's reaction. A wide, almost taunting grin appeared on his face that he didn't even try to hide. *Smug bastard.*

"So your friend wasn't lying when he said you weren't single." Scott shook his head with a sigh. "All the best ones are taken."

"Yeah. Sorry." A thought came to me, and I followed through. Cole was a fucking tease sometimes, and so it was only fair that I got to have some fun too. "It's a good thing he isn't here tonight so you didn't have to hear his off-key singing. The man can't hold a tune."

"Hey! Your boyfriend has a great singing voice," Cole interrupted, his glare reappearing.

"Great if you're tone-deaf."

"Maybe you need your ears checked," he shot back, and Scott's brows flew up. He stared at Cole for a minute, and then his expression grew calculating.

"Bro, it kind of sounds like you're into Huxley's boyfriend."

Cole studied Scott over the top of his pint, taking a casual sip. "Nah. He's not my type. But he is incredibly good-looking. Some have said he could be a model. He's funny, and clever, and a complete fucking catch, in my opinion."

"He's also completely narcissistic," I added.

"He's amazing."

"And he thinks very highly of himself."

"Okay, what's going on here?" Scott stared between us before focusing on me. "It really sounds like Cole's the one who's into your boyfriend, and you're not into him at all." He wiggled his brows. "Does this mean I have a chance?"

The fun had gone on for long enough. I shook my head. "No. Despite his many, many faults, including his huge ego, he's pretty fucking amazing, like Cole said."

"And he's really into you." Cole's eyes met mine, and my heart skipped a beat. Why did he have to look at me that way when we were around other people?

With an effort, I returned my gaze to Scott's. "I'm not interested in anyone else." Before things could get awkward, I beckoned to Tom, who had just re-entered the pub via the back exit. He was followed by Rob and Curtis. I took a moment to introduce Scott to them, and then I met Cole's eyes again, subtly indicating towards the door the band members had just come through.

When the two of us finally made it outside, at the back of the pub in the small courtyard that the van was parked in, I pulled Cole around the side of the van so we wouldn't be spotted by anyone.

"Cole. What—"

I didn't even get a chance to finish asking him what the fuck had just happened in there because he curved his fingers around my throat and yanked me into him, his mouth coming down on mine.

I responded immediately, throwing myself into his kiss, lost in the feel of his lips. Despite my worries about our parents and what people finding out would mean for our relationship, I wasn't prepared to give this up. Not yet.

TWENTY-SIX

HUXLEY

I stared down at the text on my screen. The vibration alert on my phone had broken the trance I'd fallen into in the uni library, where I was tucked into a back corner, working on my assignment.

COLE:

What happened this morning? Are you having second thoughts?

Shit. How could I explain to him what was happening in my mind? Everything had been okay after our gig, and I'd thought I'd been prepared to do what I could to keep Cole. But then, this morning, in the cold, harsh light of day, I was brought back to reality with a bang. How could we even think about being together, when the odds were stacked against us?

I'd panicked when reality had hit—there was no denying it, but even now, I couldn't see a different way that I could've reacted to the situation.

. . .

"Can you pass me the paper, love?" My dad held out his hand to Cole's mum. With a smile, she slid the newspaper across the table. It was one of our local London papers that veered between sensationalist stories and depressing London news.

"Can you believe that?" June's smile faded, and then she tutted, tapping her fingernail on the cover. "That poor family. The girl was knocked up by her stepbrother. You know the actor in that series we watched...what was its name? The one set in Brighton, with the killer who was an ice cream seller. Anyway, it was him. The whole family fell apart. Very sad."

Bile rose in my throat, and my fists clenched so tightly that I knew my nails were making crescent-shaped imprints in my palms.

"Disgusting." My dad scanned the article, his eyes moving rapidly across the page as he absorbed the text. "They should be ashamed of themselves."

I swallowed hard, trying to keep the nausea at bay. There it was. Both Cole and I had known that us being together would be a hard pill for our parents to swallow, but to hear this? We'd thought they'd be more concerned about what would happen to our family if and when our relationship fell apart and what other people would say, but it honestly hadn't crossed my mind that they'd be ashamed and disgusted with us.

"A baby!" June shook her head. "What were they thinking?"

I couldn't listen to any more. I shot up from my seat, pushing past Cole, who had just entered the kitchen, ignoring his exclamation of surprise. There was no time to say anything to him—I had to escape.

I knew I had to tell Cole what had happened. With the way I'd run out of there, he was probably confused about what was happening, and more than likely, he was jumping to conclusions.

ME:

At breakfast our parents were reading the paper and there was an article about stepsiblings who got pregnant. They said it was disgusting and the family should be ashamed and what were they thinking

COLE:

WHAT THE FUCK? Seriously?

ME:

Yeah. Here's the article

I sent him a link to the article in question and then waited for his reply.

COLE:

Fuck. They said that?

ME:

Yeah

COLE:

What does this mean? For us?

Good question. How could this go anywhere when

our own parents had such an extreme reaction to the thought of stepsiblings being together?

ME:

I don't know

COLE:

FUCK. I'm about to go into a meeting at work. I want to speak to you. Please don't make any rash decisions. Don't give up on us

I could easily picture him, tense and worried, and it brought a lump to my throat. I hated the thought of him hurting. Things would be so much easier if we didn't feel so strongly for one another. Was it better to end it now, before we'd even really started? It would save heartbreak down the line.

ME:

It's clear they disapprove but I won't make any decisions without you

I attempted to return to my coursework, losing myself in the library stacks, but my concentration was non-existent.

It was just under an hour before Cole sent me another text, and I had a total of two lines of notes in my notepad.

COLE:

OK I'm free now. Can we meet? I said I was feeling sick and they let me go

ME:

Meet at Jubilee Gardens

COLE:

Coming now

Leaving the LSU library, I headed for Jubilee Gardens, a green outdoor space close to the London Eye. It was warm for the time of year, and there were more tourists than usual. Still, it wasn't anywhere near as busy as it was in the peak summer season, and I easily found a patch of grass to slump down on. Pulling my phone from my pocket, I sent Cole a text to let him know where I was and then attempted to lose myself in my surroundings.

It was a wasted attempt. I couldn't help thinking about everything that had happened, and it led me to one conclusion that made me sick to my stomach. We'd been naive, hadn't we? How could we expect to have a relationship where we had to keep everything hidden away from every single person in our lives, a relationship where our own parents would be disgusted with us if they ever found out?

"Hi." Cole collapsed down next to me, his voice soft and hesitant. His face was so serious, and I knew we were both on the same page. A page neither of us wanted to be on, but sometimes life was fucking shit and fucked you over.

"Hi. We should talk." It was the hardest sentence I'd ever had to say, and yet somehow I managed it without my voice cracking.

Cole gave a jerky nod, which hurt way more than I thought it would. He swallowed a couple of times, his gaze intent on the grass at his feet. "I know what you're going to say. What needs to be said. Maybe this has gone as far as it should. Maybe we should end it before it goes anywhere."

"Before it properly begins. Before our parents find out." I was agreeing with his words, but inside I was fucking breaking apart.

"Yeah. Before we get in too deep. We know what our parents' feelings are. Do we want to jeopardise family relations and their happiness?"

It wasn't even a question because we both knew the answer.

"We did this once before. Maybe we shouldn't have started things back up again," I whispered.

Fuck. It hurt so much to do this. Both of us were hurting, not just me.

He glanced up at me, and I knew the look of devastation in his eyes would haunt me for a long time.

I closed my eyes so I didn't have to see his pain.

So he didn't have to see mine.

Behind my eyelids, it burned.

TWENTY-SEVEN

HUXLEY

The sixth balled-up piece of paper joined the others on my bedroom floor. This was fucking ridiculous. I was supposed to be composing a song I could share with the rest of the band at our next practice, because Tom wanted us to start gradually incorporating more of our own music to our social media. I had enough song lyrics written down that it should've been easy, but my muse had deserted me. Had left me high and dry since I'd come to that agreement with Cole. An agreement I regretted every single day, even though I told myself it was for the best.

Picking up the pen again, I closed my eyes, pushing everything else aside. Songs. Lyrics. The band.

Fucking focus, Huxley.

But all I could see behind my eyelids was him.

Was it possible to miss someone you saw every day?

I did.

It was like someone had carved a hole in my fucking

chest. Everything reminded me of him. I'd open the fridge and I'd see strawberries. I'd drive past a field, and I'd remember the day he'd taken me to the meadow, when I'd played my guitar for him. I'd go into my fucking games room and all I'd see was that moment he'd fallen asleep on me. The first time I'd consciously allowed myself to be vulnerable around him.

Fuck. I groaned aloud, letting my head fall back against the wall. My left hand rested on my guitar, and I idly plucked the strings, taking some comfort from the familiar metallic twang against the pads of my fingers, as I forced myself to count down from ten in my head.

Ten. *I wouldn't think about Cole anymore.*

Nine. *I was going to get over him.*

Eight. *Things would never work out between us. The odds had been stacked against us from the start.*

Seven. *Our parents didn't approve of stepsibling relationships.*

Six. *I needed to concentrate on the band.*

Five. *And uni. My dad would kill me if I fucked this up.*

Four. *I didn't need to be in a relationship to be happy.*

Three. *Cole was still in my life, even if we weren't together, and that was good enough for me.*

Two. *Fuck. I'd run out of ideas. Uh...I needed to concentrate on the band. No, I'd already said that. I needed to concentrate on songwriting. The band members were relying on me.*

One. *I needed to forget Cole Clarke. He was my stepbrother and my friend, and that was all he ever would be.*

I swallowed around the lump in my throat as I reached for my notebook and pen, and I locked my feelings away.

TWENTY-EIGHT

COLE

This was much worse than last time. Last time, I'd honestly thought it would be easy to get over Huxley. We barely even liked each other at the time. But now? I knew I loved him, and it was breaking my fucking heart.

I'd spent nights awake, going over everything, and I couldn't help wondering if we'd been too rash in our decision. But while I loved Huxley, I wasn't sure it was enough. I knew he had feelings for me, but *love*? I doubted it. He'd always hated me more than I hated him, and I was fairly sure that it was going to take him a lot longer to get to where I was. Not only that, but he'd changed recently. I knew now how important it was to him that he didn't fuck up his life, and with the pressures of starting uni and the band beginning to take off, I didn't want to be another problem for him to stress about.

I told myself that there was time. We could revisit things in the future, when everything had settled down, and maybe we'd have a chance to make a go of things. I wasn't interested in looking at anyone else, and I didn't think Huxley was the type to play games.

In the meantime, I just needed to stay busy and keep my head down. One day at a time.

With a sigh, I forced myself to focus on my computer screen, typing out a thank-you email for a generous donation that had been made to our charity. From its position on my desk next to my keyboard, my phone lit up with a message. My heart kicked into overdrive, but I immediately deflated when I saw the name on the screen. Not that I didn't like hearing from my cousin, but there was only one person I wanted to be texting me.

ELLIOT:

Hi Cole! I have a favour to ask you. You know you said I could use your discount on drinks at Revolve...is that still ok and is there any chance I could get some friends on the guest list?

ME:

Yeah of course. I said you can use it whenever. You'll have to come in on a day I'm working to use the discount though because I'll have to serve you. When are you thinking of coming and how many?

ELLIOT:

Friday? Not sure how many are coming but I'll let you know ASAP

ME:

Good timing. I'm down to work Friday, covering someone who's on holiday. Give me the names of everyone coming and I'll get them on the guest list. Need to check the staff rota but I think Bobby will be on the door doing the list, remember him? He should remember you, he's one of those people with a photographic memory

ELLIOT:

I remember him. I'll get the names to you ASAP. Thanks Cole. I owe you!

ME:

Family perks. Just make sure everyone spends plenty of money hahaha

ELLIOT:

Poor students, remember? *laughing emoji*

ME:

OK you can owe me

Elliot sent back a GIF of a man shaking his head in disappointment, and I smiled as I returned to typing my email. My phone lit up again, and I glanced down at the screen, expecting it to be my cousin, but it wasn't.

HUX:

How's your day going? This is mine

Huxley had attached a picture of a huge stack of books on a table, with a takeaway coffee cup next to them. In the corner of the picture, I could see part of his laptop and rows of shelves behind it. I was guessing he was in the uni library.

ME:

Mine

Fuck, I missed him. Seeing his name appear on my phone made my heart beat faster, and I wanted to prolong this moment between us for as long as possible. I was Huxley-starved, if that was a thing.

Arranging the windows on my computer screen to show both my email client and my music player, I snapped a picture and sent it to him. Grinning to myself, I waited for his inevitable reply. I'd set my Spotify to play its auto-generated playlist of my most-listened-to songs, and I knew he'd have something to say about the tracks listed on it.

Sure enough...

HUX:

WHAT THE FUCK IS THAT ABOMINATION? You have Oasis next to Cher, and the Stereophonics next to Kylie?!!!

ME:

Look at it this way. Before you I wouldn't have had either of those bands on there at all. I think the only band we both liked was The Hunna. This is all thanks to your influence

HUX:

I need to influence you more

ME:

Or maybe you should let me influence you. What about this?

I sent him a YouTube link to a rock cover of Britney Spears' "Toxic." Five minutes later, my phone rang.

"What the fuck was that, Cole?"

Fuck, I loved the sound of his voice. "There's plenty more where that came from. Maybe you could incorporate some covers into your set list. Or do a one-off gig—'the 2Bit Princes cover the queens of pop.'"

"I hate you."

"You don't."

"No, I don't." He sighed. "It might be easier if I did, though."

"No, no one wants a repeat of those days. Unless... hmm." Tapping my fingers against the desk, I cocked my head. "I am still a bit sad that I didn't get to push you into the lake at our parents' wedding. Can we pretend to hate each other and then re-enact that scene, but this time you actually end up in the water?"

"You're a dick," he said softly.

"You're a wanker," I said back, just as soft.

"Are you gonna be at home for dinner later?"

I shook my head and then remembered he couldn't see me. "No, I don't have time. I'm staying at work until six today, so I can take tomorrow off, then they want me to help set up at the club tonight. I won't have enough time to go home in between."

"Oh." There was silence for a moment, broken by the low murmur of voices around him. "We could meet if you want to and grab some food. I don't really feel like having dinner on my own with my dad and your mum."

It wasn't wise, given that we'd decided to stay apart for a second time, not to mention that I was feeling the L-word towards him, but I wasn't known for my wise decisions. I loved him, and I wanted to spend time with him. I'd take whatever he'd give me. There was nothing wrong with spending time with each other as friends, was there?

TWENTY-NINE

HUXLEY

The little coffee shop in the Aldgate area of London was nothing special. One of many. Seated next to a window on a squashy, cracked leather sofa in the corner of the shop, with Cole opposite me, I picked at my ham-and-cheese toastie. Rain pattered against the window, and every now and then, a bus would pass by outside, sending a cascade of water over the pavement. It was so good to see Cole. He looked hot as fuck, as usual, and he was asking all the right questions about uni, and the band, and life in general, knowing instinctively when to back off or change the subject. I'd never known anyone who could read me as well as he did, especially lately.

"Tom said something about the band booking a music venue. He was waiting to hear back from the promoter?"

Swallowing a bite of my toastie, I nodded. "Yeah. The Frog and Fiddle. It's a pub-slash-club that has a dedicated space for live music. One of the guys that books the bands was apparently at our gig the other day—they were saying they try to get in on the local talent before anyone else discovers them. He was interested, but we have to audition at the club, as well as send them our videos."

"Huh. I wonder why?" Cole's brows pulled together, and for some reason, I had to resist the sudden, strong urge to kiss him.

"Maybe they want to hear how we sound in their venue? I don't know, but Tom was excited when they contacted him because although it's a small venue, they have a really good reputation in the local music scene. Or so he says."

"Once they see you perform, there's no way they'll turn you down." Biting into his meatball sub, Cole met my gaze, his eyes sparkling. Once he'd finished chewing, he continued. "I'm really happy this is working out for you. It's weird how quickly things can change over one summer, isn't it?"

"Weird. Yeah."

"Weird, but good, I mean." He shot me a grin. "For both of us." His grin disappeared, and he leaned forwards in his seat, staring at me from beneath his thick lashes. I was faced with those fucking pretty eyes that haunted me in my dreams. "I wouldn't change anything, Hux. Not even the fighting. Definitely not anything that came after that. Even though things between us didn't work out the way I wanted them to, I want you to know that I don't have any regrets about you and me."

There was a sudden lump in my throat, and I had to clear my throat a few times to dislodge it. "Me neither."

When he gave me another smile that was so fucking bittersweet, misery radiating from him so strongly that I wondered how I could've missed it before now, I choked up all over again. It was obvious to me now that he'd been putting on a brave face, but he was hurting. *I* was hurting him. And myself. Yeah, there were some circumstances beyond our control, but at the end of the day, the two people at the centre of this situation were me and him, and we needed to do what was right for *us*, not for everyone else.

As I sat there, my gaze locked with his, the realisation crept up on me, until it felt like my entire body was overtaken with it. I felt as if I was vibrating out of my skin.

Right there, on a rainy Tuesday afternoon in a little coffee shop in Aldgate, I took a deep breath and welcomed the knowledge for the first time, letting myself believe what I'd tried to deny.

Now, somehow, I had to make Cole realise it. It wouldn't be enough to say the words. Our relationship had been messed up from the beginning, and it would take more than a few lines from me for him to believe how strong my feelings were.

I needed something more.

The problem was I'd never done anything like this before. I was the fuck-up—the kid that had always been in trouble at school, a disappointment to my parents, the person who'd made his new stepbrother feel so fucking unwelcome the second I'd laid eyes on him. Fuck. Who would even want me? Was it even worth trying?

With a shaky exhale, I tore my gaze away from Cole's, taking a sip of my coffee to hide my trembling lip.

"Huxley." Cole's voice was so quiet and so gentle. "Never forget just how amazing you are, okay?"

How did he know exactly what to say?

I couldn't reply. Memories filled my mind. My dad sitting me down and telling me that he'd managed to get me onto the foundation course at uni, his face full of cautious hope. Tom, Curtis, and Rob the first time we all played together and the euphoria that came from being so in tune with one another. My dad and June saying they were proud of me after the 2Bit Princes' gig. Cole taking care of me after my accident, even though we hated each other. Cole carefully bandaging my body. Cole flirting with me. Cole telling Tom how talented I was. Cole kissing me. Cole watching me play. Cole supporting me. Every. Fucking. Step of the way.

Cole. Cole. *Cole.*

I'd reached my breaking point, and I was done. I'd probably reached it way before this point, but denial was strong, especially when I was letting myself drown in outside pressure. But enough was enough. Now the blinkers had been torn away, and I could see properly. Now, I was prepared to deal with our parents, no matter what they thought. Even if they were disgusted by our relationship, I'd deal with the fallout and do my best to protect Cole.

I loved him.

And I knew something.

Cole Clarke was fucking mine, and now I needed to do something to show him just how much he meant to me.

THIRTY

COLE

"What's the matter, Cole?"

I looked up from my laptop screen to see my mum eyeing me with concern. I'd chosen to work from home today, just so I wouldn't be tempted to text Huxley and ask him to meet up with me to grab some food. It was getting too hard to put on a brave face, and although I'd accepted that things had to be this way, it didn't mean I had to like them.

"Nothing's the matter. I'm fine."

My mum took a seat at the table next to me, placing her hand on my arm. "Cole. I know you, and you haven't been yourself lately. You seem...sad. If there's something you need to talk about..." She paused, her gaze searching, before she sighed. "You know you can talk to me about anything. I promise I'll listen."

It was tempting, but no. I wasn't a kid who could go crying to their parent and they'd make it all better. Never had been, really. I had no doubt that she was genuine, but this was one secret that Huxley and I would take to the grave.

"I'm fine," I said again. "Just tired. I've had a lot of extra shifts at the club." Shifts I hadn't needed to take, but I'd volunteered because at least when I was busy, I didn't have time to think about Huxley. They also had the bonus effect of making me so tired I fell asleep easily and slept deeply, rather than tossing and turning all night, asking myself pointless questions with no answers.

Rising to her feet, she dropped a kiss to the top of my head. "I mean it, Cole. If there's anything you need to talk about—"

I forced a smile onto my face. "Thanks, but I'm fine." Like a lifeline, my phone chose that moment to buzz, and I pounced on it, glad for a chance to escape the conversation. My mum sighed again but took the hint and left me to it. When I opened my phone, I found a message from Tom waiting for me.

TOM:

Frog and Fiddle tomorrow at 7pm. BE THERE

My forced smile turned genuine.

ME:

You got on the list?

TOM:

Yeah baby! This place can make or break bands. I'm buzzing. Had the audition yesterday. The band that was opening for the main act tonight had to pull out because the lead singer got appendicitis and they asked if we wanted to fill the slot

This was *huge*. This was the venue both Tom and Huxley had spoken to me about, and the last I'd heard they were waiting for an audition spot. Now they were actually going to be playing there?

ME:

Fuck yeah! Happy for you guys

TOM:

You'd better be there. No excuses. We need our original groupie. Rob says you're a good luck charm

ME:

I'll be there

TOM:

Good. We've already put your name down on the door so you don't need to worry about getting a ticket

Hux told me to give you the set list. We're opening for another band so we're playing a short version

He sent through a picture of a piece of paper with song names written in marker pen, and I made a mental note to memorise the order.

ME:

Thanks. See you tomorrow then. Break a leg or whatever

TOM:

Cheers. Be there at 7. Don't be late. You're not going to want to miss this

---

The music venue was packed with wall-to-wall patrons. There was no way I had a chance of getting near the stage, but that didn't even matter. I was excited that the band was getting this opportunity, and all I wanted was for everyone to see how amazing the 2Bit Princes were.

Pushing my way through the crowds, I eventually made it to the bar. Once I had my drink, I moved down to the far side, where it was quieter, and tipped my pint to my lips. From here, I had a good view of the stage, and I had a bit of space from the crowds.

The last time I'd seen the 2Bit Princes perform, everything had been different. For a start, Huxley's and my parents had been there, but not only that, Huxley had been mine, even though we'd had to hide it. Now, I was adrift. I felt like the cliché of the fan hopelessly in love with the lead singer of the band, although my situation was a little different because I knew what it felt like to feel Huxley against me. To hold him all night. To share my secrets with him.

And now...now I had to watch him like he was a stranger. But I was so fucking proud of him, and of the band, for securing this gig, here in a place that was clearly so important to them.

So, yeah, it was bittersweet.

The band launched into their first song, and my heart swelled with pride and love. My smile was helpless. They were amazing.

The crowd seemed to agree, their cheers almost deafening, singing along with the lyrics. I lost myself in the music of the 2Bit Princes, also singing along with all the songs, all the way until the end of their set list.

Then something changed.

Huxley stepped up to the microphone. There was a sheen of sweat across his head, and his eyeliner was smudged around his eyes, as it often was when he performed. He grinned at the crowd, and the way his smiles came so easily now, compared to when I'd first known him, gave me fucking butterflies. His fingers curled around the hem of his T-shirt, and as I watched, mesmerised, he lifted it over his head, throwing it to the floor and baring his torso to everyone in the venue.

Screams and whistles sounded as my mouth went dry at the sight. I was envious, and jealous, and wishing that this was a private performance, just for me. I'd seen his gorgeous body before, but now *everyone* was seeing it.

Wait a fucking minute. I blinked, and then blinked again, as I stared at my stepbrother.

Huxley had his nipples pierced.

Fuuuck. Since when?

Two barbells gleamed on his pecs, and fucking *wow*.

He was the hottest fucking thing I'd ever seen in my life.

"Hi." He dipped his head to the microphone, and the screams from the crowd were deafening. A shy smile crossed his face at the cheers. "Thanks...I...uh...I want to dedicate the next song to someone special. Someone that means a lot to me."

My breath caught in my throat. Nothing else mattered except for the beautiful bleached blond boy on the stage, his shirtless, pierced torso bared to the world, a glossy black electric guitar strapped across his body, and his gaze...his gaze found mine in the crowd, fixing on me for a brief moment before he glanced down at his guitar, but that moment was everything.

"This next song means a lot." Huxley's hand visibly shook around the mic, but his face was resolute. I watched as he visibly gathered himself, straightening his shoulders, and his grip on his mic steadied. His gaze arrowed straight back to mine. "I let someone think they weren't the most important thing in my world, and I want them to know that they are."

What?

His tongue slid across his lips in a slow, sensuous glide, his fingers moving across his guitar strings, ready for the song to begin.

Then the band began to play "Never Enough" by The Hunna, and I was completely fucking lost.

The song was about falling in love, and the way Huxley fixated on me while he was singing, never once looking away, left me with no doubt in my mind.

I blinked back the sudden tears that sprang to my eyes, watching the boy I loved sing his heart out. *To me.*

When the final notes of the song died away, Huxley

held out his hand. The whole time he'd been singing, he hadn't looked away from me.

The emotions rising up in me made the breath catch in my throat. I felt like I couldn't fucking breathe as I moved forwards on wobbly legs, barely even aware of the eyes on me as I made my way through the crowd and up to the edge of the stage.

"Hi," Huxley murmured, still holding out his hand. Reaching out, I curled my fingers around his and let him pull me onto the stage.

"Hi." We were so close I could feel his breath on my skin, could see the anxiety in his eyes beneath the veneer of confidence. "Hux. What's...what's happening?"

He swallowed hard, his Adam's apple bobbing. His gaze darted to the crowd behind me and then back again, a hint of a smile tugging at his lips as he exhaled shakily. "I think I just told you how I feel about you in front of two hundred and fifty strangers."

"Is it true?" My mind couldn't process this, trying to reconcile the Huxley I knew with this man who would share his feelings with the world.

One trembling hand cupped my cheek. "Yes," he whispered, and I couldn't take it anymore, winding my arms around him, guitar and all, and slanting my mouth across his.

It felt like coming home. It felt like everything.

"I love you," I rasped against his lips, so choked up that it was a struggle to get the words out. "I love you, and I want to be with you. Even if it's hard. Even if no one understands."

"I don't care what anyone thinks anymore, Cole. This

is worth it." He kissed me again, so fucking soft and sweet, and it was like a moment in a film or something, right up until the loud screech of feedback from a mic had us springing apart.

"Congratulations to our lead singer and his boyfriend." Tom smirked at us both. "You can thank me later for making sure that Cole was here tonight to witness this moment." He raised his voice, speaking directly into the mic. "Now...who wants to hear "Never Enough" again?"

"Me." I raised my hand, and I suddenly became aware of the audience behind me, almost deafening in volume. How had I missed that? Slowly spinning on my heel, I turned to face them, dropping into a deep bow. Out of the corner of my eye, I could see Huxley's bright smile, even as he rolled his eyes at my antics, and it warmed me all the way through.

Moving to the side of the stage, I held out my hand to the band, and the cheers sounded even louder. Tom shook his head, grinning, before he indicated that the song should begin. As Huxley started singing, his hand caressing the stand like a lover, his lips so fucking close to touching the mic, my mind was blown all over again that he'd just done that, right in front of *two hundred and fifty people*. And now those people were excited for more from him, and from his band.

I was so proud of him.

I wanted to let him know how much I loved him.

THIRTY-ONE

HUXLEY

The roads were quiet. There was something peaceful about being out late at night in London, away from the busy centre, after all the tourists had disappeared. Winding our way higher, out of the city, I navigated the streets with ease, my confidence finally restored after that first drive since the crash. I'd never have made it through that drive without Cole. Now, he was a reassuring presence at my side, strong and supportive, and I was hit all over again with the knowledge that he was so fucking special to me and I was so lucky to have him.

I'd purposely brought my car tonight instead of travelling to the gig with the rest of the band. When I'd taken my idea to Tom and the rest of the band, they'd been all for it, but none of us had known how it would turn out. There were no guarantees, and I'd still been half expecting Cole not to show up, even though Tom had assured me he would. Tonight had been a gamble, but it had paid off, and now here I was, with Cole next to me.

"I didn't mean it. What I said to that guy in the pub

before about you being tone-deaf." The words fell from my lips without thought. I wanted there to be no more barriers between us.

Cole chuckled softly, his hand sliding onto my thigh and briefly squeezing, making my breath hitch. "I know. Want to tell me about your pierced nipples?"

"You know I wanted to get them done." Lifting one hand from the steering wheel, I brought it down to slide my fingers between his. "Thinking of how much you'd like them was an extra incentive."

He squeezed my hand as the familiar sign appeared up ahead, almost glowing in the beams of my headlights. Flipping the indicator on, I turned off the road and onto the bumpy track that led to the small car park on this side of the hillside park. When I switched the headlights off, the darkness enveloped us, with the city lights stretched out below, sparkling in the night.

Turning off the engine, I unclipped my seat belt and opened my door, swinging my legs around to the outside of the car. Across from me, I could hear Cole doing the same thing, and when his door slammed, I exited the car.

Cole met me around the front of the car, reclining against the bonnet with one leg casually crossed over the other. "They're sexy as fuck." When my gaze shot to his, he smiled. "Your piercings."

"Glad you approve." I came to a stop next to him, leaning back against the front of the car and staring straight ahead. "What do we do now?"

He was silent for a while, and then his hand slid across the car bonnet, his fingers covering mine. "We have to tell our parents. But before that..." There was a long pause, and then he said, "Hux?"

I turned my head to his. "Yeah?"

His eyes were dancing with humour as he glanced around the empty car park. "You didn't bring me to a dogging spot, did you?"

"What? *Fuck*, no."

Now he was openly grinning at me. "Just checking. You know. Late night...a deserted car park...seems a bit suspicious. I keep expecting someone to turn up and shine a light in my face or something."

There was only one way to shut him up. "Believe me." I curved my fingers around his jaw, bringing his face in line with mine. "I don't want an audience for this. Just you."

Then I kissed him.

We ended up with me standing between his spread thighs, my arms wrapped around him, while he gave me the softest, deepest kisses I'd ever had in my life, his hands cupping my face.

"Fuck, I love you," he murmured when we paused for breath. "It's fucking insane that of all the people in the world, it's you, but you're..." Stroking his thumb over my jaw, he shook his head. "You're everything that I never knew I wanted."

I wasn't used to big declarations, or any declarations, for that matter, and I thought I'd used my lifetime's quota during the gig earlier. But here in the dark, with Cole's body a warm, solid weight against mine, it was easy. Fixing my gaze on his, I spoke the words aloud for the first time.

"I love you, Cole."

He melted against me, his head coming down to rest in the crook of my neck as he held me even tighter. "We did everything the wrong way, didn't we? We fucked before we'd even done anything else. We split up, even when we weren't properly together. But we can do it properly now, can't we?"

"Yeah." I felt him place a soft kiss to my throat, and I returned it with a press of my lips to his head. "We'll have to work out what to say to our parents, but I don't want to be without you anymore."

His head shot up, his eyes meeting mine. "Boyfriends?"

I nodded, and he smiled, so wide and happy. "We'll face them together. We'll make them see that this is important to us."

"I'm not looking forward to that conversation."

"Me neither. Fucking hell, can you imagine how they'll react? It'll be worth it, though. You're mine, Huxley, and now I have you, I'm not letting you go again."

He was right. It would be worth it, and there was no way I was letting him go again. As much as I was his, he was mine.

Headlights cut through the darkness, slashing straight through our moment. *Fuck.* We instantly sprang apart. Another set of lights followed the first, blinding us both, before they started flashing. Covering his eyes with his arm, Cole scrambled off the car and dived for the passenger door.

"I'm not being caught in a dogging spot, Hux!" he shouted. "Get us out of here, now!"

My heart was pounding as I slid behind the wheel, starting the engine and spinning the wheels like we were in a car chase. "I didn't know! Maybe they're just people who wanted to see the view like we did."

Cole glanced back through the rear windscreen, the cars disappearing from sight as I got us out of there. "They were flashing their headlights. That means they wanted to know if we were doggers."

"You seem to know a lot about it for someone who's so adamant that they don't want to have public sex." I couldn't help grinning at him, and he finally laughed, shaking his head.

"Fucking hell. No. No shame to anyone who wants to, if you're into all that. It could be hot, I guess. But I think we have enough to worry about with telling our parents we're together. Getting caught at a local dogging spot isn't on my to-do list."

"Nor mine." I reached over and squeezed his thigh. "Want to pick some music?"

He smiled to himself as he began fiddling with my stereo. Predictably, terrible pop music filled the car, but I bit down on my lip and forced myself not to say anything. I'd let my ears bleed. For him. Only ever for him.

"Huh. You do love me." Leaning over, he pressed a soft kiss to my cheek before settling back in his seat. "But this relationship is all about give and take. You and me are equal partners in this. I think you'll like the next song."

"Equal partners. You know I don't have much experience with relationships. Or any. I might fuck up. It's probably a given."

"Me neither, and I know I'll fuck up too. But we're good together, aren't we?"

The familiar opening notes of "Somewhere Only We Know" sounded through the car speakers as I nodded. "Yeah. We are."

THIRTY-TWO

COLE

Next to me, Huxley swallowed hard, his thigh bouncing against mine as we sat in the lounge, our parents across from us and a coffee table with a Trivial Pursuit board separating the two sofas. Tonight was the night. We'd done a lot of talking, and we'd worked out what we wanted to say to them. Now we just had to hope they'd listen and understand just how much we wanted to be together.

Fuck, how should we even begin?

"Okay, who wants the purple wheel?" When no one replied, David glanced up from the board, his gaze going first to my mum, then to me, and finally to Huxley. A frown pulled his brows together. "We're doing this now, are we?" he muttered under his breath, and Huxley and I exchanged panicked looks. What was happening?

Clearing his throat, he took my mum's hand. "June? Do you want to take the lead?"

My mum nodded, straightening up in her seat. "We know about the two of you."

"What?" My heart fucking stopped, my mouth dropping open. Next to me, Huxley's entire body went rigid.

"Cole." Her voice softened. "I could tell when I saw the way you looked at Huxley when he was singing that Oasis song. It was clear to me that you were in love with him."

Fucking hell. I buried my face in my hands, feeling my cheeks heat against my palms.

She continued speaking. "What I didn't know at the time was if Huxley felt the same way about you. I spoke to David, and, well, we weren't sure if it was one-sided. We decided to hold off, to let you tell us in your own time, if there was anything to tell." From behind my hands, I heard the sound of rustling fabric, and then I felt her arm slip around my shoulders as she perched on the arm of the sofa next to me. "Then lately... You were so sad, Cole. I was worried about you."

I lowered my hands, even though I could still feel my cheeks burning, and took a shaky breath. "I told you I was fine."

"You weren't fine."

From my other side, Huxley swallowed again, and then, slowly, cautiously, he placed his hand on my leg, palm up. Just as slowly, I placed my hand over his, our palms meeting and our fingers sliding together.

"Bill from accounts sent me a link last night." David spoke up, his gaze flicking to our joined hands before he turned his attention to his son. "I'd spoken to him about your band, you see. He enjoys the sort of music you play, and, well, I was proud of you, so I wanted him to know."

"What was the link?" Huxley whispered. I stroked my thumb over his skin, gently squeezing his hand.

"It was a video clip from your band's performance at...what's the name of the place... The Frog and Fiddle."

Oh. *Oh.*

"It wasn't one-sided," my mum murmured.

Huxley's grip on my hand tightened. "No, it wasn't."

David took over the conversation again. "Your mother and I spent a long time talking about things. If you're truly serious about each other, if you're prepared to invest in this relationship, then you have our support. But I want you both to promise us that you'll work at this. The last thing we want is to see our family torn apart if you decide it's not worth the effort."

I was completely speechless, blindsided by the fact that they already knew about us, and somehow, they were prepared to support our relationship. Huxley shook his head next to me. "I—I don't understand. We thought...we didn't think you'd approve. And what about the article?"

"What article?"

"The article with the stepsiblings that got pregnant. You...you said it was disgusting and they should be ashamed of themselves."

David's frown disappeared. "No. That wasn't—it was the situation, not their relationship. The family should have supported them."

"We can't help who we fall in love with." My mum looked over at David, and he gave her a fond smile. "Yes, we do have our concerns, and if it hadn't been clear that you're in love with each other, then, yes, we would have had some issues. Not because the two of you are related by marriage—you didn't choose that. Our concerns relate to our family. David and I have been trying our hardest to get things right this time around, and if things between the two of you were to sour, then there's a big risk that it could tear apart the family unit we've been trying to build."

I finally found my voice. "Hux and I have spoken about this. We know it's a risk, and we know there's no guarantees. But we love each other." Turning to Huxley, my words just for him, I added softly, "He means everything to me."

Same, he mouthed, his sapphire eyes brimming with emotion. Fuck, I loved him so much.

Clapping his hands, David cut through the moment. "Right. So we're all in agreement. The two of you will work at this, and in return, your mother and I will give you our support. There are a few house rules, though."

Please don't mention sex.

"What you do together—no. I can't. June? Do you want to go through the rules?"

"Not really, David," she said dryly, making us all laugh nervously. "Just...stay safe and be sensible. I'm sure we don't need to spell it out."

This was so fucking awkward.

"Trivial Pursuit!" My voice came out way too loud and weirdly high-pitched. "I'll have the purple wheel. Mum, you want the orange one?"

"I need a drink before we begin," my mum said with a grimace, rising from the arm of the sofa. "Anyone else?"

"*Yes*. Scotch. I'll get them. You boys want anything?" David leapt to his feet, probably glad for a reason to leave the awkward atmosphere. I didn't blame him. If I could, I'd be leaving the room and dragging my boyfriend with me. But since we had to sit through this forced bonding time, aka family games night, then alcohol would have to help ease the way. I couldn't be too annoyed about it, though. Our parents had given us their blessing, and we hadn't even had to give them any of our rehearsed speech. I was still in shock.

"I'll take a beer. Cole? Want one?" Huxley nudged me, bringing me back into the present.

"What? Oh, yeah. Beer. Thanks."

Our parents disappeared into the kitchen, and I slumped back onto the sofa, pulling Huxley into me. He huffed out a breath against the side of my face, his arm sliding around my waist. "I'm not dreaming, am I? Did our parents really just give us their support?"

My arm tightened around him as I tugged him even closer, savouring the feel of his warm body against mine. I didn't have to hide my affection for him anymore, and I'd never been so happy. "Yeah. I think they did."

THIRTY-THREE

HUXLEY

"That's the last of it." My dad placed a large cardboard box onto the pile at the side of the room. The flat wasn't much to look at—small and neutrally decorated, with an open-plan living space, a bedroom, a bathroom, and a storage closet in the hallway. But the lounge had its own tiny balcony that was big enough for a little table and two chairs, and best of all, the entire flat had been soundproofed by a previous owner. That meant we wouldn't be disturbing the neighbours with my music...and other things.

We were only renting the space, but it was somewhere that was just ours. Now we were properly together as a couple, Cole and I had agreed that it was too weird and awkward living under the same roof as our parents, and I was sure they agreed, so we'd started looking for a place to live. It had taken a while to find somewhere because we needed to be within easy reach of Cole's job at the charity, LSU, and our parents' house. Eventually, though, this flat had come up, and we'd taken it as soon as we'd seen it.

My dad had insisted on subsidising our rent on the proviso that we would have dinner or a catch-up with him and June at least once a fortnight, preferably more often, and of course, we had our family bonding nights. We'd agreed to his conditions without question. I was a student and would be for a long time to come. Neither of Cole's jobs paid particularly well, and he was still only part-time at the charity, so it was good to have the financial support of our family. We were lucky to have it.

"Thanks, David." Grinning, Cole clapped him on the back. "If you ever feel like a career change, you could easily get a job in removals."

It was weird to see him teasing my dad like this, but ever since the truth about our relationship had come out, it felt like everyone had relaxed more around each other. My dad and I were still trying to navigate this new dynamic, but the abrasiveness that had always been between us was gone for good.

June entered the room, carrying two familiar chipped blue-and-white mugs. "Something old and something blue." She raised the mugs into the air. "The flat is a rental, so that covers the something borrowed part, and it also covers the new part because it's your new home."

"Mum, you do realise we're not getting married, right? We're just moving out." Cole shook his head with a grin, taking the mugs from her and heading over to the kitchen cupboards.

"Indulge me. Our children are leaving the nest." Her

words were joking, but her lip trembled, and Cole was instantly there, wrapping his arms around her.

"We're only a few tube stops away, and we'll be coming over to eat all your food as often as we can. I'll text you every day, okay?"

"You'd better." She sniffed as he released her and then gave me a wobbly smile. "Huxley?"

I wasn't a hugger, but I also wasn't a heartless bastard. Not anymore. Crossing the room, I gave her a brief hug, and then my dad wanted to hug me, and my hug quota for the year was used up in about five minutes.

When they'd left us alone, Cole collapsed onto his back on the sofa, holding out his arms. He smirked at me. "Are you all hugged out, or can your sexy, amazing boyfriend get a turn?"

"I don't know. When's he getting here?"

"Ha ha. Come here, Hux."

Before I went to him, I took a minute to cue up some music on my phone, linking it to the Wi-Fi speakers. That done, I left my phone on the table and let my sexy, amazing boyfriend pull me down on top of him.

"Mmm. There you are." He smiled up at me, sliding his hands down my back and onto my ass. "I love this song."

"Yeah?" I knew he did. It was "Never Enough" by The Hunna, the song I'd sung to him in front of all those people to show him I loved him.

"Yeah." His lips met mine. "I really, really love it."

"Do you love it more than Katy Perry? Or Cher? Or Beyoncé?"

"Hmm. Let's not go that far." He shot me a grin

before kissing me again. "I think we should complete the most important part of the moving-in tradition now."

"Which is?"

"Christening the flat. Let's start with the sofa."

After christening the lounge, bathroom—which was like playing Tetris, trying to fit all our body parts into one tiny space—and bedroom, there was only one place left. The balcony.

"It's my turn to fuck you." Cole pressed me up against the wall of the lounge, next to the balcony door, his mouth at my throat and his fingers lightly tugging at one of the barbells in my nipples. To say he was obsessed with them was an understatement. It was a win-win situation for both of us because when he played with them, it felt so fucking good.

We were both completely naked, and I could see this becoming a habit. Cole was fucking hot, so why wouldn't I want to take every opportunity I could to look at his body?

Dipping down, I gave into my urges and dragged my tongue over his abs, stopping at the top of the dark trail of hair that led to his rapidly stiffening erection. Glancing up at him through my lashes, I felt him shiver against me, his hand coming down to run his fingers through my hair.

"Suck me on the balcony, Hux. Then I'm gonna fuck you."

There was a reason we'd left the balcony until now. It was too exposed. Out there, there were other buildings overlooking ours. The traffic noises below. The sounds from other flats around us.

But now it was dark, it was late at night, and we hadn't switched the outside lights on.

"Come on, then." I kissed up the inside of his thigh, enjoying the tease. His cock grew even more erect, making my mouth water. When he spread his legs, I tapped his thighs, tearing my gaze away from his dick. "Outside."

He groaned, gripping the base of his thick cock, precum glistening at the tip. "Fuck, yeah. I want your mouth on me."

When we made it onto the balcony, there was a chill in the air, but I ignored it. Cole Clarke was naked, with a deliciously hard cock, and he was about to fuck me with it.

After I'd sucked him.

I backed him up against the balcony wall, kissing him until we were both breathless, and then I dropped to my knees. Leaning forwards, I curled my fingers around his cock, my mouth skating across the tip as I stared up at him.

"Fuck, Huxley. Suck me." His fingers tangled in my hair, his hips arching forwards, sending his dick down my throat. "So good."

Lowering my head, I took him into my mouth. He groaned long and low as I took him further in. The taste of him against my lips...the hard, heavy weight of his erection, the musky deliciousness of his scent, and the soft skin covering his rock-hard length...it was so fucking hot.

He felt so good against my tongue, but I wanted more. I wanted him to fuck me like he said he would.

Releasing his dick from my mouth, I rose to my feet, licking my lips. I met Cole's heavy-lidded gaze, watching as his blown pupils focused on my lips, then dropped to scan my body before returning to meet my eyes. He smiled, slow and sexy, both of us on exactly the same page. After hand jobs and blow jobs and rim jobs, interspersed with food and talking and just enjoying being together in our own place, here we were. Ready for this moment.

Curving my fingers around the railing that ran along the top of our balcony wall, I spread my legs, arching my back. I was facing away from Cole, but I was close enough to hear the hitch in his breath, to feel the heat of his body against the backs of my thighs.

"Hux. You're so fucking hot." His finger began to circle my hole, open and lubed from the prep we'd already done. When I arched my back further, he pressed inside, so fucking slowly.

"Cole. I need your cock. All of it."

"Yeah. Hold still, baby." He kept pressing forwards at the same slow, torturous pace, until he was all the way in. Fuck. It was so good, feeling him bare inside me, skin to skin. The first time he'd fucked me in our parents' kitchen, it had been hard and fast and just what we both needed at the time, but here on our tiny balcony, with his thick cock stimulating my prostate with every slow thrust, it was so fucking perfect.

"Yeah! Give it to him!"

I gasped, my gaze flying to the building across from

us, seeing two guys hanging out of a window, waving their fists at us and cheering. *Shit.*

"Fucking hell." Cole collapsed over my back, burying his face in the nape of my neck, although he kept up his slow, thrusting movements. "Is this really happening? Our first day here, and we've already got a reputation with the neighbours." A pained groan tore from his throat. "I don't even care. It feels too good to stop."

"Keep going." Fuck it. They'd already seen us.

"I knew you took me to that dogging place on purpose. You like this, don't you?"

Releasing the railing with one hand, I reached back, sliding my fingers through his damp hair. "Fuck. Right there, Cole. Do that again." A shout of encouragement from across the road sounded again, and Cole raised his middle finger at the noise. Pushing back against him, I moaned. He felt so good inside me. "Maybe I don't care because I'm getting the best dick of my life."

"Good point." His fingers curved around my cock as he pressed a kiss to the back of my neck. "They can't see much, anyway. The balcony wall's blocking their view, and we didn't turn the lights on."

It was surprisingly easy to tune out our two voyeurs as Cole began to stroke my cock in time with our movements, and I lost myself in the feeling, surrounded by his body, letting the pleasure build until I came, covering his hand and the balcony wall in my release.

"Cole," I moaned, and he swore under his breath, his hips jerking.

"Fucking...milking my dick...so fucking good," he ground out, emptying himself inside me. We both slumped forwards, the wall taking most of our combined weight, breathing hard, sweating and sated.

"Yeah!"

At the sound of the noise from across the road, Cole chuckled tiredly, raising his hand in acknowledgement. "Thanks for the encouragement," he called, receiving a laugh in reply.

"Thanks for the show!"

"Can you shut up? Some of us have to get up early for work in the morning!" came another shout from somewhere to our left.

"Sorry!" The guys opposite us finally closed their window, leaving us alone.

"So. That was an interesting welcome to the neighbourhood," Cole said eventually. He slowly withdrew from me and then placed a soft kiss to my shoulder. "Wanna clean up in the shower, then cuddle in our bed all night long? I'll even make you breakfast in the morning."

"Yeah." I straightened up with a groan. "That wall isn't comfortable."

"Next time, we'll do it in our nice soft bed."

"*Without* an audience."

When I turned around, Cole was smiling at me, his gorgeous brown eyes shining in the soft glow from the streetlights. "Just us."

"Just us."

We sealed our agreement with a kiss.

COLE

EPILOGUE

FIVE YEARS LATER

From my vantage point in the wings at the side of the stage, I surveyed the crowd. It went back for as far as the eye could see, thousands of people here for the weekend to enjoy music and soak up the atmosphere. This was a huge moment for the 2Bit Princes. After five years of building up a reputation in London, as well as online, they'd been invited to play at a big festival, performing on a stage dedicated to talented local artists that deserved the exposure. Part of their set was even due to be filmed by the BBC for a compilation show that would be broadcast the following week.

I was so fucking proud that I could burst. My boyfriend and his bandmates had worked incredibly hard to get to this point, and it had paid off. Now all these people here at the festival would get to see just how amazing they were.

My phone buzzed in my pocket, and I swore under

my breath as the sudden vibration jogged my memory. I knew there was something I'd forgotten to do—put my out-of-office reply on. After the festival weekend was over, I was taking Huxley for a surprise week away, flying to Amsterdam, then on to Berlin, where I'd booked us tickets to see one of his favourite bands.

Pulling up my work email, I set my out-of-office reply. I couldn't help the smile that crossed my face when I added my email signature. *Cole Clarke, Director of Development*. Yeah, my hard work was paying off too. I loved my job at Mind You, where I got to do something I was passionate about every day, and at nights, I got to come home to my incredible boyfriend. And every weekend was spent with the two of us exploring London, or me watching Hux play, or us hanging out with our family and friends. Oh, and our now-monthly Trivial Pursuit nights with our parents, of course, which we took turns hosting.

The one downside? The more famous the band got, the more people there were that wanted a piece of Huxley Granger. But at the end of the day, he was mine. No one else knew him like I did. No one else got to see his soft, vulnerable side, to hold a piece of his heart, to be there to support him through the bad days as well as the good.

Also, no one else got to experience his fucking delicious body. No one else got to map out all his lines and contours and his beautiful dick with their hands and mouth, to make him gasp, to see his face when he came—

Okay...that wasn't strictly true. One or two of the residents in the buildings around us *might* have seen his face, y'know, if they happened to be looking when we were in the mood for fun times on our huge balcony with its view of the Thames in the distance. Whoever designed those balconies with glass walls was clearly an exhibitionist. At least in our first flat, we'd had a solid brick wall.

But the most important point was he was the one I came home to every night. He was the one I loved, deeply and irrevocably, and I knew he loved me the same way in return. Even if we'd had nothing, as long as we were together, then we'd be happy.

Shoving my phone back into my pocket, I moved closer to the stage, smiling as the 2Bit Princes launched into their first song—one that was climbing the downloads charts, slowly but surely. The exposure they'd be getting here would give it a huge boost, and I couldn't wait. Huxley and Tom had written the song together, and there was a line about pretty brown eyes that referred to me, which was very fucking cool. It also mentioned the singer—Huxley—being in love with a man, and fuck, yeah. We needed more gay representation in music. It was one of a thousand different things that made me so proud of the band. I couldn't even put it into words, but I was buzzing for them.

Halfway through the set list, when the band had finished one song and were taking a moment to swig from the bottles of water the backstage crew had provided, Huxley clasped the microphone stand. He looked over at me, and a smile curved over his lips.

"The next song isn't one of ours, but it'll always be special to me. Those of you who've been following us will know the background, but if you're new to the 2Bit

Princes...this song is how I first told my boyfriend I loved him. He's the love of my life, and yeah...I want to dedicate this song to him."

Then the band launched into "Never Enough" by The Hunna, and I did my best to swallow around the sudden lump in my throat. I sang along with the lyrics, unable to tear my eyes away from Huxley, up there performing on that huge stage for thousands of people, making a statement about his love for me.

When the song finished, my legs moved of their own accord. I ran to him, grabbing him around the back of his neck and pulling him into a hard kiss, right there onstage.

"I love you so much. So. Fucking. Much." I pressed my forehead against his, trying to catch my breath, so overwhelmed with everything I felt for this man who was my whole world. "You're everything to me. Everything."

"I love you, too." He kissed me again. "And now everyone knows it."

"Yeah. And everyone knows I love you." Shifting my gaze to the crowds, I smiled. "Finish your set list. I'll be waiting."

With a wave, I jogged off the stage to the roar of the crowds, shooting a wink at Tom, who shook his head at me with a grin on his face.

Then I watched the man I loved blow the audience away with his talent, leaving them begging for more.

And after that? Let's just say it involved the two of us, plenty of strawberries, and a balcony with a clear glass wall...

THE END

THANK YOU

Thank you so much for reading Cole and Huxley's story! Are you interested in reading more from some of the other characters? Check out the following:

Blindsided (Liam & Noah)

Sidelined (Ander & Elliot)

If you want to know what's coming next, sign up to my newsletter for updates or come and find me on Facebook or Instagram.

Check out all my links at https://linktr.ee/authorbeccasteele Feel free to send me your thoughts, and reviews are always very appreciated ♥

Becca xoxo

ACKNOWLEDGMENTS

There are so many people I need to thank. First and foremost, I need to mention the bands and artists that played such a huge part in Cole and Huxley's story. Yes, they'll never read this, but they deserve recognition. And along those lines, I need to thank A and T, because watching my best friend's boyfriend's band play local gigs on Saturday nights was a big inspiration for this book.

To Kelly, for the video that inspired the events that led to Cole and Huxley's relationship changing. You're awesome!

A special thank you to Krystal for naming Huxley's band. The 2Bit Princes are forever grateful!

I can't write these acknowledgements without thanking my amazing betas. Jenny, Sue, and Amy, you are so appreciated! And of course, my ride or die, Claudia, for supporting me every step of the way.

Finally, thank you to my brother, who put up with my five million questions about guitars and chords and music, and many other little facts that never even made it into the book, but still helped to shape the story.

Okay, I lied. I do have more thanks to give—to my awesome editor, Sandra, my amazing proofreader, Rumi, and all the PR, bloggers, readers, and everyone else who believed in Cole and Huxley.

One more thank you!!!

To you. Thank you for reading this. I appreciate it more than you will ever know.

Becca xoxo

ALSO BY BECCA STEELE

LSU Series

(M/M college romance)

Collided

Blindsided

Sidelined

The Four Series

(M/F college suspense romance)

The Lies We Tell

The Secrets We Hide

The Havoc We Wreak

*A Cavendish Christmas (free short story)**

The Fight In Us

The Bonds We Break

The Darkness In You

Alstone High Standalones

(new adult high school romance)

Trick Me Twice (M/F)

Cross the Line (M/M)

*In a Week (free short story)** (M/F)

Savage Rivals (M/M)

London Players Series

(M/F rugby romance)

The Offer

London Suits Series

(M/F office romance)

The Deal

The Truce

*The Wish (a festive short story)**

Other Standalones

Cirque des Masques (M/M dark circus romance)

Reckless (M/M soccer romance - standalone as part of the Unlucky 13 multi-author series)

*Mayhem (M/F Four series dark spinoff)**

*Heatwave (M/F summer short story)**

Boneyard Kings Series (with C. Lymari)

(RH/why-choose college suspense romance)

Merciless Kings

Vicious Queen

Ruthless Kingdom

Box Sets

Caiden & Winter trilogy (M/F)

(The Four series books 1-3)

**all free short stories and bonus scenes are available from https://authorbeccasteele.com*

***Key - M/F = Male/Female romance*

M/M = Male/Male romance

RH = *Reverse Harem/why-choose (one woman & 3+ men) romance*

ABOUT THE AUTHOR

Becca Steele is a USA Today and Wall Street Journal bestselling romance author. She currently lives in the south of England with a whole horde of characters that reside inside her head.

When she's not writing, you can find her reading or watching Netflix, usually with a glass of wine in hand. Failing that, she'll be online hunting for memes or making her 500th Spotify playlist.

Join Becca's Facebook reader group Becca's Book Bar, sign up to her mailing list, check out her Patreon, or find her via the following links:

- facebook.com/authorbeccasteele
- instagram.com/authorbeccasteele
- bookbub.com/profile/becca-steele
- goodreads.com/authorbeccasteele
- patreon.com/authorbeccasteele
- amazon.com/stores/Becca-Steele/author/B07WT6GWB2